Future Smart Grid Systems

Future Smart Grid Systems

Editors

Michael Short
Tracey Crosbie
Maher Al-Greer

MDPI • Basel • Beijing • Wuhan • Barcelona • Belgrade • Manchester • Tokyo • Cluj • Tianjin

Editors

Michael Short
School of Computing,
Engineering and Digital
Technologies
Teesside University
Middlesbrough
United Kingdom

Tracey Crosbie
School of Computing,
Engineering and Digital
Technologies
Teesside University
Middlesbrough
United Kingdom

Maher Al-Greer
School of Computing,
Engineering and Digital
Technologies
Teesside University
Middlesbrough
United Kingdom

Editorial Office
MDPI
St. Alban-Anlage 66
4052 Basel, Switzerland

This is a reprint of articles from the Special Issue published online in the open access journal *Energies* (ISSN 1996-1073) (available at: www.mdpi.com/journal/energies/special_issues/Future_Smart_Grid).

For citation purposes, cite each article independently as indicated on the article page online and as indicated below:

LastName, A.A.; LastName, B.B.; LastName, C.C. Article Title. *Journal Name* **Year**, *Volume Number*, Page Range.

ISBN 978-3-0365-1336-2 (Hbk)
ISBN 978-3-0365-1335-5 (PDF)

Contents

About the Editors

Michael Short

Michael Short is a professor of control engineering and systems informatics at Teesside University and leads the multidisciplinary Centre for Sustainable Engineering. He holds a BEng degree in electronic and electrical engineering (1999, Sunderland) and a PhD degree in real-time robot control (2003, Sunderland). Michael's research interests encompass aspects of control engineering and systems informatics applied to smart energy systems. He has authored over 150 reviewed publications in high-quality conferences and journals, has received over 1300 citations and has won five best paper awards at international conferences. He currently has an h-index of 21 and an i-10 index of 37. He has supervised six PhD completions and is the (co-)investigator on numerous completed and ongoing funded research projects. He is an associate editor for *Energies* (Basel), a full member of the IET, a fellow of the HEA and a sub-committee member of the IEEE Technical Committee on Factory Automation.

Tracey Crosbie

Dr. Tracey Crosbie is a leader in sustainability in the built environment at the School of Computing Engineering and Digital Technologies at Teesside University. She is a transdisciplinary academic with degrees in the social and technical sciences. Dr. Crosbie has researched issues associated with sustainability for over 20 years. She has vast experience in leading the development of successful collaborative research proposals and in managing work in national and international interdisciplinary research and innovation projects. Dr. Crosbie has over 40 academic journal publications in a diverse range of disciplines.

Maher Al-Greer

Maher Al-Greer Received his BSc degree in electrical engineering and his MSc degree in computer engineering from the University of Mosul, Mosul, Iraq, in 1999 and 2005, respectively, and his PhD degree in electrical and electronic engineering from Newcastle University, Newcastle upon Tyne, UK, in 2012. In 2016, he joined the Electrical Power Research Group at Newcastle University. He is currently a senior lecturer of electrical power engineering at Teesside University, Middlesbrough, UK. His main research interests include battery systems management, estimation and control. Currently, he is an associate editor for the *IET Power Electronics* journal, a topic editor for MDPI *Energies* and a guest member of the staff at the School of Engineering at Newcastle University.

Preface to "Future Smart Grid Systems"

Decarbonisation, digitalisation and decentralisation are transforming energy systems across the globe and supporting the transition to a sustainable future. Within this context, smart grids are electrical grids that include a variety of interoperable communication and control devices to optimally (or near-optimally) facilitate the production and distribution of electricity. Smart grids allow for better integration and management of volatile renewable energy sources, flexible transmission resources, energy storage devices, electric vehicles, microgrids and controllable loads; they are seen as key enablers in the decarbonisation of both the industry and society. This book focuses on the analysis, design and implementation of future smart grid systems. This book contains eleven chapters, which were originally published after rigorous peer-review as a Special Issue in the *International Journal of Energies* (Basel). The chapters cover a range of work from authors across the globe and present both the state-of-the-art and emerging paradigms across a range of topics including sustainability planning, regulations and policy, estimation and situational awareness, energy forecasting, control and optimization and decentralisation. This book will be of interest to researchers, practitioners and scholars working in areas related to future smart grid systems.

Michael Short, Tracey Crosbie, Maher Al-Greer
Editors

Article

Towards Self-Sustainable Island Grids through Optimal Utilization of Renewable Energy Potential and Community Engagement

Marko Jelić [1,2,*], Marko Batić [2], Nikola Tomašević [2], Andrew Barney [3], Heracles Polatidis [3], Tracey Crosbie [4], Dana Abi Ghanem [4], Michael Short [4] and Gobind Pillai [4]

1 School of Electrical Engineering, University of Belgrade, 11120 Belgrade, Serbia
2 Institute Mihajlo Pupin, University of Belgrade, 11060 Belgrade, Serbia; marko.batic@pupin.rs (M.B.); nikola.tomasevic@pupin.rs (N.T.)
3 Department of Earth Sciences, Uppsala University Campus Gotland, S-621 57 Visby, Sweden; andrew.barney@geo.uu.se (A.B.); heracles.polatidis@geo.uu.se (H.P.)
4 School of Computing, Engineering and Digital Technologies, Teesside University, Middlesbrough TS1 3BX, UK; t.crosbie@tees.ac.uk (T.C.); d.abighanem@tees.ac.uk (D.A.G.); m.short@tees.ac.uk (M.S.); g.g.pillai@tees.ac.uk (G.P.)
* Correspondence: marko.jelic@pupin.rs; Tel.: +381-(11)-677-4024

Received: 25 May 2020; Accepted: 16 June 2020; Published: 1 July 2020

Abstract: Solving the issue of energy security for geographical islands presents a one-of-a-kind problem that has to be tackled from multiple sides and requires an interdisciplinary approach that transcends just technical and social aspects. With many islands suffering in terms of limited and costly energy supply due to their remote location, providing a self-sustainable energy system is of utmost importance for these communities. In order to improve upon the status quo, novel solutions and projects aimed at increasing sustainability not only have to consider optimal utilization of renewable energy potentials in accordance with local conditions, but also must include active community participation. This paper analyzes both of these aspects for island communities and brings them together in an optimization scenario that is utilized to determine the relationship between supposed demand flexibility levels and achievable savings in a setting with variable renewable generation. The results, specifically discussed for a use case with real-world data for the La Graciosa island in Spain, show that boosting community participation and thus unlocking crucial demand flexibility, can be used as a powerful tool to augment novel generation technologies with savings from flexibility at around 7.5% of what is achieved purely by renewable sources.

Keywords: renewable energy; sustainability; island communities; demand flexibility; energy management; optimization

1. Introduction

Geographical islands and island communities in general present a very distinct problem from the perspective of electric energy infrastructure planning due to a number of unique contributing factors. In most cases, these island systems are heavily dependent on the mainland energy markets since there is a limited supply of energy and transporting it to the island itself can be very costly. With many islands relying mostly on either high-capacity underwater power cable connections with the nearest land mass or huge amounts of diesel or similar fuel to be shipped to the island in order to meet demand, providing energy security and stability to the islands presents a key task with significant repercussions on both the environment and the economic sustainability of the island.

Furthermore, these island electric energy supply systems generally have a single point of failure which, in the unfortunate case of a malfunction in the supply circuits or temporary delay in fuel shipment due to adverse weather conditions, may leave several hundreds or thousands of users without power or, at the very least, relying on a limited storage capacity. For this redundancy to even be available, the utilization of environmentally unfriendly and expensive lithium-ion batteries is often required. Unlike power grids onshore where, when need be, the energy supply can be obtained from multiple sides and sources such as different neighbouring countries or different energy generation technologies, the generation and import capacities of islands is commonly much less diverse and much more limited. This increases the vulnerability of island communities to power outages and their impacts, which disproportionately affect older residents, resulting in negative health and well-being outcomes [1,2]. Island communities can also be at risk of energy poverty if sufficient attention is not applied to self-sustainability as an energy strategy [3]. Having all these points in mind, many island communities have been looking into and actively implementing different strategies in order to be less reliant on polluting technologies and fuels as well as conditions dictated by events on the mainland.

Key factors in providing self-sustainability and increasing energy efficiency in island electric power grids are the optimal utilization of locally available clean renewable energy in accordance with the appropriate potentials of each island and community involvement through participation in interactive load management and demand response (DR) programmes [4,5]. With these two combined, they would unlock a certain flexibility from the demand side and allow for the demand to be moulded and adjusted to match the intermittent profile of generation typical for contemporary renewable energy sources as a result of their dependency on uncontrollable meteorological (photovoltaic panels, wind turbines, wave generators) and geological conditions (geothermal production sources). With demand-side flexibility provided by end users, an optimization algorithm can be utilized in order to provide the best match between supply and demand in order to minimize the costs of running the (energy) system, and maximize the exploitation of locally available renewable sources. With minor modifications, such algorithms can also be employed in cases where storage systems are present, and used to monitor a wide variety of economic factors as well as indicators of grid stability which could be used as key performance indicators that should be managed in high renewable penetration systems.

This paper aims to explore the two aforementioned aspects, community engagement and renewable energy potential, crucial to self-sustaining island grids, and propose a simulation scenario that includes both of them and is capable of assessing the potential savings that can be achieved with varying levels of energy generation from renewable sources and demand flexibility. The theoretical presentation of the methodology will be accompanied with a use case demonstration for which real-world data for the La Graciosa island in Spain is used.

The remaining part of the paper is structured as follows: First, Section 2 analyzes related work with regards to different aspects that are noteworthy for improving self-sustainability of island grids and methods of testing their joint impact. Here, Section 2.1 goes into detail on various key contributing aspects that influence community participation in programmes essential to sustainable energy systems, Section 2.2 looks into common practices with respect to assessments of renewable energy potential while Section 2.3 provides an overview of relevant papers in terms of planning and optimization implementations with demand flexibility. Afterwards, Section 3 presents an efficient simulation methodology using linear programming capable of assessing long-term behaviour of energy systems with different levels of renewable generation and demand flexibility while Section 4 provides an overview of data for La Graciosa, used to instantiate the optimization scenario and calculate the required parameters. Then, Section 5 provides the results of the conducted simulations and illustrates the effects of demand flexibility on total costs savings. A specific focus in result presentation was placed on decoupling the effects that renewable generation and demand flexibility have on the savings in order to be able to analyze them separately, as well as to allow for their mutual relationship to be inferred. A sensitivity analysis is also included to illustrate the stability of the results.

Finally, Section 6 provides concluding remarks regarding the proposed methodology and summarizes the results provided in this paper.

The methodology that follows utilizes a set of different variables. Their labels, units and descriptions are summarized in Table 1.

Table 1. Variable nomenclature overview.

Label	Unit	Description	Label	Unit	Description
$P_{\text{in}}(t)$	[kW]	Imported power variable vector to the Energy Hub	f_L	[%]	Load flexibility level
$Q_{\text{in}}(t)$	[kW]	Charge/discharge variable vector from input storage of the Energy Hub	J, J_{opex}	any, [EUR]	Criterion function and operational costs as a criterion function
$P_{\text{cin}}(t)$	[kW]	Power inflow variable vector to the conversion stage of the Energy Hub	$n_h(t)$, $n_m(t)$	[%]	Hourly and monthly noise levels
$P_{\text{cout}}(t)$	[kW]	Power outflow variable vector from the conversion stage of the Energy Hub	η_t, η_i	[%]	Grid transformer and invertor efficiency
$P_{\text{exp}}(t)$	[kW]	Exported power variable vector from the Energy hub	B_i	[EUR]	Initial (equipment) acquisition cost
$Q_{\text{out}}(t)$	[kW]	Charge/discharge variable vector from output storage of the Energy Hub	γ_i	[a]	Expected (equipment) lifetime
$L(t)$	[kW]	Load variable vector of the Energy Hub	δ	[%]	Monthly discount rate
$L_{\text{min}}(t)$, $L_{\text{max}}(t)$	[kW]	Lower and upper load flexibility margins	X_i	[EUR]	Monthly expenditure
$L_{\text{base}}(t)$	[kW]	Predicted load level	C_{total}	[EUR]	Total costs of running the system

2. Different Aspects of Self-Sustainable Island Grids

2.1. Factors Influencing Willingness to Participate in Sustainable Energy Systems

The success of any solution that would manage renewable sources and user demand depends on island communities becoming active parts of the energy system changing their consumption patterns through DR and smart grid (SG) technologies [6,7]. To develop effective strategies to encourage active participation and financial investment in sustainable energy systems it is important to understand which factors influence the willingness to participate [8,9].

2.1.1. Community Identity and Trust

Studies have recognized community identity and trust as key to peoples' active participation in sustainable energy systems [10,11]. Community identity can mobilise action and shift the interests of individuals from being self-oriented to being community oriented. Community identity can be summarised as: "Feelings of attachment to the community, taking pride in the community, and having friends within the community" [12] (p. 797). The shared intention to make the community a "better place" can be an important factor for the success of community energy projects [10].

Trust, a fundamental concept of interpersonal relationships and collaboration, is positively related to volunteering which is a basic type of participation and is shown to be crucial for economic decision making, such as financial investments [10,13]. Trust between local people and stakeholders that take projects forward is essential [14], especially if information is handled with transparency and accuracy throughout all stages of the project [15]. Communities that enjoy high levels of collective trust and have trust in their community-based institutions and organisations are thus more likely to succeed, as members would be willing to invest their time or money in projects that they believe are beneficial and of good value to the community.

2.1.2. Social Norms

Citizen participation is also influenced by social norms, which, in the context of sustainable energy projects, can be thought of as peers' expectations regarding to energy issues. The effect of social norms on environmentally related behaviours has been analysed in [16,17], and the impacts of social norms on engagement with sustainable community energy projects has also been given attention [18]. Social norms have been shown to exert a powerful influence on people's behaviour and consequently one's intention to contribute to a sustainable community energy project [18]. Social norms often rely on a person's perception of social pressure to perform or not perform the behaviour under consideration [10,16]. Furthermore, cooperation is central to many community sustainable energy projects and that itself is influenced by social norms [19].

2.1.3. Environmental Attitudes

Environmental reasons, e.g., climate protection and sustainability, climate awareness, have been found to be among the motivations for collective sustainable energy projects. People involved in community energy projects are generally more receptive to ethical and environmental commitment and question their behaviour with respect to energy consumption and carbon emissions reduction [13,20,21]. With respect to sustainable community energy projects, environmental attitudes are an influential positive factor and an important motivator for collective energy action [22–24].

2.1.4. Economic Benefits

Economic benefits which can be generated by sustainable community energy projects are identified as another incentive to get involved with a sustainable energy project. Community-owned means of production can generate income locally, through returns on investment, the sale of generated energy in the form of electricity or heat, tax revenues and the creation of employment during the construction and maintenance of renewable energy (RE) installations [15,20,21]. Distribution of financial benefit for shareholders and/or the community is another leading motivation. Financial benefits made by community institutions or organisations, whether local authorities or non-governmental organisations as a result of implementing DR technologies, can have wider benefits to the community if these are shared through a transparent process. Accordingly, identifying collective economic benefits as well as individual household economic savings [25] are both important considerations.

2.1.5. Island Communities and Sustainable Community Energy Projects

In the case of islands, community-based energy interventions are not uncommon. Research presented in [26] indicated the role that external elements such as the regulatory framework and national guiding visions and plans, as well as internal elements such as the strong sense of community and identity have in ensuring the success of community energy projects. A study conducted on the Reunion island [27] indicates that the security of supply is a major factor that drives island communities to take part in projects based on new innovations.

Paper [28] reports on the success of DR projects on Magnetic Island off Queensland, Australia, suggesting that engaging residents as members of a community was one of the main reasons for the DR project's success. This, however, indicates that whilst community identity can be strong, it should not be taken for granted when mobilising for DR implementation. The same study examined changes in energy-related behaviour. Equally, the effort and support expended by the utility company helped increase the level of behavioural change [28], achieved through effective marketing and communication efforts. Accordingly, the different actors involved in a reactive solution to increase self-sustainability have an important role to play not only in recruiting the households to take part in the project demonstration but to engage positively with the community to build needed levels of trust and acceptance of DR innovations

2.1.6. Sociotechnical Implications

As mentioned previously, DR is seen as a key enabling technology for the integration of renewable energy to support island self-sustainability. However, it has been well reported that one of the major drawbacks to successful engagement in DR programs by consumers, for example in some recorded cases in the US, relates to little or diminishing levels of user participation from residential customers enrolled in DR programmes [29,30]. It has also been suggested that, since DR programmes can be broadly classified into two sub-types (price-based and incentive-based), with each requiring a different levels of home automation, consumers with little or diminishing user participation should be identified as early as possible and new ways sought to encourage participation [31]. This can be achieved, for example, by switching from a voluntary Time-of-Use (ToU) incentive-based programme to a Direct Load Control (DLC) price-based programme by increasing the penetration of home automation, to allow automatic remote switching of residential loads [31]. Although techniques designed for online implementation can be employed to identify participation levels of residential customers in incentive-based DR programmes (e.g., [32]), the use of surveys such as that described in this section can reveal preliminary insights into likelihood to participate in different kinds of DR programmes for island communities. It is suggested that such preliminary information could be combined with appropriate assessment tools (e.g., [30,33]), and simulation models (e.g., [31]) to assist with the planning and execution of DR activities, programmes and use cases to support energy transitions for island communities.

2.1.7. Policy Barriers

There are a number of regulatory and policy barriers and enablers that are significant in relation to the development of autonomous and semi-autonomous energy islands. One key issue in this respect is the ability of independent aggregators to operate on energy markets. This is because "through aggregation the value of flexibility (DSR, storage and embedded generation) can be enhanced by bringing together providers who would be too small to participate in the markets individually due to specified load sizes" [34]. The ability of independent aggregators to operate on energy markets varies across the EU and associated countries [34–36]. Recent research [34] ranked the eight EU and associated countries that are leading in respect to the ease with which independent aggregators can operate on energy markets. In descending order these are: France, Switzerland, Ireland, Great Britain, Belgium, Finland, Germany and Denmark. Overall, it is fair to say that the regulatory and policy framework in Europe for demand response is progressing, but further regulatory and policy improvements are needed [34].

2.2. Renewable Energy Potential

Another key contributing factor to the diversification of the energy supply of islands, as a prerequisite to achieving self-sufficiency and decreasing the reliance on traditional energy supply sources, are the implementation efforts of clean renewable energy generation solutions. Keeping in mind the meteorological conditions to which islands are exposed to are very specific when compared to use cases on the mainland, and that they vary significantly based on the location and the microclimate of the considered island, each renewable generation deployment effort must be preceded with an appropriate analysis of what the potentials of the island are with respect to different generation technologies.

2.2.1. Methods for Production Estimation

The assessment method of renewable energy potentials varies by the energy type being evaluated. For wind power, the mean annual wind speed (usually expressed in [m/s]) at a given height serves as common indicator of potential [37]. For solar power, the global horizontal irradiance (GHI) (usually expressed in [W/m^2]), which is equal to the total solar radiation on a horizontal surface, can serve the same purpose for some solar technologies [38]. Geothermal potential can be assessed, in part, by its surface heat flow (usually expressed in [mW/m^2]). Hydro-electrical potential is based on the head and the long-term flow statistics, where head can be estimated looking at topographical maps while flow is much more complicated to estimate though models do exist such as described in [39]. Less mature renewable resources like wave and tidal can be evaluated from annual mean wave power density within each wave (usually expressed in [kW/m]) and tidal range (usually expressed in [m]) as well as tidal currents (usually expressed in [m/s]), respectively.

2.2.2. Data Availability

In all cases of different generation technologies, it is preferable to have long term, high quality measurement data of the energy type to be exploited. Unfortunately, in many cases, such data does not exist in a readily accessible form for any number of reasons. In those cases where specific locational data does not exist, the values for potential assessment can be sometimes obtained from several local and international official sources. Local sources can include nearby official meteorological towers such as those common near airports or from national resource surveys. International sources include services like the Global Solar Atlas provided by the World Bank Group [40], the New European Wind Atlas partly funded by the European Commission [41], as well as from a number of corporate sources. Some energy modelling software, such as RETScreen developed by Natural Resources Canada, provide climate data from both local official as well as international sources [42]. The values received from governmental and international sources are generally useful for the initial assessment of an area or site's potential while the corporate services can often gather actual measurement data at a proposed location.

The onsite data gathering for wind power includes the erection of a measurement mast with a variety of different anemometers, normally installed at varying heights, to gather wind speed data. Alternatively, laser and sound detection technologies can be used that do not require the erection of a measurement tower to obtain the wind speeds at different heights [43]. The process for measuring solar irradiation uses pyranometers and pyrheliometers at the site to obtain global horizontal irradiation (GHI) as well as direct irradiance [38]. Testing of geothermal potential onsite is done by drilling a geothermal well or wells to gather required measurements [44]. Hydro electrical potential can be evaluated onsite by review of actual head potentials and flow rates using measurements of a river that are convertible into flow data. Evaluation of wave energy onsite can be done using a floating buoy that moves along the surface of the ocean and records its vertical displacement that allows for the calculation of wave height and period [45]. Tidal range is determined through measurement of the site's high tide and low tide points while tidal flow can be assessed using acoustic Doppler equipment [46].

2.3. *Planning and Optimization*

Planning is a crucial aspect when discussing improvements to island grid systems. Among others, [47] discusses the dependency of Mediterranean islands on fossil fuels and the underutilization of renewable energy source (RES) potential and discusses economic feasibility of RES. Long-term cost-benefit analyses are the main focus of [48] which also analyzes the potential effects of RES installations for small islands. A Caribbean use case with high RES penetration is elaborated by [49]. On the other hand, [50] looks at energy poverty and energy planning with local energy policies in mind for the Canary Islands. Long-term analyses are expanded with the introduction of battery storage in [51] where a comparison between small-scale urban systems and island grids is made.

Some authors place a special focus on discussing the effects of optimization, but very few consider geographical island scenarios specifically. Simulations on this topic are explored by [52] where multiple different energy sources such as photovoltaics, wind turbines, diesel generators and many others are integrated in a comprehensive scenario. Flexibility of the power system in general is discussed in [53] where a European use case is analysed in a stepwise methodology. Regarding optimization with demand flexibility, [54] provides a scenario with real-time pricing, but focuses specifically on heating systems. Once again, demand flexibility is explored by [55] where participation in DR programmes as sources of load elasticity is analysed.

This paper aims to extend the state of the art by considering demand flexibility and renewable potential as the two previously mentioned key factors that influence the ability of achieving self-sustainability of island energy grids. By using proactive user participation as a driver to unlock crucial demand flexibility and renewable energy generation as chosen in accordance with respective meteorological conditions and practical constraints, the methodology that follows will bring these two aspects together in a cohesive simulation scenario. The proposed methodology will then be implemented on a specific use case of the La Graciosa island in Spain using real-world data, followed by the results that analyze savings that can be achieved by using demand flexibility on top of renewable energy generation. By doing so, this paper aims to explore the goals that an island community eager to achieve self-sustainability can accomplish.

3. Simulation Methodology

In order for the full potential of an island electric grid that utilizes demand flexibility and renewable sources to be assessed, an optimization procedure has to be conducted in order to couple the intermittent supply with the demand-side flexibility so that optimal performance can be achieved. By doing so, traditional rule-of-thumb methods can be augmented using computer simulations, thus utilizing precise numerical data that depicts variable pricing schemes for both energy import and export, variable weather conditions distinguishable through renewable production profiles and constraints related to user comfort with demand flexibility in mind. This section presents a methodology capable of conducting such simulations which is later applied to a use case in order to investigate the impact of user demand flexibility in island renewable energy systems.

3.1. Optimization Procedure

In order for the optimization procedure to be conducted, a set of scenarios has to be first defined. Each scenario has a variable demand flexibility level and a variable amount of renewable generation. Using the data that will be laid out in the following sections, each scenario is instantiated with an adequate pricing tariff, renewable production curve and demand levels with flexibility.

As depicted in Figure 1, every one of the scenarios are individually optimized using a model for energy management formulated as an appropriate optimization problem and a predefined criterion that will be deliberated on in the following text. Besides this criterion, other key values of each of the optimized scenarios will be monitored and logged for later analysis. Once all the cases are optimized for, the relationship between demand flexibility, renewable energy generation and the monitored output(s) will be analyzed using the output data.

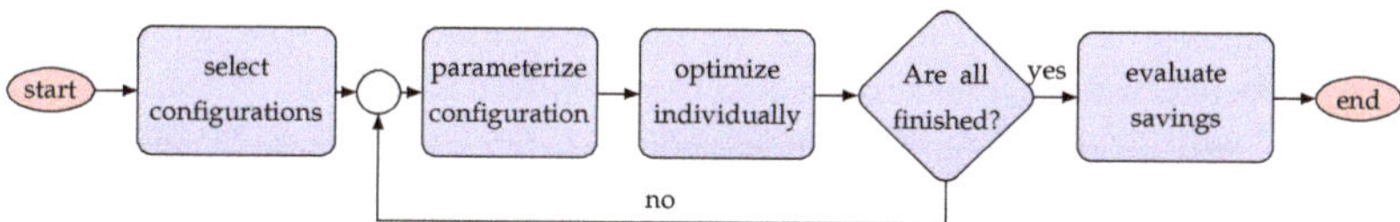

Figure 1. Simplified flowchart illustration of the proposed optimization procedure.

3.2. Operational Optimization Model

The operational optimization is based on a model of energy transmission and transformation called the Energy Hub, conceptualized in [56] and instantiated as a sequential structure presented in Figure 2. This model is used to simplify the grid of the island as follows: The Hub is supplied with input energy $P_{\text{in}}(t)$ which, as other variables in the model, represents a vector of instantaneous values of a certain physical quantity, in this case power coming either from the submarine cable or local generation. In this model, each type of carrier can be limited in a different way. For example, the import from a cable connection can be considered to be a variable that has a value between zero and the maximum power transmission capacity of the cable. On the other hand, all energy produced by the renewable sources has to be imported and either stored in the input stage storage ($Q_{\text{in}}(t)$) or the output stage ($Q_{\text{out}}(t)$) for later use, exported as $P_{\text{exp}}(t)$ or used to fulfill the load $L(t)$. In the path from the import stage to the load, the energy passes through several stages: input transformation (defined by the F_{in} block), conversion (defined by the C block) and output transformation (defined by the F_{out} block). This allows for the modelling of ways in which different energy carriers can be mixed together or depiction of losses associated with necessary transformations in this process.

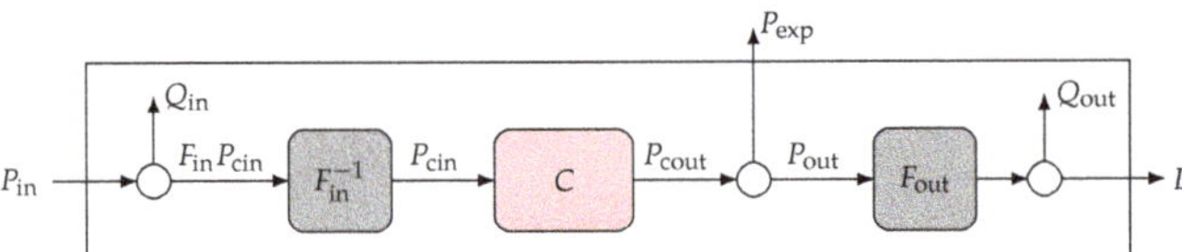

Figure 2. Illustration of the structure of the Energy Hub instance.

Owing to its modular and flexible nature, the Energy Hub can be used to simultaneously model fundamentally different types of carriers such as electric and thermal energy as in [57]. However, since the target of this paper are improvements to the electric energy supply, a focus will be placed solely on the electric aspect. By specifying different layouts of blocks F_{in}, C and F_{out}, different layouts of the grid can be modelled and adapted to the considered scenario with an adequate number of carriers of which each one corresponds to one generation technology (photovoltaic, wind turbine, etc.) or supply type (submarine cable connection, power generated from burning diesel, etc.).

3.3. Load Flexibility Model

In this methodology, the load $L(t)$ in the Hub is considered as a variable that can be optimized between a lower $L_{\text{min}}(t)$ and an upper $L_{\text{max}}(t)$ limit, constructed around some baseline load value $L_{\text{base}}(t)$. Different levels of user willingness to adjust their loads are expressed by varying the width of the band between $L_{\text{min}}(t)$ and $L_{\text{max}}(t)$. If f_L is assumed to be the flexibility of the demand, the aforementioned limits can be constructed as

$$L_{\text{min}} = (1 - f_L)L_{\text{base}}(t) \quad \text{and} \quad L_{\text{max}} = (1 + f_L)L_{\text{base}}(t).$$

Figure 3 illustrates an example that shows how the lower and upper limits relate to the baseline consumption and what an optimal profile between these two limits might look like for a given time period. This type of load flexibility can sometimes be utilized as a replacement for storage capacities. Namely, by introducing a flexibility margin, a power reserve is introduced in the system whereby it can utilize the region between the minimum and maximum allowed load value in order to dynamically match the supply with the demand in the same way as a storage solution would allow this with adequate charge and discharge rates. As the results will later show, if such flexibility is displayed by the users, observable savings can be achieved even in cases without any storage facilities. However, in reality, the levels of reserve that such flexibility can provide may not always be sufficient in which cases storages are a necessity. Still, having any flexibility whatsoever can go a long way towards reducing the overall costs as will be illustrated later.

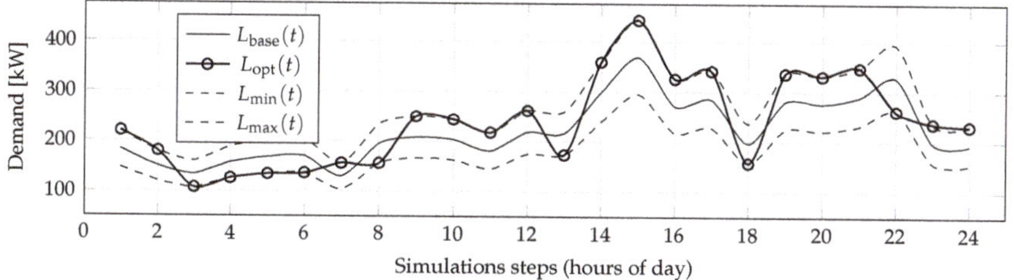

Figure 3. Illustration of the demand flexibility band and the optimized profile.

Any optimization engine that has to optimize for lowest price (which is a common practise) and has the ability of utilizing load flexibility margins will tend to minimize the load wherever possible. In order to avoid this effect, as it would significantly impact user comfort and not be realistic, an integral energy constraint is added, stating that the total energy consumed within a 24 h period must be the same in the case of the optimal profile and the baseline profile. Such a constraint can be expressed as

$$(\forall k \in \{1, 2, ..., 365\}) \quad \sum_{t=24(k-1)+1}^{24k} L(t) = \sum_{t=24(k-1)+1}^{24k} L_{\text{base}}(t).$$

3.4. Implementation

By laying out all the constraints of the model with regards to energy transmission and transformation as well as load management in matrix form

$$A_{\text{eq}}x = b_{\text{eq}} \quad \wedge \quad A_{\text{ineq}}x \leq b_{\text{ineq}} \quad \wedge \quad l_b \leq x \leq u_b$$

and an objective function defined as $J = f^T x$, the Energy Hub can be optimized as a linear programming (LP) problem using any number of solvers (CPLEX, Gurobi, MatLab's linprog, lpsolve, etc.) very efficiently and key performance indicators can be extracted in order to summarize the performance of the system over a long time span, allowing for a long-term (multi-year) assessment to be performed. Furthermore, by analysing individual variables of the Hub that are integrated in the vector x where each variable of the model at a given time instance in the simulation is assigned a corresponding position in x, additional parameters regarding the performance of the grid, as given in [58], can be easily extracted from the optimized outputs.

4. Use Case Definition

In order for the joint impact of social engagement and renewable generation potential to be assessed using the previously defined simulation methodology, a specific use case should be assumed. Therefore, the following subsections will present a selection of available data, gathered as part of the EU Horizon 2020 REACT project [59–61]. The main goal of the project is to improve the self-sufficiency of island energy supply systems and it considers eight islands in total, of which three (La Graciosa in Spain, San Pietro in Italy and the Aran Islands in Ireland) are used as pilot islands, i.e., islands where the project's final research solutions will be piloted. The following sections place the main focus on data for La Graciosa as this data is used to generate a simulation scenario, but also utilize results from other islands for reference.

4.1. End User Engagement

Given the priorities of island communities, the factors which influence the willingness of the island residents to engage with the project and become active parts of the energy system changing their consumption patterns through DR and SG technologies are summarised in Table 2. Along with them, some key findings from a survey conducted on the La Graciosa island with a sample size of roughly 13% of the total number of households, were also presented. The survey assessing the communities' perception of SGs and DR at each of the REACT project's pilot islands has some very encouraging results related to island community's motivation to be part of sustainable energy systems.

Table 2. Motivations for participation in island sustainable energy projects in island communities and findings from a survey conducted in La Graciosa.

Motivation	Description	Relevance to Island Communities	Focus Case Island La Graciosa
Community identity	An intention and desire to make the community better for everyone; benefits oriented to the community and away from the self.	Securing supply for the island and becoming more energy autonomous. Financial savings from the community energy project being re-directed to other community projects/investments; integrating with the community and having shared experiences.	The majority of respondents are 'motivated' or 'strongly motivated' by environmental and altruistic values suggesting a strong community identity. For example, 75% of respondents from La Graciosa, report that they are motivated to use smart grid technologies by a desire to reduce CO_2 emissions.
Financial	Save on energy bills; receive financial rewards from energy supplier.	Lower energy costs due to better security of supply.	74% of respondents in La Graciosa indicate that energy saving is very important to them.
Environmental	Reducing CO_2 emissions; increasing the sustainability of the island community	Taking action on CO_2 reductions empowers island communities in the face of climate change.	80% of the respondents from La Graciosa report that the use of renewable energy technologies is very important to them.
Social norms	Following the example of others in the community; being sensitive to the opinion of others.	The size and closeness of communities driving a sense of togetherness and responsibility, generating positive social norms.	Social norms pertaining to sustainability and high-end technologies are reported as 'strongly motivating' with 71% of respondents in La Graciosa being interested in attaining a sustainable character for their homes.

By analyzing the survey results from the rightmost column of Table 2, it can be observed that utilizing renewable energy, achieving energy savings and being environmentally friendly are all important aspects of participation in the project. Therefore, it is assumed that an overwhelming majority of the residents will be open to the idea of adapting their demand so as to achieve these goals. In order to get detailed insight into achievable savings in different scenarios, the demand elasticity that would be unlocked by active user engagement is modelled in accordance with specifications from Section 3.3 using a varying level of flexibility which is used to reflect different levels of community participation. 25 different demand flexibility levels were chosen, from 0 to 60% (inclusive) with 2.5% increments.

4.2. Renewable Energy Generation

In order to first illustrate how different conditions impact the renewable generation potential of islands in different locations, data that indicates the relative energy potentials for REACT pilot islands in terms of wind power, solar power and wave power is collected and summarized in Table 3. Both La Graciosa and San Pietro have been identified as having a potential for geothermal development [62,63] but specific data for neither island was available. No specific data for hydro electrical or tidal potential was available for the islands.

Table 3. Wind, solar and wave power potentials on the islands of La Graciosa, San Pietro and Aran.

Island	Annual Average Wind Speed at 10 m Height [m/s]	Annual Average Hourly Global Horizontal Rrradiance [W/m^2]	Average Annual Wave Power Density [W/m]	Sources
La Graciosa	4.1	205.7	31.0	[45,64,65]
San Pietro	6.1	171.5	N/A	[66]
Aran Islands	7.0	128.6	20.8	[67,68]

None of the values presented in the Table 3 are for a specific location on any of the three islands and provide only a general assessment of the resources on each. While La Graciosa may appear to have the lowest average wind speed there may be locations on the island where higher wind speeds can be achieved, such as on the island's mount Mojón, and conversely lower speeds may also be present within Caleto del Sabo, the island's capital. Irradiance is also impacted by local conditions, such as shading from buildings, which can reduce the islands' potentials. That noted, the Aran Islands and San Pietro clearly have significant wind resources at the relatively low height of 10 m, far below the hub height of any commercial wind turbine, while La Graciosa's resource is more modest. Conversely, La Graciosa has a significant solar resource while the Aran Islands have somewhat more modest values typical for northern Europe. Wave power estimates also indicate significant potential around La Graciosa and moderate potential near the Aran Islands.

Looking specifically at La Graciosa as the focus case for this paper, the island's land and sea-based energy production capabilities are both quite limited despite the potentials. In 1986, the local government designated most of the island as a regional Natural Park and in 1993 the entire island was declared a UNESCO Biosphere Reserve to protect the island's environment and biodiversity [62]. In 2003, ownership of the majority of the island's territory was granted to the Spanish agency for national parks. Additionally, the island is part of the Maritime-Terrestrial Chinijo Archipelago Natural Park and all nearby waters are protected [69]. These designations onshore and off essentially limit the types and sizes of RES production to those onshore technologies acceptable within the island's two small, populated areas. Similar limitations on RES production exist for other islands, like San Pietro and the Aran Islands.

An additional limitation to keep in mind is that not all countries have a regulatory framework that allows for the full application of demand response. For the specific case of La Graciosa, the Spanish electricity market is not entirely open to explicit demand response activities and doesn't allow for aggregation of demand-side resources [36]. These regulatory limitations create an environment where demand response on La Graciosa would be difficult to implement, though recent legislation indicates this situation may be changing [70].

Due to mentioned regulatory limitations and the illustrated potentials, this paper will focus on a theoretical photovoltaic generation deployment scenario in the capital of La Graciosa, Caleta de Sebo (at the specific location of LAT $= -29.232°$ and LON $= -13.502°$). In order to illustrate the potential of solar energy for this site, interfaces for an international official database [71] were used and hourly meteorological data depicting irradiance components and temperature was obtained. A summary of the data is illustrated in Figure 4. As can be observed, the direct irradiance has a noticeable seasonality, peaking during the summertime while, on the other hand, diffuse irradiance is relatively constant. The temperature on the island appears to be moderate with monthly mean values between 17 °C and 22.5 °C which can be beneficial as the photovoltaic production is sensitive to overheating issues.

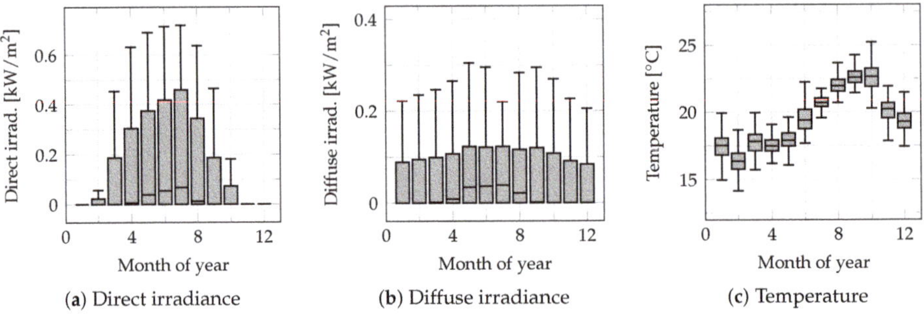

(**a**) Direct irradiance (**b**) Diffuse irradiance (**c**) Temperature

Figure 4. Seasonal variability in hourly irradiance components (**a**,**b**) and temperature (**c**) measurements for La Graciosa illustrated using box plots in monthly steps.

Following the mentioned remarks regarding renewable production, a two-carrier supply is given to the Energy Hub with one of them corresponding to the imported power from the submarine cable (with a maximum capacity of 1030 kW equal to the existing cable between La Graciosa and Lanzarote) and the other corresponding to the energy provided by the photovoltaic (PV) array. Using the previously presented data sources as well as the methodology associated with [72], the expected output of a non-tracking PV generation system with a reference capacity of 1 kW, 10 % losses, tilted 35° around the horizontal axis from the horizontal plane, facing south and located in the capital of La Graciosa, was generated. In order to simulate the output of differently sized PV generation plants, the unit production is scaled up linearly. For this purpose, a set of 21 different rated powers of PV generation was selected, from 0 to 1000 kWp (inclusive) with 50 kWp increments, as this range is thought to roughly correspond to what would theoretically be possible if the available roof space of buildings in the capital of La Graciosa was used.

4.3. Demand Profile

Since the total electricity demand of the entire island of La Graciosa cannot be obtained from publicly available sources, this baseline curve must be somehow constructed. Using hourly demand measurements from ten houses that are supplied by Fenie Energia (Spanish retailer for electric and thermal energy), averaged normalized working day and weekend day profiles are derived for each month and their variations between different months are illustrated using box-plots in Figure 5.

However, since total monthly consumption data is available, and is depicted in Figure 5 also, the hourly data can be scaled up with the appropriate monthly consumption divided by the total number of days in each month so that an estimated consumption curve can be obtained. In order to make this process more realistic, random unit white noise samples ($\mu = 0$, $\sigma = 1$) are scaled into the range $(-10,\ 10)\%$ to model hourly noise $n_h(t)$ and into $(-4,\ 4)\%$ to model monthly noise $n_m(k)$ and are superimposed to the 100% of the demand that would have been calculated without the given noise levels. The resulting demand profile is illustrated in Figure 6 and this profile is used as the baseline load $L_{\text{base}}(t)$, mentioned previously, around which a flexibility band will be formed.

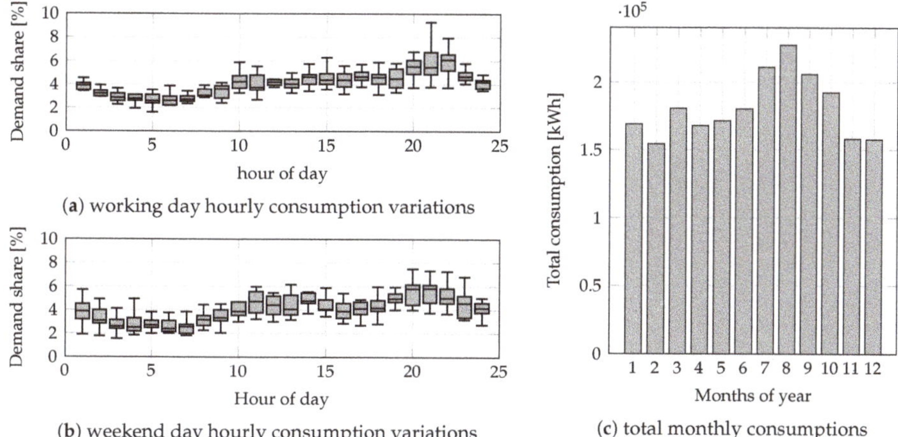

Figure 5. Monthly seasonal variations in hourly electric energy consumption share for working days (**a**), weekend days (**b**) and total monthly electric energy consumptions (**c**) based on data provided by Fenie Energia.

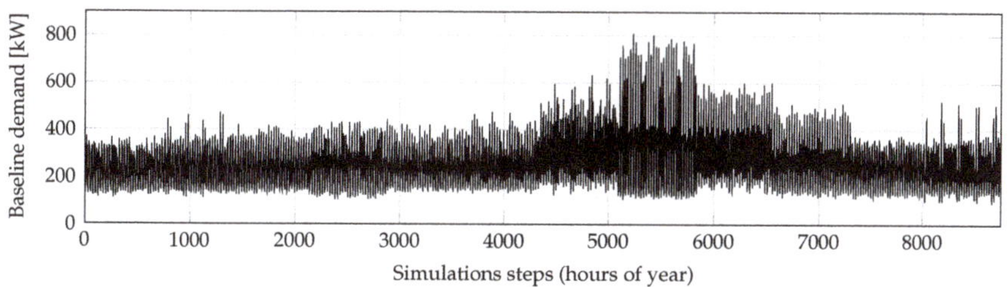

Figure 6. Estimated electric demand data for La Graciosa.

4.4. Pricing Profile

Survey data [60] indicates that 558 out of 613 (approximately 91%) grid-connected users in La Graciosa are contracted to a single rate tariff (with maximum power draw of 10 kW) with a fixed price of 0.162 EUR/kWh, so this value is assumed as the cost at which power can be imported from the submarine cable. In a scenario where renewable generation would be community owned, besides the cost of acquiring the system and its maintenance, the energy it generates comes without direct costs, and the excess energy can be sold on the wholesale market. Using the interfaces provided by OMIE [73] (Due to the crises related to COVID-19 ongoing at the time of writing this paper and the changes in the API provided by OMIE which now only provides the last year of hourly wholesale prices, historical (previously acquired) data was used for 2018 in order for the price profile not to include the months of February, March and April of 2020 which saw significant price fluctuations attributed to the effects of the pandemic that do not depict general trends), the hourly price profile for the year 2018 can be constructed. Namely, for each month and each first working and weekend day of that month, hourly prices have been obtained. These values are then normalized for each month into hourly profiles for a typical working and weekend day so that the average of both profiles equals one and are illustrated in the left side of Figure 7. Then, for each day of the year, the average wholesale price was obtained and is depicted in the right side of Figure 7. Afterwards, day by day, these daily values are used as a scale factors that are applied to the corresponding working or weekend day price profile. By concatenating these profiles together, the estimated hourly wholesale price profile is derived which is assumed to be the export price of electric energy.

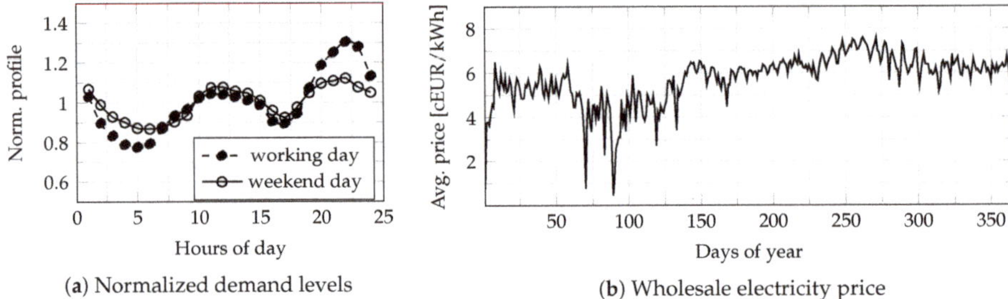

(**a**) Normalized demand levels

(**b**) Wholesale electricity price

Figure 7. Estimated averaged and normalized price profiles for working and weekend days (**a**) and reported average daily prices (**b**) for the Spain wholesale electric energy market in 2018.

4.5. Optimization Setup

With the Energy Hub having two carriers (cable imports and PV generated electricity) and only one load type (electric demand), the transformation stages can be defined by the matrices

$$F_{\text{in}}^{-1} = \begin{bmatrix} 1 & 0 \\ 0 & 1 \end{bmatrix}, \quad C = \begin{bmatrix} \eta_{\text{t}} & 0 \\ 0 & \eta_{\text{i}} \end{bmatrix} \quad \text{and} \quad F_{\text{out}} = \begin{bmatrix} 1 & 1 \end{bmatrix}$$

where η_{t} = 98% represents the typical grid transformer efficiency and η_{i} = 95% represents the typical invertor efficiency. All other losses are neglected. The combination of selected PV generation and demand flexibility increments provides a total of 525 configurations for simulation. These ranges are assumed as theoretical and the results at the end of the paper focus on a selection that is considered most realistic. The baseline load is instantiated using the estimated year-long hourly curve derived previously, while the criterion of the optimization model is set to equal the yearly operating cost of the model $J_{\text{opex}} = f^T x$ with regards to the assumed energy import and export prices, as is achieved by setting the appropriate values of f to the corresponding prices.

However, the optimization scenario should also be able to considered the costs associated with the assumed renewable installation. The upfront investment regarding the acquisition of solar panels is based on [74] with a cost of 1115 EUR/kWp. However, due to the fact that the PV panels come with a typical warranty and a matching expected lifetime of around 25 years, the cost associated with using them is split into yearly installments. The expected operational lifetime of PV panels may be longer that the mentioned duration. However, after the guaranteed amount of time had passed, the efficiency of the panels is expected to have dropped off significantly which, in turn, means that a replacement is in order. For a piece of equipment labeled i with the initial acquisition cost B_i, an estimated lifetime of γ_i years, one month's equated installment (equivalent to rent) would be calculated as

$$X_i^{\text{EMI}} = B_i \delta \cdot \frac{(1+\delta)^{12\gamma_i}}{(1+\delta)^{12\gamma_i} - 1}$$

where δ = 0.42% is the monthly discount rate which is assumed to be constant. With the monthly maintenance costs of each device given as X_i^{maint} (assumed to equal, in total, 2% of acquisition cost over lifetime), the total yearly cost of running the system would equate to

$$C_{\text{total}} = J_{\text{opex}} + 12 \sum_{i \in E} \left(X_i^{\text{EMI}} + X_i^{\text{maint}} \right)$$

where J_{opex} is the aforementioned cost determined by the optimization output through its criterion function. With the considered use case limited to only photovoltaic panels, the set of all considered equipment is given purely as $E = \{PV\}$. However this same methodology can also be implemented in other cases where either different energy sources are used or storage systems are utilized.

5. Results and Discussion

5.1. Simulation Outputs

With each simulation scenario being defined by a given PV rated power and demand flexibility level, by running individual optimization procedures, a set of results is obtained whereby a selection of monitored values is derived for each of the considered parameter combinations. Of these monitored values, the main focus of further analysis will be placed on total operating costs. Keeping in mind that, through previously completed projects, the island of La Graciosa has already installed around 50 kWp of solar generation capacity, this scenario could be considered as a baseline for further expansions of renewable production. However, for the sake of simplicity in this theoretical analysis, a baseline is actually assumed to be the case with no on-site renewable generation and no demand flexibility (i.e., all demand is met using energy imported from the submarine cable), as would be the case on an island with no existing renewable generation. For reference, the simulations show that the aforementioned existing implementation of PV generation saves a little over 2.5% in operating costs on a yearly basis (with monthly installments and maintenance accounted for) against the theoretical baseline with no renewable generation.

The first evaluated effect relates only to the potential savings that can be achieved from the implementation of renewable generation (i.e., without any demand flexibility). In order to derive this, only scenarios with zero flexibility are extracted and their operational costs are compared to the baseline scenario with no renewable generation. The resulting savings are shown on the left side of Figure 8.

Afterwards, each of the scenarios with renewable configurations is considered with different levels of demand flexibility in the range defined previously. When the initial savings achieved only due to the installation of renewable generation are subtracted, the additional (additive) saving levels can be calculated, and the appropriate values obtained in this way are presented in a three-dimensional space in Figure 8 where one dimension represents the considered rated power levels of PV generation, the second are the considered demand flexibility levels and the third are additional savings that can be achieved over the sole contribution of the renewable generation by utilizing demand flexibility. In other words, the total savings over baseline operational costs that can be achieved using both renewable generation and demand flexibility can be calculated by first determining the isolated impact of renewable generation given in the two-dimensional graph in Figure 8 and then adding an additional impact of demand flexibility as presented in the surface plot in Figure 8. By utilizing such a presentation, the impacts of renewable generation and demand flexibility are decoupled and can be analyzed separately while, at the same time, allowing for the total savings to be easily determined. Furthermore, more insight into the relation between additional savings, renewable generation capacity and the supposed flexibility is given by visualizing projections of the surface plot from Figure 8, as is presented in Figure 9.

By analyzing all three graphs, it can be clearly deduced that there is a notable increase in the performance that can be achieved by utilizing demand flexibility when there is ample renewable generation as the load can be more drastically shifted and aligned with renewable generation. An interesting point to note is that the impact from the demand flexibility only becomes apparent when the rated power capacity of RES becomes large enough (compared to the baseline demand levels) which, according to the results of the conducted simulations, is between 400 kWp and 450 kWp, at the point where the lines from the projections in the left of Figure 9 begin to diverge. Having in mind the fact that the efficiency of PV generation is calculated very conservatively with noticeable losses, it should be mentioned that this point of divergence may occur with less renewable generation if a more efficient system were to be considered. Also, the second projection graph from Figure 9 shows that for each renewable generation capacity, the returns tend to level off after a certain amount of flexibility with the value of that point increasing as the capacity of the generation increases.

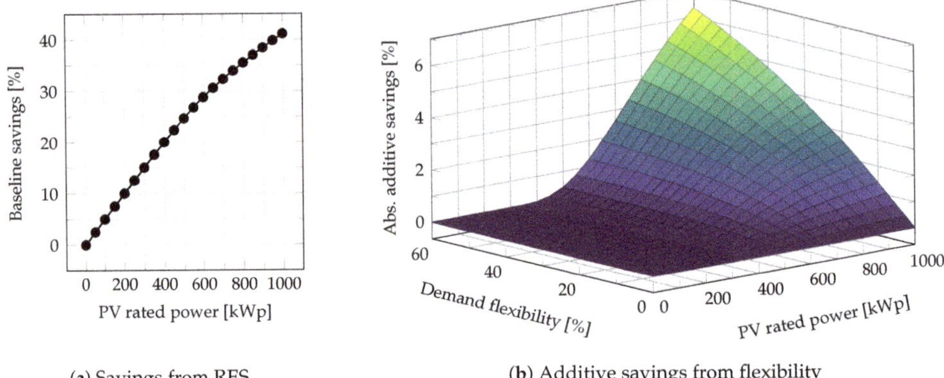

(**a**) Savings from RES (**b**) Additive savings from flexibility

Figure 8. Results of the simulations showing the (baseline) total operating costs savings attributed only to the deployment of renewable energy (i.e., without demand flexibility) (**a**) and the additional savings that can be achieved with different levels of demand flexibility (**b**). All savings are calculated against the baseline with no demand flexibility and no renewable generation.

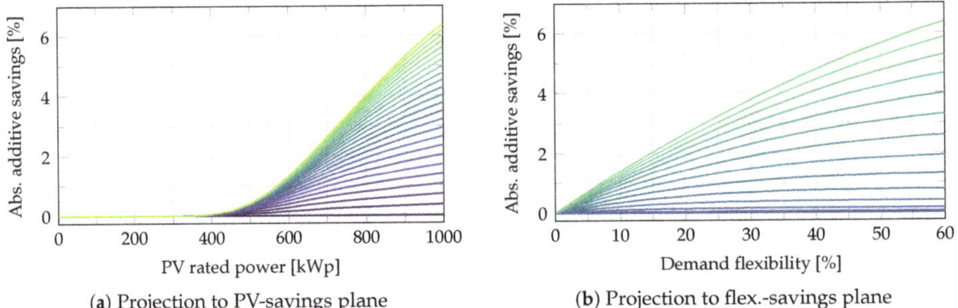

(**a**) Projection to PV-savings plane (**b**) Projection to flex.-savings plane

Figure 9. Two-dimensional projections of the additional achievable savings to the PV rated power-savings plane (**a**) and flexibility-savings plane (**b**). The color shift from blue to green corresponds to the increase of flexibility on the left graph and the increase in PV rated power on the right graph.

In order to illustrate the relationship between renewable generation capacity and demand flexibility, this theoretical analysis considers very large supposed amounts of demand flexibility which are difficult to achieve in real-world practice [75] while, on the other hand, results from Section 4.1 show promising potential in terms of demand flexibility for a motivated island community. Therefore, a selection of results that are more likely to be achieved is extracted from the experiments and presented in Table 4. This selection is coupled with several PV installations of different capacities, with the two medium sized capacities being considered potentially realistic, considering land restrictions limiting RES expansion to inhabited areas and the RES potential described and specifically analyzed for La Graciosa in Section 4.2. As can be observed from these results, for example, even with realistically achievable levels of supposed flexibility (i.e., 30%), the addition of demand flexibility can result in a notable decrease of operational costs. In the case with a PV installation capacity of 800 kWp, the decrease is 2.66%, which equates to 7.5% of the impact achieved solely by the utilization of that same renewable source with no flexibility. Also, Table 4 shows that some scenarios with lower renewable generation and higher demand flexibility can perform better than those with higher generation and lower flexibility. This indicates that considerable savings can be achieved by boosting social engagement rather than investing in equipment.

Table 4. A summary of achievable additive savings (expressed in percent of baseline operational costs) over the baseline savings achieved using only renewable generation and ratio of these additional savings when compared with baseline savings with the same PV power (in percentage, given in in parentheses) with realistic demand flexibility levels and supposed high renewable penetration.

dem. flex. [%]	PV Rated Power [kWp]			
	400	600	800	1000
10	0.02 (0.1)	0.39 (1.4)	0.80 (2.2)	1.06 (2.6)
15	0.03 (0.2)	0.69 (2.4)	1.51 (4.3)	2.04 (5.0)
20	0.04 (0.2)	0.84 (2.9)	1.93 (5.4)	2.65 (6.4)
25	0.04 (0.2)	0.96 (3.3)	2.31 (6.5)	3.23 (7.8)
30	0.04 (0.2)	1.06 (3.7)	2.66 (7.5)	3.77 (9.2)
35	0.04 (0.2)	1.13 (3.9)	2.96 (8.3)	4.29 (10.4)

In the end, it is worth mentioning that the considered model supposes a constant level of demand flexibility. However, by utilizing variable pricing schemes, DR programmes and others means of boosting community engagement, the users can be influenced to be more adaptive, especially in periods of peak demand, which could further increase the potential savings.

5.2. Sensitivity Analysis

In order to ascertain the stability of the obtained result and get a deeper understanding of how variability of different input parameters affects the savings, a limited sensitivity analysis was conducted. Namely, for the previously considered realistic scenario with 800 kWp of PV and 30% demand flexibility, each sample of the year-long hourly demand and predicted PV production was varied in the range between 95% and 105% of the corresponding nominal value, as dictated by a pseudo-random uniform distribution. These modified inputs were independently generated 200 times which correspond to 200 instances of optimizations that were conducted in this sensitivity analysis. By analyzing the achieved savings from demand flexibility in the same manner as was done in the previous part of this section, a distribution of the achievable additional savings due to demand flexibility can be observed in the form of a histogram, as given in Figure 10. Furthermore, using this data, an approximation of the cumulative distribution function (CDF) can be obtained which can also be used to reach the same conclusion.

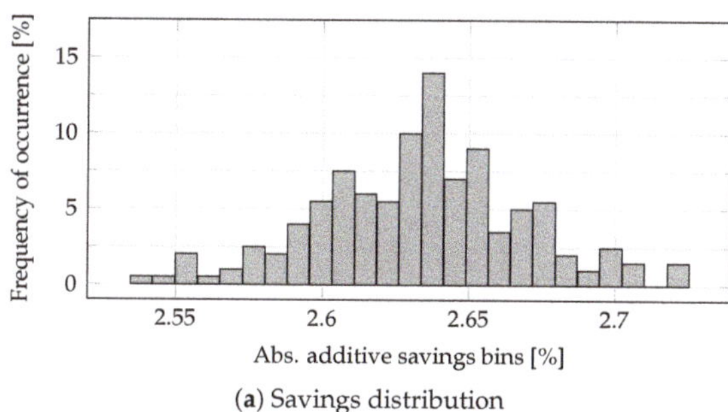

(**a**) Savings distribution

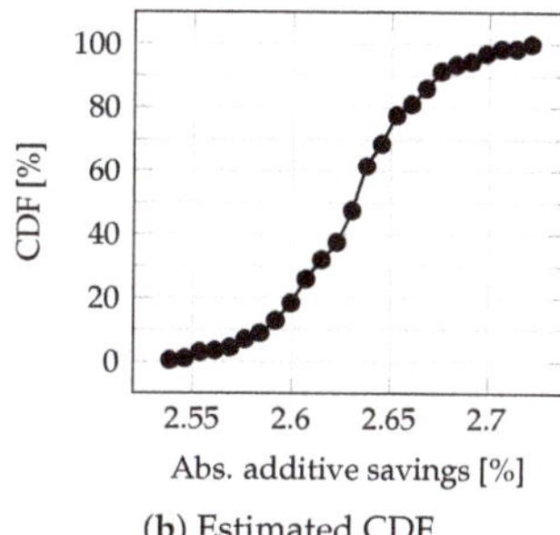

(**b**) Estimated CDF

Figure 10. Histogram illustrating the distribution of the absolute additive savings in 25 bins for the considered realistic use case (**a**) and the corresponding estimate of the cumulative distribution function (**b**).

As clearly shown by the data, almost all of the instances have resulted in the additive savings between 2.55% and 2.7%, with the nominal value previously calculated to be 2.66%. More precisely, around 78% of the outputs of 200 instances belong to a range between 2.6% and 2.7%. By comparing the mentioned four range limits with the nominal savings percentages from Table 4 for the same 800 kWp installation, but with adjacent flexibility levels (i.e., 2.31% savings for 25% flexibility and 2.96% savings for 35% flexibility), it can be concluded that even with a moderate amount of variability in the inputs, the results appear to be relatively stable since the limits do not overlap with the other savings.

6. Conclusions

This paper presents a view into ways in which the unique problem of increasing self-sustainability levels of geographical islands in terms of electric energy supply can be achieved when crucial flexibility in user demand is unlocked, allowing for it to be adjusted to an intermittent energy supply provided by clean on-site renewable generation. Renewable energy supply and demand flexibility are integrated in a comprehensive simulation procedure that determines the optimal energy management strategy, and in doing so, the relationship between the installed capacity of renewable generation, supposed demand flexibility and the expected savings is derived and discussed.

The results simulating 525 different combinations of PV generation capacity and demand flexibility levels show that, although significant savings in yearly operational costs can be achieved using only renewable generation and no demand flexibility, notable additional savings can be obtained with community participation through users' willingness to adjust their load. In some cases, even with realistic flexibility levels, the impact of load elasticity on the total savings was large enough for a given scenario to be more cost-effective than the next few larger renewable installations with less flexibility. In summary, as the survey results in Section 4.1 suggest high interest in sustainability issues, strong community identity and willingness to use renewable technologies among residents, different motivation factors (such as incentive-based (ToU), DR programmes) should be effective for island users. Moreover, the simulation results presented in Section 5 show that significant economic benefits will be leveraged if residents display flexibility. Concretely, with the increase in installed renewable capacity, the achievable savings attributed purely to demand flexibility go from negligible (between 0.02% and 0.04%) at 400 kWp to noticeable (between around 1% and 4.3%) at 1000 kWp. Furthermore, depending on the level of this flexibility, the increase for 10% and 35% boosts these savings 2.9 fold for 600 kWp of PV capacity and 3.7 fold for 800 kWp.

Overall, this indicates that an incentive-based programme will be both accepted by residents and effective at leveraging wider benefits, in turn giving some preliminary support to the combined approach of residential surveys and calibrated simulations. Given these results, it can be concluded that demand flexibility without significant disruptions to user comfort can be utilized as a powerful mechanism to augment some and replace other technologies while returning noticeable savings in operational costs, but also increasing the supply security and paving the way for a more resilient island energy supply network. In cases where the production and demand can technically be balanced using existing solutions, demand flexibility can be utilized as a means to ease this process but also lower costs and reduce the need for expensive and environmentally unfriendly storage solutions. Community user participation, however, remains key to achieving the benefits of this mechanism.

The limiting factors of this study have been isolated to be policy barriers for novel incentive-based programmes that would drive community participation and motivate the users to display flexibility as well as local regulatory constraints preventing the expansion of RES installments. The latter is especially crucial with regards to geographical islands as they are often the objects of strict laws designed to protect their unique environments. Further efforts in this field should be aimed at looking into novel and ecologically friendly generation technologies that would comply with local regulation, but a focus should also be placed on expanding DR and other similar programmes into the residential sector as contemporary efforts in this regard are often aimed solely at the industry.

Author Contributions: Conceptualization, N.T., M.B., H.P., T.C.; methodology, M.J., H.P., T.C. and D.A.G.; software, M.J.; validation, M.J.; formal analysis, M.J., A.B., C.T. and D.A.G.; investigation, H.P., A.B. and T.C.; data curation, M.J.; writing—original draft preparation, M.J., H.P., A.B., T.C. and D.A.G.; writing—review and editing,

N.T., M.B, H.P., A.B., T.C., M.S. and G.P.; visualization, M.J.; project administration, M.J., N.T. and M.B. All authors have read and agreed to the published version of the manuscript.

Funding: The research presented in this paper is partly financed by the European Union (H2020 REACT project, Grant Agreement No.: 824395) and partly by the Ministry of Education, Science and Technological Development of the Republic of Serbia.

Acknowledgments: The authors would like to especially express their gratitude towards the team from Fenie Energía for the energy consumption data that was kindly provided regarding the La Graciosa use case.

Conflicts of Interest: The authors declare no conflict of interest. The funders had no role in the design of the study; in the collection, analyses, or interpretation of data; in the writing of the manuscript, or in the decision to publish the results.

References

1. Ghanem, D.A.; Mander, S.; Gough, C. "I think we need to get a better generator": Household resilience to disruption to power supply during storm events. *Energy Policy* **2016**, *92*, 171–180. [CrossRef]
2. Christine, D.; Kathryn, L.; Sarah, J.; Kazuhiko, I.; Thomas, M. Health Impacts of Citywide and Localized Power Outages in New York City. *Environ. Health Perspect.* **2018**, *126*. [CrossRef]
3. Uche-Soria, M.; Rodríguez-Monroy, C. Energy planning and its relationship to energy poverty in decision making. A first approach for the Canary Islands. *Energy Policy* **2020**, *140*, 111423. [CrossRef]
4. International Renewable Energy Agency (IRENA). *A Path to Prosperity: Renewable Energy for Islands*; International Renewable Energy Agency: Bonn, Germany, 2016.
5. Smart Islands Initiative. *Smart Islands Declaration: New Pathways for European Islands*; Smart Islands Initiative 2017. Available online: www.smartislandsinitiative.eu (accessed on 22 May 2020).
6. Gangale, F.; Mengolini, A.; Onyeji, I. Consumer engagement: An insight from smart grid projects in Europe. *Energy Policy* **2013**, *60*, 621–628. [CrossRef]
7. Verbong, G.P.J.; Beemsterboer, S.; Sengers, F. Smart grids or smart users? Involving users in developing a low carbon electricity economy. *Energy Policy* **2013**, *52*, 117–125. [CrossRef]
8. Claudy, M.C.; Michelsen, C.; O'Driscoll, A.; Mullen, M.R. Consumer awareness in the adoption of microgeneration technologies: An empirical investigation in the Republic of Ireland. *Renew. Sustain. Energy Rev.* **2010**, *14*, 2154–2160. [CrossRef]
9. Palm, J.; Tengvard, M. Motives for and barriers to household adoption of small-scale production of electricity: Examples from Sweden. *Sustain. Sci. Pract. Policy* **2011**, *7*, 6–15. [CrossRef]
10. Kalkbrenner, B.J.; Roosen, J. Citizens' willingness to participate in local renewable energy projects: The role of community and trust in Germany. *Energy Res. Soc. Sci.* **2016**, *13*, 60–70. [CrossRef]
11. Bauwens, T. Explaining the diversity of motivations behind community renewable energy. *Energy Policy* **2016**, *93*, 278–290. [CrossRef]
12. Vugt, M.V. Central, Individual, or Collective Control?: Social Dilemma Strategies for Natural Resource Management. *Am. Behav. Sci.* **2002**. [CrossRef]
13. Hammami, S.M.; chtourou, S.; Triki, A. Identifying the determinants of community acceptance of renewable energy technologies: The case study of a wind energy project from Tunisia. *Renew. Sustain. Energy Rev.* **2016**, *54*, 151–160. [CrossRef]
14. Walker, G.; Devine-Wright, P.; Hunter, S.; High, H.; Evans, B. Trust and community: Exploring the meanings, contexts and dynamics of community renewable energy. *Energy Policy* **2010**, *38*, 2655–2663. [CrossRef]
15. González, A.M.; Sandoval, H.; Acosta, P.; Henao, F. On the Acceptance and Sustainability of Renewable Energy Projects—A Systems Thinking Perspective. *Sustainability* **2016**, *8*, 1171. [CrossRef]
16. Dwyer, P.C.; Maki, A.; Rothman, A.J. Promoting energy conservation behavior in public settings: The influence of social norms and personal responsibility. *J. Environ. Psychol.* **2015**, *41*, 30–34. [CrossRef]
17. Gadenne, D.; Sharma, B.; Kerr, D.; Smith, T. The influence of consumers' environmental beliefs and attitudes on energy saving behaviours. *Energy Policy* **2011**, *39*, 7684–7694. [CrossRef]
18. Mulugetta, Y.; Jackson, T.; van der Horst, D. Carbon reduction at community scale. *Energy Policy* **2010**, *38*, 7541–7545. [CrossRef]
19. García-Valiñas, M.A.; Macintyre, A.; Torgler, B. Volunteering, pro-environmental attitudes and norms. *J. Socio-Econ.* **2012**, *41*, 455–467. [CrossRef]

20. Walker, G.; Cass, N. Carbon reduction, 'the public' and renewable energy: Engaging with socio-technical configurations. *Area* **2007**, *39*, 458–469. [CrossRef]
21. Brummer, V. Community energy—Benefits and barriers: A comparative literature review of Community Energy in the UK, Germany and the USA, the benefits it provides for society and the barriers it faces. *Renew. Sustain. Energy Rev.* **2018**, *94*, 187–196. [CrossRef]
22. Bomberg, E.; McEwen, N. Mobilizing community energy. *Energy Policy* **2012**, *51*, 435–444. [CrossRef]
23. Dóci, G.; Vasileiadou, E. "Let's do it ourselves" Individual motivations for investing in renewables at community level. *Renew. Sustain. Energy Rev.* **2015**, *49*, 41–50. [CrossRef]
24. Van der Schoor, T.; Scholtens, B. Power to the people: Local community initiatives and the transition to sustainable energy. *Renew. Sustain. Energy Rev.* **2015**, *43*, 666–675. [CrossRef]
25. Christensen, T.H.; Friis, F.; Bettin, S.; Throndsen, W.; Ornetzeder, M.; Skjølsvold, T.M.; Ryghaug, M. The role of competences, engagement, and devices in configuring the impact of prices in energy demand response: Findings from three smart energy pilots with households. *Energy Policy* **2020**, *137*, 111142. [CrossRef]
26. Sperling, K. How does a pioneer community energy project succeed in practice? The case of the Samsø Renewable Energy Island. *Renew. Sustain. Energy Rev.* **2017**, *71*, 884–897. [CrossRef]
27. Bouly de Lesdain, S. The photovoltaic installation process and the behaviour of photovoltaic producers in insular contexts: The French island example (Corsica, Reunion Island, Guadeloupe). *Energy Effic.* **2019**, *12*, 711–722. [CrossRef]
28. Morris, P.; Vine, D.; Buys, L. Critical Success Factors for Peak Electricity Demand Reduction: Insights from a Successful Intervention in a Small Island Community. *J. Consum. Policy* **2018**, *41*, 33–54. [CrossRef]
29. Berger, L.T.; Iniewski, K. *Smart Grid Applications, Communications, and Security*; Wiley: Hoboken, NJ, USA, 2012.
30. Anuebunwa, U.R.; Rajamani, H.S.; Pillai, P.; Okpako, O. Evaluation of user participation capabilities in demand response programs for smart home applications. In Proceedings of the 2017 IEEE PES PowerAfrica, Accra, Ghana, 27–30 June 2017; pp. 483–488. [CrossRef]
31. Wang, Z.; Paranjape, R.; Sadanand, A.; Chen, Z. Residential demand response: An overview of recent simulation and modeling applications. In Proceedings of the 2013 26th IEEE Canadian Conference on Electrical and Computer Engineering (CCECE), Regina, SK, Canada, 5–8 May 2013; pp. 1–6. [CrossRef]
32. Zhao, S.; Ming, Z. Modeling demand response under time-of-use pricing. In Proceedings of the 2014 International Conference on Power System Technology, Chengdu, China, 20–22 October 2014; pp. 1948–1955. [CrossRef]
33. Crosbie, T.; Broderick, J.; Short, M.; Charlesworth, R.; Dawood, M. Demand Response Technology Readiness Levels for Energy Management in Blocks of Buildings. *Buildings* **2018**, *8*, 13. [CrossRef]
34. Bray, R.; Woodman, B. *Barriers to Independent Aggregators in Europe (EPG Working Paper: EPG1901)*; University of Exeter: Exeter, UK, 2019.
35. Crosbie, T.; Vukovic, V.; Short, M.; Dawood, N.; Charlesworth, R.; Brodrick, P. Future Demand Response Services for Blocks of Buildings. In *Smart Grid Inspired Future Technologies*; Lecture Notes of the Institute for Computer Sciences, Social Informatics and Telecommunications Engineering; Hu, J., Leung, V.C.M., Yang, K., Zhang, Y., Gao, J.; Yang, S., Eds.; Springer International Publishing: Cham, Switzerland, 2017; pp. 118–135. [CrossRef]
36. Smart Energy Demand Coalition (SEDC). *Smart Energy Demand Coalition: A Demand Response Action Plan for Europe*; Smart Energy Demand Coalition (SEDC): Brussels, Belgium, 2017.
37. Elliott, D.L.; Holladay, C.G.; Barchet, W.R.; Foote, H.P.; Sandusky, W.F. Wind Energy Resource Atlas of the United States; 1987. Available online: rredc.nrel.gov (accessed on 29 April 2020).
38. Reno, M.; Hansen, C.; Stein, J. *Global Horizontal Irradiance Clear Sky Models: Implementation and Analysis*; Sandia National Laboratories: Albuquerque, NM, USA, 2012. [CrossRef]
39. US Departmant of Energy. *New Stream-Reach Hydropower Development*; US Departmant of Energy: Washinton, DC, USA, 2014.
40. World Bank Group. *Global Solar Atlas*; World Bank Group: Washington, DC, USA, 2019.
41. NEWA. About the New European Wind Atlas, 2019. Available online: map.neweuropeanwindatlas.eu (accessed on 4 April 2020).
42. Natural Resources Canada. *RETScreen*; Natural Resources Canada: Ottawa, Canada, 2010. Available online: www.nrcan.gc.ca (accessed on 28 April 2020)

43. Barthelmie, R.J.; Wang, H.; Doubrawa, P.; Pryor, S.C. *Best Practice For Measuring Wind Speeds and Turbulence Offshore Through In-Situ And Remote Sensing Technologies*; report; eCommons: Boston, MA, USA, 2016. [CrossRef]
44. van Wees, J.-D.; Boxem, T.; Calcagno, P.; Dezayes, C.; Lacasse, C.; Manzella, A. GEOELEC Project Deliverable 2.1: A Methodology for Resource Assessment and Application to Core Countries, 2013. Available online: www.geoelec.eu(accessed on 29 April 2020).
45. Sierra, J.P.; González-Marco, D.; Sospedra, J.; Gironella, X.; Mösso, C.; Sánchez-Arcilla, A. Wave energy resource assessment in Lanzarote (Spain). *Renew. Energy* **2013**, *55*, 480–489. [CrossRef]
46. Salvatore, F.; Lidderdale, A.; Dick, C.; Richon, J.; Porter, K.; Maisondieu, C.; Johnston, C. MARINET Project Deliverable 4.11: Report New Instrumentation and Field Measuring Technology for Tidal Currents; 2015. Available online: www.marinet2.eu (accessed on 29 April 2020).
47. Cassar, D.; Erika, M.; Evangelos, R.; Christoforos, P.; Pfeifer, A.; Groppi, D.; Krajacic, G.; Garcia, D.A. A Methodology for Energy Planning in Small Mediterranean Islands, the Case of the Gozo Region. In Proceedings of the 2019 1st International Conference on Energy Transition in the Mediterranean Area (SyNERGY MED), Cagliari, Italy, 28–30 May 2019; pp. 1–6. [CrossRef]
48. Szabó, S.; Kougias, I.; Moner-Girona, M.; Bódis, K. Sustainable Energy Portfolios for Small Island States. *Sustainability* **2015**, *7*, 12340–12358. [CrossRef]
49. Dominković, D.F.; Stark, G.; Hodge, B.M.; Pedersen, A.S. Integrated Energy Planning with a High Share of Variable Renewable Energy Sources for a Caribbean Island. *Energies* **2018**, *11*, 2193. [CrossRef]
50. Uche-Soria, M.; Rodríguez-Monroy, C. Special Regulation of Isolated Power Systems: The Canary Islands, Spain. *Sustainability* **2018**, *10*, 2572. [CrossRef]
51. Li, X.; Chalvatzis, K.J.; Stephanides, P. Innovative Energy Islands: Life-Cycle Cost-Benefit Analysis for Battery Energy Storage. *Sustainability* **2018**, *10*, 3371. [CrossRef]
52. Wu, F.; Yang, L. Simulation of the integrated energy system for isolated island. In Proceedings of the 2017 China International Electrical and Energy Conference (CIEEC), Beijing, China, 25–27 October 2017; pp. 527–531. [CrossRef]
53. Poncela, M.; Purvins, A.; Chondrogiannis, S. Pan-European Analysis on Power System Flexibility. *Energies* **2018**, *11*, 1765. [CrossRef]
54. Finck, C.; Li, R.; Zeiler, W. Optimal control of demand flexibility under real-time pricing for heating systems in buildings: A real-life demonstration. *Appl. Energy* **2020**, *263*, 114671. [CrossRef]
55. El Geneidy, R.; Howard, B. Contracted energy flexibility characteristics of communities: Analysis of a control strategy for demand response. *Appl. Energy* **2020**, *263*, 114600. [CrossRef]
56. Favre-Perrod, P. A vision of future energy networks. In Proceedings of the 2005 IEEE Power Engineering Society Inaugural Conference and Exposition in Africa, Durban, South Africa, 11–15 July 2005; pp. 13–17. [CrossRef]
57. Batić, M.; Tomašević, N.; Beccuti, G.; Demiray, T.; Vraneš, S. Combined energy hub optimisation and demand side management for buildings. *Energy Build.* **2016**, *127*, 229–241. [CrossRef]
58. Salom, J.; Marszal, A.J.; Widén, J.; Candanedo, J.; Lindberg, K.B. Analysis of load match and grid interaction indicators in net zero energy buildings with simulated and monitored data. *Appl. Energy* **2014**, *136*, 119–131. [CrossRef]
59. Barney, A.; Polatidis, H. REACT Project Deliverable 2.1: Assessment of RES potential at pilot sites. REACT: Catalonia, Spain, 2019, submitted.
60. Jelić, M.; Tomašević, N.; Pujić, D.; Biosca, L.M.; Fabres, J.; Totschnig, G. REACT Project Deliverable 2.2: RES/storage enabled infrastructure planning. REACT: Catalonia, Spain, 2019, submitted.
61. Ghanem, D.A.; Crosbie, T. REACT Project Deliverable 4.1: Criteria and framework for participant recruitment. REACT: Catalonia, Spain, 2020, submitted.
62. Canary Islands Technological Institute (ITC). *Plan de Acción Insular para la Sostenibilidad Energética para La Graciosa (2012-20)*; Canary Islands Technological Institute (ITC): Las Palmas de Gran Canaria, Spain, 2013. Available online: www.datosdelanzarote.com (accessed on 29 April 2020.).
63. Consiglio Nazionale dei Geologi. *Geotermia*; Consiglio Nazionale dei Geologi: Rome, Italy, 2019. Available online: www.cngeologi.it (accessed on 29 April 2020).
64. Luis Manuel Santana Perez. El clima de La Graciosa, 2016. Available online: www.agrolanzarote.com accessed on 28 April 2020).

65. SISIFO. *DataInput—Simulación Fotovoltaica*; SISIFO: Madrid, Spain, 2019. Available online: www.sisifo.info (accessed on 28 April 2020).
66. Comune di Carloforte—Isola Ecologica del Mediterraneo. *Piano d'azione per l'energia Sostenibile*; Comune di Carloforte—Isola Ecologica del Mediterraneo. 2012. Available online: https://mycovenant.eumayors.eu/docs/seap/3594_1339169680.pdf (accessed on 28 April 2020).
67. Global Modeling and Assimilation Office (GMAO). *Energía Geotérmica—CanariWiki*; GMAO: Greenbelt, MD, USA, 2015.
68. Ecar Energy Ltd.; ESB International and Mullany Engineering Consultancy Ltd. *All Electric Aran Islands Concept: A Design for a Wind and Ocean Powered System Prepared for the He Sustainable Energy Authority of Ireland & Department of Arts, Heritage and the Gaeltacht*; Ecar Energy Ltd., ESB International and Mullany Engineering Consultancy Ltd.: Dublin, Ireland, 2015.
69. Ministerio para la Transición Ecológica y el Reto Demográfico. La Graciosa, 2019. Available online: www.miteco.gob.es (accessed on 28 April 2020).
70. Comisión Nacional de los Mercados y la Competencia. Resolución de 11 de diciembre de 2019. BOE-A-2019-18423, 307, 139647-139668, 2019. Available online: www.boe.es (accessed on 10 June 2020).
71. Gelaro, R.; McCarty, W.; Suárez, M.J.; Todling, R.; Molod, A.; Takacs, L.; Randles, C.A.; Darmenov, A.; Bosilovich, M.G.; Reichle, R.; et al. The Modern-Era Retrospective Analysis for Research and Applications, Version 2 (MERRA-2). *J. Clim.* **2017**, *30*, 5419–5454. [CrossRef] [PubMed]
72. Pfenninger, S.; Staffell, I. Long-term patterns of European PV output using 30 years of validated hourly reanalysis and satellite data. *Energy* **2016**, *114*, 1251–1265. [CrossRef]
73. OMIE. *Electricity Market | OMIE*; OMIE: Madrid, Spain, 2020.
74. International Renewable Energy Agency (IRENA). *Renewable Power Generation Costs in 2018*; International Renewable Energy Agency: Bonn, Germany, 2019.
75. European Environmental Agency. *Achieving Energy Efficiency through Behaviour Change: What Does It Take?* European Environmental Agency: Copenhagen, Demark, 2013.

Article

Life Cycle Costing Analysis: Tools and Applications for Determining Hydrogen Production Cost for Fuel Cell Vehicle Technology

Martin Khzouz [1,2,*], Evangelos I. Gkanas [2], Jia Shao [3], Farooq Sher [4], Dmytro Beherskyi [4,5], Ahmad El-Kharouf [6] and Mansour Al Qubeissi [2,4]

1. Department of Systems Engineering, Military Technological College, Al Matar Street, Muscat 111, Oman
2. Institute for Future Transport and Cities, Coventry University, Priory Street, Coventry CV1 5FB, UK; ac1029@coventry.ac.uk (E.I.G.); ac1028@coventry.ac.uk (M.A.Q.)
3. Faculty Research Centre for Financial and Corporate Integrity, Coventry University, Priory Street, Coventry CV1 5FB, UK; ac3679@coventry.ac.uk
4. Faculty of Engineering, Environment and Computing, Coventry University, Coventry CV1 2JH, UK; ad0040@coventry.ac.uk (F.S.); Ad4509@coventry.ac.uk (D.B.)
5. Departmet of Automobiles and Transport Technologies, Zhytomyr Polytechnic State University, 10005 Zhytomyr, Ukraine
6. Centre for Hydrogen and Fuel Cell Research, School of Chemical Engineering, University of Birmingham, Birmingham B15 2TT, UK; a.el-kharouf@bham.ac.uk

* Correspondence: marcin.khzouz@gmail.com

Received: 15 March 2020; Accepted: 15 July 2020; Published: 23 July 2020

Abstract: This work investigates life cycle costing analysis as a tool to estimate the cost of hydrogen to be used as fuel for Hydrogen Fuel Cell vehicles (HFCVs). The method of life cycle costing and economic data are considered to estimate the cost of hydrogen for centralised and decentralised production processes. In the current study, two major hydrogen production methods are considered, methane reforming and water electrolysis. The costing frameworks are defined for hydrogen production, transportation and final application. The results show that hydrogen production via centralised methane reforming is financially viable for future transport applications. The ownership cost of HFCVs shows the highest cost among other costs of life cycle analysis.

Keywords: hydrogen economy; cost analysis; life cycle costing; methane reforming; water electrolysis; centralised hydrogen production

1. Introduction

The phrase 'Hydrogen Economy' is not a recent concept. John Bockris first introduced it in 1976, where hydrogen was identified as clean energy carrier. In a Hydrogen Economy, the lightest of all gases must be processed as any other market commodity. Hydrogen is to be produced, packaged, transported, stored and transferred to the end-user [1], where it can be converted to electricity by the usage of fuel cells or other conversion devices [2].

Hydrogen can be produced from conventional fossil fuels, but also from more environmentally friendly and renewable resources. The total annual world production of hydrogen was around 368 trillion m^3 [3]; 48% of hydrogen was produced from natural gas, about 30% from oil, 18% from coal and 4% via water electrolysis [4]. Eighty percent of the produced amount was mainly consumed by the chemical industry and by petrochemical refineries [5]. The remaining amount was utilised in various processes including situations that hydrogen used as energy carrier.

In addition, the global demand for hydrogen in 2010 was 43 Mtons and the forecast is to reach around 50 (or more) Mtons by 2025, majorly affected from the demand for ammonia production, methanol and petroleum refinery operations [6,7]. The hydrogen consumption (in million Tons) is shown in Figure 1. Asia and Pacific are the world's leaders in consuming hydrogen (almost 1/3 of the global consumption), followed by North America and Western Europe [6,7]. The hydrogen as an alternative fuel can be used as potential energy carrier at the transportation sector for Fuel Cell Electric Vehicles (FCEVs) [8].

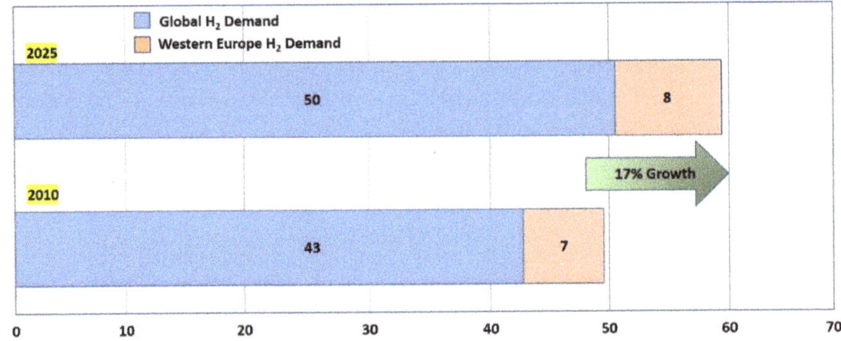

Figure 1. Hydrogen consumption (in millions of Tons) for the year 2010 (below) and the forecast for the year 2025 (up).

The hydrogen production via methane steam reforming and water electrolysis can take place in centralised or decentralised facilities [9]. For the case of centralised hydrogen production, hydrogen is distributed to the area of application via tank trailers in liquefied or gaseous form [10]. For the case of decentralised production, hydrogen is produced and stored in the location of usage, by normally utilising hydrogen fuelling stations [11].

The viability of the hydrogen technology as alternative fuel for transportation applications depends on several factors; the current and future cost of hydrogen, the technological advantages that employs hydrogen as fuel when utilising fuel cells, the long-term restrictions on greenhouse gases emission and the cost of competitive technologies, such as batteries and super capacitors [12]. Therefore, the cost analysis for hydrogen production is a very important an crucial aspect to identify the economic feasibility of using hydrogen in the transportation sector, regardless the technological obstacles at the current time, such as the hydrogen storage capacity and the specifications to meet the future high demand for transportation [13,14]. The introduction and implementation of using the life cycle cost analysis is one method that can be deployed in order to identify and decide the feasibility of using hydrogen as alternative fuel [15].

Accurate evaluation techniques for decision-making are required for economic, social, and environmental aspects. Various models have been developed such as [16]; the E3-database model in Germany and France, the H2A model developed by U.S Department of Energy Hydrogen, G4-ECONS methodology to estimate the levelled unit energy cost, and, finally, the HEEP model, which applies analysis and feasibility studies related to hydrogen production using nuclear energy [16–18]. The models mentioned above can be classified according to the tools and methods deployed for the cost estimation. The classification can be performed based on the following criteria; life cycle energy analysis models including energy flows and environmental assessment criteria [19,20], infrastructure development models and future benefit [21,22], social life cycle infrastructure and vehicle market models [23–25], and finally, energy economy models including hydrogen production and environmental assessment [26]. Table 1 summarises the most recent studies and techniques for hydrogen economy evaluation for hydrogen mobility applications. The current article focuses on the study of the cost analysis for hydrogen production for fuel cell vehicles applications. The costing analysis is applied for four hydrogen production routes. The analysis includes a framework and sensitivity analysis to compare costing results.

Table 1. Recent studies of hydrogen economy for mobility applications.

Scope of Study	Framework of Study	Area of Investigation	Ref
Social cost-benefit analysis framework for fuel cell vehicle versus internal combustion engine vehicle.	Cost benefit analysis framework was introduced; considering economic comparison, external costs estimation and social–economic comparison.	Cost Benefit Analysis of German Market based on previous published study for fuel cell electric vehicle, including externality costs in Europe for society benefits analysis.	[27]
Techno-economical characterization for Alkaline water electrolysis.	Alkaline water electrolysis life cycle was studied focusing on the metrics' approach and the approaches to specify realistic projections of sensitive technical and economic parameters, such as investment cost or future electricity cost.	The weighted average cost of capital for alkaline water electrolysis was used for analysis for cost estimation and financial analysis on three different production sites.	[28]
Techno-economic modelling of future hydrogen supply chains with spatial resolution.	Study evaluates all parts of the supply chain, from hydrogen production to refilling, the case of Germany for the target year 2050, considering a spatial resolution regarding costs, primary energy demand and CO_2 emissions. It also optimizes potential for hydrogen distribution.	Simulation approach of each step of the supply chain using optimization method.	[29]
Techno-economic model of future hydrogen supply chains.	Investigates the application area of different hydrogen supply chain architectures through a point-to-point analysis based on the methodology of the lowest-cost hydrogen delivery mode investigated by [30], full supply chain from hydrogen production by electrolysis, large-scale storage, transportation and fuelling.	Well-to-tank analysis to estimate greenhouse gas emissions for conditioning a transportation fuel	[31]
Cost of hydrogen applications are compared with conventional energy supply.	Five scenarios have been developed to compare the cost of using hydrogen with conventional energy sources, taking into account the cost of CO_2 emissions.	Methodology of life cycle cost is employed to conduct the cost of hydrogen production and application for islands and specific applications.	[32]
Economic model is developed to evaluate the investment and operational cost.	Build an economic evaluation model that describes the investment cost and operational cost, making the mathematical optimization to reach a minimum total annual cost of Hybrid Battery/Hydrogen Storage.	Optimizing the life cycle capital was formulated to determine the optimal configuration of a hybrid renewable energy.	[33]
Techno-economic analyses and life cycle assessments of four hydrogen production technologies using natural gas as a feedstock.	Understanding of the techno-economic and life cycle environmental performance of a set of emerging hydrogen production technologies.	Technical and financial conditions under which each technology is expected to be attractive are explored for carbon price.	[34]
Well-To-Wheel (WTW) analysis for hydrogen.	Global emission model for integrated systems (GEMIS) interacted with greenhouse gases, regulated emissions, and energy use in transportation (GREET) for Portugal.	Greenhouse gases emissions (GHGs) compared with the gasoline vehicle from different hydrogen production routes.	[35]
Integrate multi-objective optimization with principal component analysis to address the environmentally conscious design of hydrogen networks.	A Framework has been proposed for optimizing hydrogen supply chains in accordance with several environmental indicators.	Multi-objective mixed-integer linear program (MILP) is formulated that takes into account the simultaneous minimization of the most significant impacts of life cycle assessment (LCA) for Spain.	[36]
Life cycle including the effect on environment.	Demonstrate the costs of every step and to discuss their relationship for coal hydrogen production.	The minimum cost of each production step was analysed and focused on strategic selection for China.	[37]

Table 1. *Cont.*

Scope of Study	Framework of Study	Area of Investigation	Ref
Techno-economic analyses and life cycle assessment (LCA).	Two main gasification processes for producing hydrogen from biomass showing minimum hydrogen selling price.	Evaluate and compare the impact of gasification technology on the techno-economic and life cycle environmental performance of hydrogen production from biomass.	[38]
Life cycle costing analysis framework for hydrogen production and utilization.	Apply the method of life cycle cost for hydrogen production and utilization via analysis the feasible economic tools for forecasting hydrogen cost by developing an economic model framework including sensitivity analysis criteria.	Engineering economic tools that can be applied in general/universal form for estimating feasible hydrogen production systems, considering major cost breakdown structure to simplified life cycle model using Microsoft Excel as the tool.	Current study

Analytical tools must be standardised in order to develop a decision-making tool for hydrogen end-users and policy makers. In the current study, the life cycle costing model is proposed and introduced specifically to investigate an in-depth analysis in the Hydrogen Economy. It is a systematic analytical process for the evaluation of various alternatives with the objective of choosing the most suitable alternative. The main objective of this work is to apply the concept of life cycle cost analysis and explore the feasibility of various hydrogen production systems and techniques utilising this methodology. A life cycle costing concept in the energy field by defining a possible system boundary for various hydrogen sources is investigated and implemented. The proper life cycle costing framework for hydrogen production is proposed and used to determine the most cost effective and economically feasible hydrogen source as alternative fuel. The analysis focuses in small to medium-scale hydrogen production for FCVs. The proposed life cycle model also investigates the impact of changing several technological parameters on the hydrogen cost through a sensitivity analysis.

In the present work, essential economic evaluations have been identified in order to estimate the hydrogen cost based on life cycle methodology. Using the life cycle analysis principles, where the feasible and simplified costing framework structure is developed, a simple procedure and a general way to estimate the hydrogen cost production, transportation and utilization is introduced. The costing breakdown structure has been identified according to the boundaries of the hydrogen system and it is linked to the developed engineering economic model to simulate the feasible hydrogen cost using Microsoft Excel as a simulation tool. The estimated cost is then applied, using sensitivity analysis for a fair and feasible alternative selection in a simplified manner. This estimation is conducted without influence of the environmental assessment aspects or advanced energy selection modelling software. In such a way, the current approach can offer a more general and universal form to estimate the hydrogen production cost and it is not limited by the regional factor.

2. Methodology

2.1. The Life Cycle Costing Model

The concept of life cycle cost includes the total cost of the product from the early stages (development and manufacturing), mid stages (storage and transport) to the final stage where the product reaches the end-user. The life cycle costing is a management cost method which can be used for all sorts of products. However, the nature and objective of the analysis depends on the product itself.

For the needs of the current analysis, the Life Cycle Costing (LCC) model studies the cost-effective activities during hydrogen production and distribution. The feasible system of hydrogen technology is used to develop the LCC model, which defines a common hydrogen cost breakdown structure. The life cycle model is defined to estimate the hydrogen cost based on several hydrogen resources. The model defines various cost categories involved in hydrogen technology. Figure 2 presents the proposed life cycle costing model structure and strategy for hydrogen fuel costing analysis. The framework includes sensitivity analysis of feedstock price, vehicle cost, change on demand and capacity of hydrogen production. Both technical and economical parameters are included during the life cycle costs analysis.

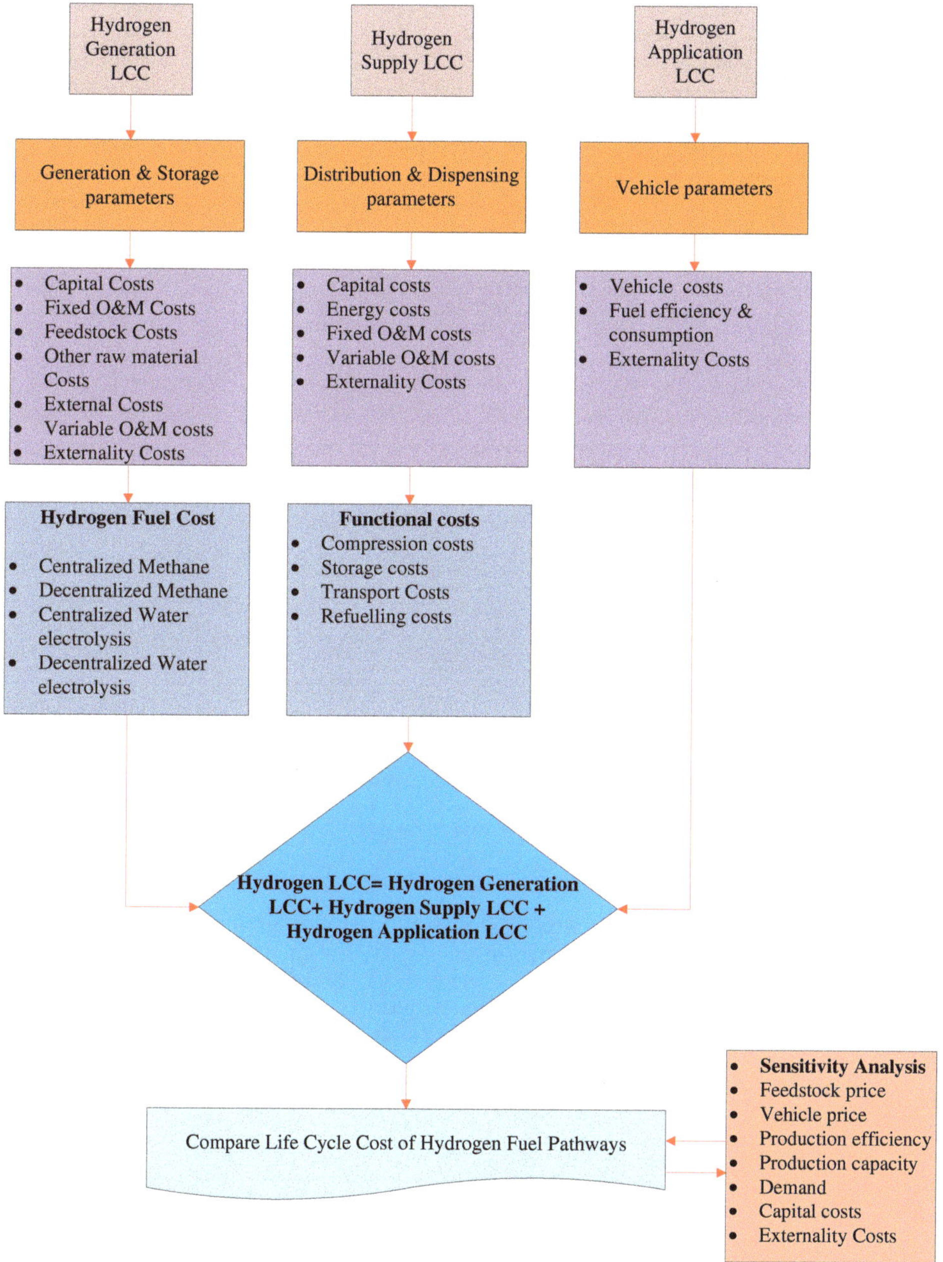

Figure 2. Proposed life cycle costing model structure and strategy for hydrogen fuel costing analysis.

Figure 3 presents the cost categories taken into account in the current work, in terms of hydrogen production, hydrogen distribution and usage. The capital costs consist of construction, preparation and cost for equipment. The running costs include: raw and other materials, primary energy usage, utilities, labour and other variable operating costs. The disposal costs consist of wastewater and CO_2 treatment. Finally, other costs take into account any costs not included in the previously mentioned cost categories that can have potential effects on the analysis. The technical data that are used to perform the life cycle analysis are presented in Table 2. The basic requirements to estimate the life cycle costing is to generate accurate cost data. Hydrogen production depends on: process efficiency, capacity and availability factor and hydrogen storage methods onsite. Hydrogen supply includes mode of transportation, dispensing components and supply capacity. The hydrogen utilization cost depends on the vehicle type and system. Several cost estimations techniques are used, such as the bench marking technique, the parametric approach, and estimating costs from first principles [39–41].

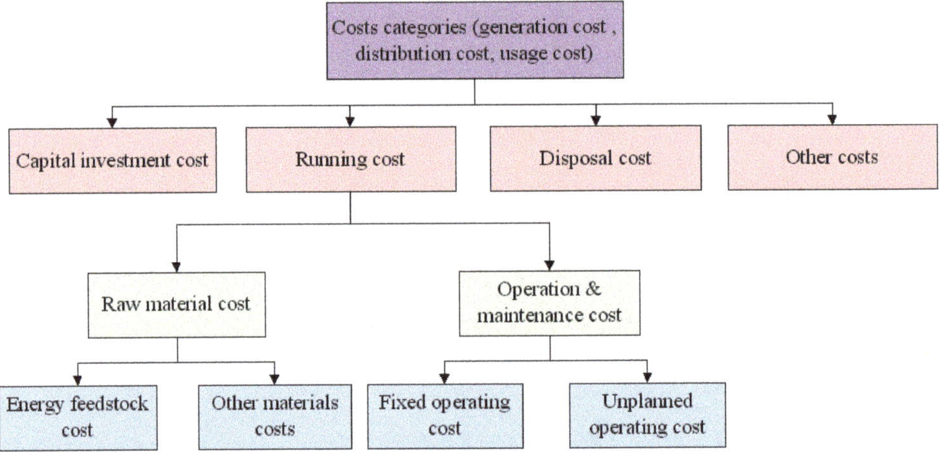

Figure 3. The major cost categories for life cycle costing analysis.

Table 2. Technical data and parameters for life cycle analysis for economic data identifications.

Hydrogen Production	Hydrogen Supply	Hydrogen Use
• Hydrogen production plant • Process efficiency • Capacity and availability factor • Annual hydrogen production rate • Number of generation units • Maximum amount of utilities (raw material) required for process • Other non-direct raw material amounts required • Hydrogen storage methods	• Mode of transportation • Transportation distance • Type of transportation vehicles • Transportation capacity • Vehicle driving distance • Dispensing components (storage and compression)	• Vehicle capacity • Vehicle type· Vehicle driving distance • Average speed of vehicle

2.2. Economic Analysis

For the needs of the current study, the economic comparison between alternatives is the main objective of the life cycle cost analysis. The equations applied in this study are listed in Table 3. The operation period (lifetime) is considered as 40 years for the centralised hydrogen production and as

20 years for the decentralised production. The data can be validated regarding analytical model outputs using the cause-effect relationship, data treatment and comparison with similar production process.

Table 3. Economic analysis equations used for life cycle analysis.

Equation Number	Equation	Abbreviation
(1)	$NPV = \sum_{t=0}^{t=N} \frac{PV}{(1+i^*)^t}$	NPV : *Net Present Value* PV : *Present Value* N : *Study period* i^* : *After Tax Nominal IRR*
(2)	$\% \, C_j = \frac{NPV \, C_j}{Total \; NPV}$	C_j: *is j cost value*
(3	$Hydrogen \; cost \; contribution = Hydrogen \; LCC \times \% \, C_j$	C_j: *is j cost value*
(4)	$i^* = ((1+i) \times (1+f)) - 1$	i^* : *After Tax Nominal IRR* i : *After tax Real IRR* f : *Inflation Rate*
(5)	$IIF = (1+f)^{(AY-SY)}$ $IF = (1+f)^{(SY-RY)}$	AY : *Actual Year* SY : *Start up Year* RY : *Refernce Year* IIF : *Inflation Increase Factor* IF : *Inflation Factor* f : *Inflation Rate*
(6)	$DCC = DDCC + IDCC$ $CI = DCC + NDCC$ $Inflated \; DCC = -DCC \times IF \times IIF \times \% \, CI \; at \; start \; up$ $Inflated \; NDCC = -NDCC \times IF \times IIF$	DCC : *Depreciable Capital Cost* $DDCC$: *Direct Deprciable Capital Costs* $IDCC$: *Indirect Deprciable Capital Costs* CI : *Capital Investment* $NDCC$: *Non Depreciable Capital Costs* IIF : *Inflation Increase Factor* IF : *Inflation Factor*
(7)	$Inflated \; other \; NDCC = -NDCC \times IF \times IIF$	$NDCC$: *Non Depreciable Capital Costs* IIF : *Inflation Increase Factor* IF : *Inflation Factor*
(8)	$Inflated \; Replacement \; Costs =$ $-Replacement \; Costs \times IF \times IFF$ $Inflated \; FC = -FC \times IF \times IIF$	FC : *Fixed Cost* IIF : *Inflation Increase Factor* IF : *Inflation Factor*
(9)	$Feed \; Cost =$ $-Inflated \; Feedstock \; Cost \times Annual \; H_2 \; Produced \times IIF$ $MC = -inflated \; MC \times IF \times IIF$	MC : *Material Costs* IIF : *Inflation Increase Factor* IF : *Inflation Factor*
(10)	$Other \; VOC =$ $-IF \times Other \; Feedstock \times Actual \; H_2 \; Produced \times IIF$	VOC : *Variable Operating Costs* IF : *Inflation Factor* IIF : *Inflation Increase Factor*
(11)	$WC = \% \, WC \times (FC + Feed \; Costs + MC + VOC)_t -$ $WC(FC + Feed \; Costs + MC + VOC)_{t-1}$	WC : *Working Capital* VOC : *Variable Operating Costs* FC : *Fixed Cost*
(12)	$SV = \%TCI \times IIF$	SV : *Salvage Value* TCI : *Inflated Total Capital Investment at start up year* IIF : *Inflation Increase Factor*
(13)	$DC = Infalted \; DCC \times IIF$	DC : *Decommissioning Costs* DCC : *Depreciable Capital Cost* IIF : *Inflation Increase Factor*
(14)	$R = H_2 \; Nominal \; LCC \times IIF \times Annual \; H_2 \; Produced$	R : *Revenue* IIF : *Inflation Increase Factor*
(15)	$D_t = \frac{DCC-SV}{n}$ $B_t = DCC - t\left(\frac{DCC-SV}{n}\right)$	DCC : *Depreciable Capital Cost* SV : *Salave Value* B_t : *Book value at the year t* D_t : *Depreciation charge during year t* n : *estimated life of the asset*
(16)	$TI = Pre \; Depreciation \; Income + D_t$	TI : *Taxable Income*

Table 3. *Cont.*

Equation Number	Equation	Abbreviation
(17)	$Total\ Taxes = Tax\ Credit - (TI \times Tax\ Rate)$ $Pre\ Depreciation\ Income =$ $R + SV + FC + DC + Feed\ Costs + MC + Other\ VOC$	R : *Revenue* SV : *Salave Value* FC : *Fixed Cost* DC : *Decommissioning Costs* MC : *Material Costs* VOC : *Variable Operating Costs*
(18)	$After\ Tax\ Income =$ $Pre\ depreciation\ Income + Total\ Taxes$	
(19)	$CFBT = DCC + Replacement\ Cost + WC + NDCC +$ $Pre\ Depreciation\ Income$ $CFAT = DCC + Replacement\ Cost + WC + NDCC +$ $Pre\ Depreciation\ Income + Total\ Taxes$	$CFBT$: *Cash Flow Before Tax* $CFAT$: *Cash Flow After Tax* DCC : *Depreciable Capital Cost* WC : *Working Capital* $NDCC$: *Non Depreciable Capital Costs*
(20)	$Actual\ H2\ Produced =$ $Plant\ Design\ Capacity \times Capacity\ Factor$	
(21)	$Annual\ H2\ Produced = Actual\ H2\ Produced \times 365$	

3. Case Study

Hydrogen is currently produced from various resources via steam reforming process and water electrolysis [42,43]. The proposed model will be based on hydrogen that is produced from natural gas steam reforming and water electrolysis (Tables A1–A6) Hydrogen can be produced by following two paths: large-scale centralised production plants (centralised generation) or small-scale distributed production plants (decentralised generation). The analysis for the produced hydrogen at centralised form includes the stage of the production pathway, starting from the preparation of feedstock (raw materials). The central production equipment, distribution preparation equipment and the necessary storage equipment have to be considered. The stage of the distribution pathway starts from the gate of the centralised plant and ends at the gate of hydrogen refuelling station, including the hydrogen transmission equipment. The dispensing pathway stage includes all the processes and equipment within the refuelling station, such as hydrogen compression and hydrogen storage processes. The analysis for hydrogen produced at decentralised form includes the production pathway stage, including the preparation of raw materials and onsite raw material conversion to hydrogen. The dispensing pathway stage includes the processes within the refuelling station, such as hydrogen compressing and hydrogen storage.

3.1. Natural Gas Steam Reforming

Hydrogen production via methane steam reforming can be achieved in both centralised and decentralised facilities as illustrated in Figure 4. In the case of centralised production, hydrogen should be distributed to the area of the application via tank trailers in liquefied or gaseous form. During the decentralised production, hydrogen is produced and stored in the location of usage.

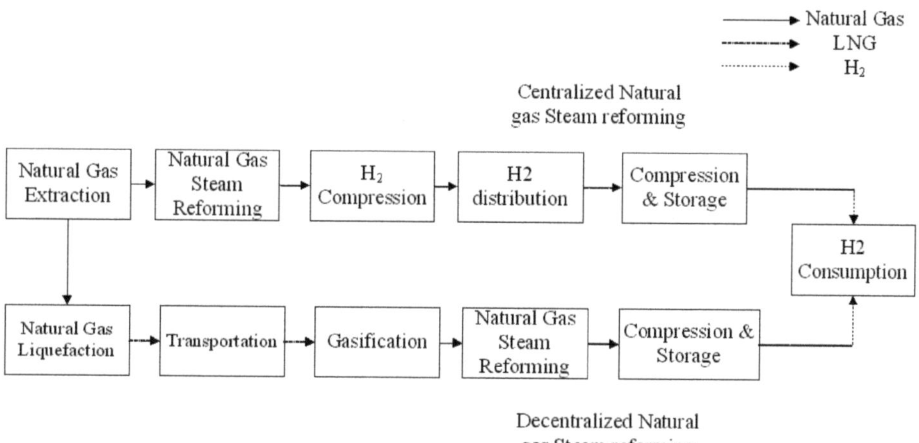

Figure 4. Hydrogen production process via natural gas steam reforming, centralized and decentralized forms.

3.2. Water Electrolysis

The conversion of pre-treated water to hydrogen and oxygen is known as water electrolysis. For the needs of the current study, decentralised hydrogen production through electrolysis is considered as small-medium scale hydrogen refuelling stations are available on the market representing the decentralised form of hydrogen production for fuel cell vehicles. This process is represented in Figure 5. The centralised process is a large-scale hydrogen production operation that produces hydrogen on-site and requires hydrogen transportation and distribution.

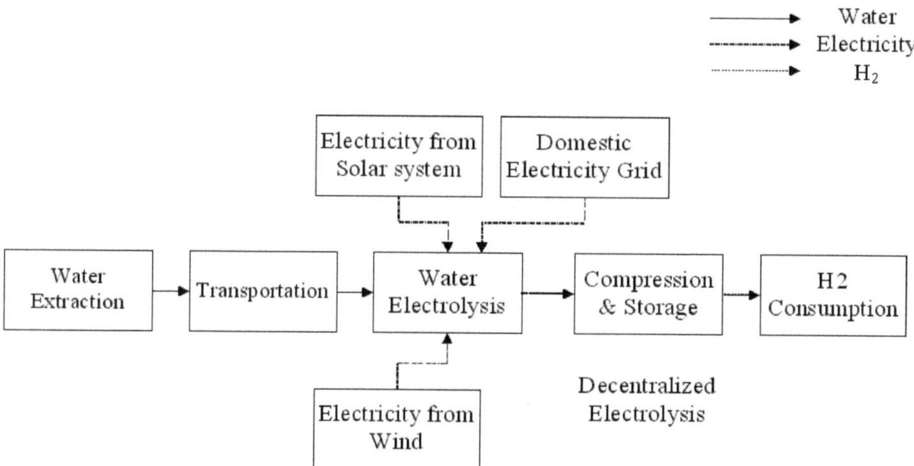

Figure 5. Hydrogen production process via water electrolysis, representing decentralized form of hydrogen refuelling station.

4. Results and Discussion

4.1. Hydrogen Production and Storage Life Cycle Costs

The outcome of the life cycle model presents a minimum rate of return of investment. Table 4 shows that centralized methane reforming achieved the lowest hydrogen costs through the life cycle span (0.90 USD/kg). The most expensive process on the life cycle analysis for hydrogen production and storage was found to be the decentralized electrolysis with a value of 4.30 USD/kg. The major cost parameters contributing to the life cycle results are: the feed cost, the cost for raw materials and the capital costs. Figure 6 presents the contribution of the cost parameters individually to the hydrogen cost for each production method analysed. It can be extracted that the feed cost for the centralised methane reforming, the centralised electrolysis and the decentralised electrolysis has the lions share in the total cost of hydrogen production. For the case of decentralised methane hydrogen production, the capital costs, the fixed operating costs, the feed cost and the raw material cost are almost equally contributing to the final cost of hydrogen. Finally, for the decentralised electrolysis, besides the contribution of the feed cost, the capital cost and the raw material cost also affect the hydrogen cost.

Table 4. Life cycle costs of hydrogen production and storage, minimum rate of return of investment.

Hydrogen Alternative	Life Cycle Cost of Generation and Storage (USD/kg)
Centralized methane reforming	0.90
Decentralized methane reforming	3.83
Centralized electrolysis	2.92
Decentralized electrolysis	4.30

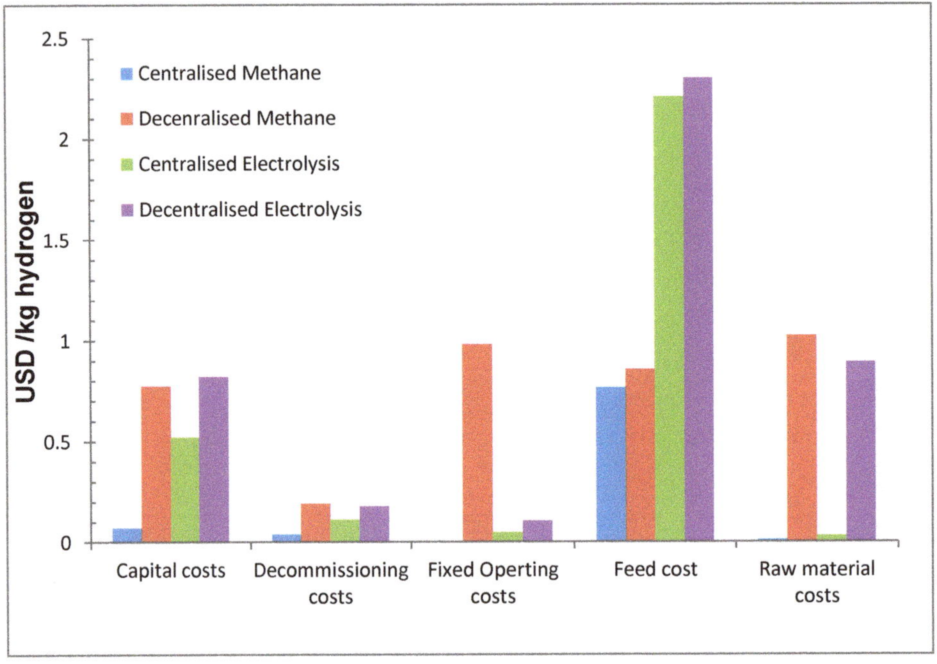

Figure 6. Hydrogen cost contribution for each hydrogen production life cycle.

4.2. Life Cycle Cost for Hydrogen Transportation and Dispensing

Hydrogen is produced in centralised forms and usually transported to the application area immediately. The life cycle model for the hydrogen transportation and dispensing applied for both the cases of centralised methane reforming and centralised electrolysis showed that the case of centralised methane reforming had lower minimum rate of return of investment compared to the case of centralised electrolysis production as presented in Table 5. The cost for the hydrogen transportation and dispensing depends on the capacity and demand of the produced hydrogen. The hydrogen cost contribution for the transportation and dispensing for the centralised methane and centralised electrolysis production is presented in Figure 7. The major cost contributor in the hydrogen transportation model is the cost of the fuel required for the transportation, where for both the examined cases the contribution is equivalent. For the case of the centralised electrolysis, the capital costs and the raw material cost are also contributing towards the final cost. The life cycle cost resulted from electrolysis resulted in the highest cost as the transportation of hydrogen produced from the electrolysis method depends on the size and capacity of the centralized electrolysis plant, which is normally smaller in production capacity compared to the centralised methane steam reforming. In addition, the dispensing cost of high-pressure hydrogen gas for the methane reforming production contributed towards lowering the cost of energy required for dispensing process compared to the case of hydrogen production via centralised electrolysis. Thus, the compression and dispensing cost for high pressure and large hydrogen production capacity is economically more viable compared to a low pressure/or low hydrogen production capacity.

Table 5. Life Cycle Costs of Hydrogen Transportation.

Hydrogen Alternative	Life Cycle Cost of Transportation and (USD/kg)
Centralized methane reforming	0.41
Centralized electrolysis	0.92

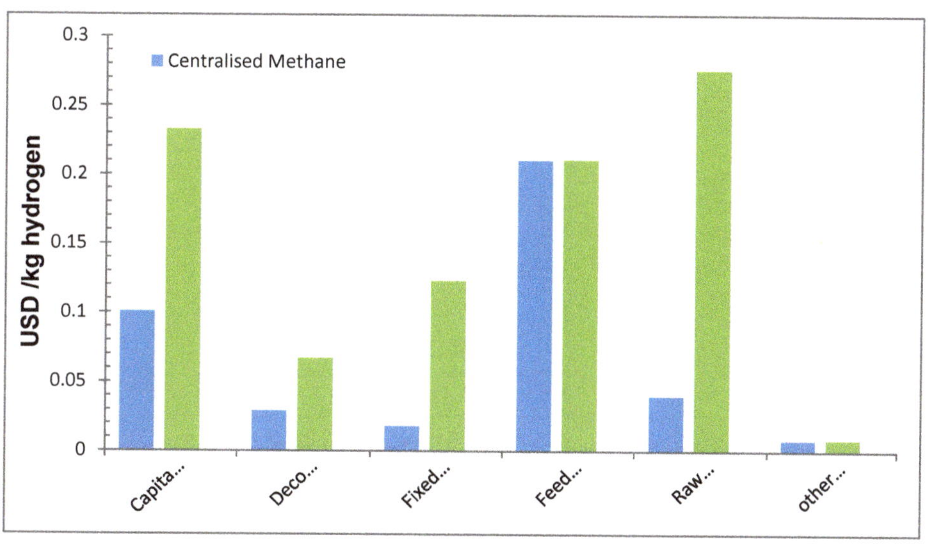

Figure 7. Hydrogen cost contribution for hydrogen transportation and dispensing life cycle.

4.3. Hydrogen Application Life Cycle Costs

The produced hydrogen can be used as fuel to feed Fuel Cell Vehicles (FCVs). The cost of hydrogen from the previous life cycle analysis is used to identify and evaluate the total entire usage cost of hydrogen in FCVs during the life span. The investment cost of FCVs is the main cost contributor for hydrogen life cycle applications. The capital investment showed 77% of the total life cycle cost of the applications, 19% was for hydrogen as fuel cost and 4% for fuel cell vehicle maintenance.

4.4. Sensitivity Analysis

The uncertainty of data cannot be eliminated. Uncertainty refers to the costs at which the probability of occurrence is unknown. Sensitivity analysis is the most used technique to deal with uncertainty. The approach is to find and identify the critical assumptions that can affect the cash flow analysis. The purpose of this analysis is to study high costs data items that may affect the future cost. The 10–20% of changing the cost will identify 60–80% of the total cost. A sensitivity analysis was applied for the hydrogen production process and it was majorly focused on the capacity factor of production, the feedstock cost and the capital cost changes. For hydrogen mobility applications, the contribution of the capital cost was compared. For the analysis of the hydrogen transportation and dispensing, there was a drawback regarding the availability of data for the simple case introduced; thus, further investigation is required for future forecasting analysis.

In general, hydrogen production cost was found to be affected from the capacity factor as shown in Figure 8a. For the case of centralised methane reforming was the effect of the capacity factor is almost negligible, as the designed production plant is compatible for high demand requirements. For the cases of centralised/decentralised electrolysis and the decentralised methane reforming, the shape of the hydrogen nominal cost when the capacity factor increases is almost hyperbolic and tends to reach the minimum cost at the maximum capacity factor.

The effect of increasing the feedstock costs showed that hydrogen production via electrolysis was very sensitive compared to the methane source, as the slope for both the centralised and decentralised cases was found to be larger compared the methane steam reforming cases, as presented in Figure 8b. The cost of electricity used for electrolysis is dependent on the grid supply, which is directly connected to the fossil fuel cost. It was difficult to predict the electricity generation cost from renewable sources, and the present model assumed the contribution of fossil fuel-based electricity sources only. In addition, the water price is increasing, which adds further higher cost into the vehicle cost price electrolysis hydrogen production route.

For the case of hydrogen fuel cell vehicle usage, the cost of the vehicle is the main issue for the current technology. Figure 8c presents the effect of the vehicle cost reduction on the present value of hydrogen application. The fuel cell vehicle cost should be reduced. In the current study, the cost of the vehicle is reduced up to 60%. The reflection of this into total life cycle cost was 57% for capital cost and 35% for hydrogen fuel cost. This indicates that even with a high reduction in the cost of FCVs, the total cost of using such technology today will remain relatively high. For the entire life span, the fuel cost is a good option if it is compared with internal combustion engine cars.

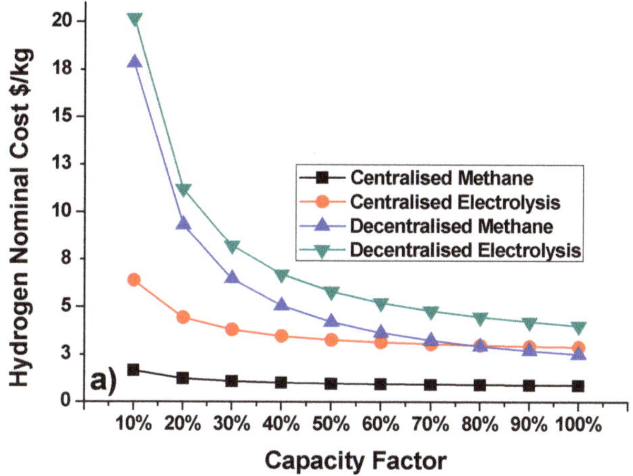

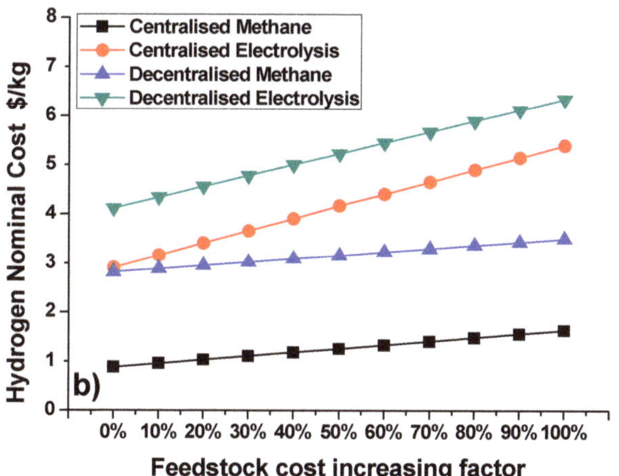

Figure 8. *Cont.*

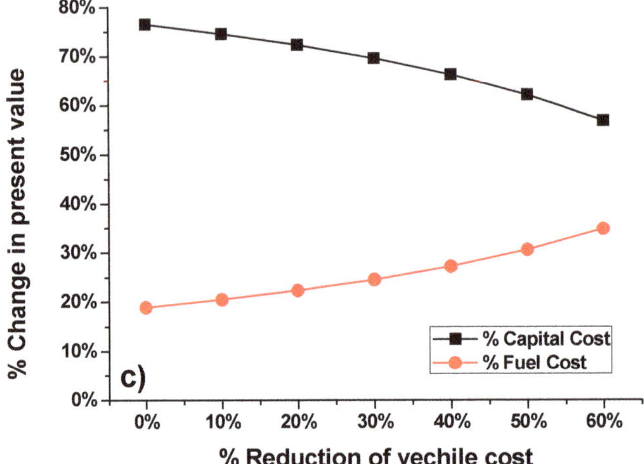

Figure 8. Effect of increasing the capacity factor for hydrogen production to the hydrogen production life cycle cost (**a**), effect of increasing the feedstock cost to the hydrogen production life cycle cost (**b**) and effect of the vehicle cost reduction on the present value of hydrogen application (**c**).

5. Conclusions

The cost estimation of hydrogen technology is essential for the acceptance of a future Hydrogen Economy, especially in the transportation sector. The main objective of this study was the definition and adoption of the life cycle costing method regarding hydrogen production for hydrogen utilization in fuel cell vehicles. The simulation results of the hydrogen production and storage showed that the hydrogen production via centralised methane steam reforming is the most economically feasible alternative amongst the rest production routes at current study. Further investigation on the hydrogen transportation and dispensing model has been performed and the outcomes showed that the centralised production via methane reforming is still the most prominent alternative compared to the other decentralized production methods. The FCV cost is a drawback for adapting this technology in the near future, due to the high cost of vehicle. Sensitivity analysis investigated the effect of changes of capacity factor and feedstock cost in hydrogen price where the effect of changes was obvious for hydrogen production via electrolysis. The challenges for hydrogen costing analysis—such as changes in technology, changes in renewable energy acceptance, and changes in material costs—can be added into the costing framework to increase the forecasting reliability of hydrogen. The framework costing structure for hydrogen production and data analysis suggested at current work can be used for stationary applications.

Author Contributions: M.K.: Conceptualization, Investigation, Methodology, Writing–original draft; E.I.G.: Data curation, validation; review & editing J.S.: Visualization; F.S.: Data curation; D.B.: Resources; A.E.-K.: Writing–review and editing; M.A.Q.: Writing–review & editing, Funding acquisition. All authors have read and agreed to the published version of the manuscript.

Funding: This research was funded by the EU Commission KA107 project, grant number: 2018-1-UK01-KA107-047386.

Conflicts of Interest: The authors declare no conflict of interest.

Appendix A

Table A1. Hydrogen production and storage input data.

Data	Centralised Methane Reforming	Decentralised Methane Reforming	Centralised Electrolysis	Decentralised Electrolysis	Units
Number of Hydrogen Units (Assumed)	1	2	1	1	
Plant Design Capacity (Typical available plant)	380,000	1500	52,300	1500	kg/day
Capacity Factor (assumed)	90%	85%	95%	95%	
Efficiency of the Process [44]	80%	75%	75%	70%	
Hydrogen Storage Pressure (Typical available storage system)	70	35	70	35	MPa
Hydrogen Storage Capacity [45]	98,589	49,294	49,294	49,294	kg
Hydrogen Compressor Power [45]	74,472	64,223	74,472	64223	kWe
Plant Capital Cost, corrected to year 2018 based on reference [44]	52,673,000	640,000	29,234,000	840,000	USD
Indirect Depreciable Costs (calculated)	7,374,220	70,400	2,923,400	92,400	USD
Non Depreciable Costs (calculated)	50,000	25,000	50,000	25,000	USD
Installation Costs [46], (Forecasted to 2018)	21,069,200	64,000	5,846,800	84,000	USD
Feedstock Usage Calculated (Lower Heating Value of Hydrogen % Lower Heating Value of Feedstock) % Conversion Efficiency Price of electricity (0.05370 USD/kWh)	4.1 ($Nm^3/kg\ H_2$)	4.4 ($Nm^3/kg\ H_2$)	44.5 ($kWh/kg\ H_2$)	47.7 ($kWh/kg\ H_2$)	
Raw materials Usage Water consumed for process production,	12.5	16.3	11.0	11.0	$l/kg\ H_2$
Labour Costs (assumed for typical industry)	25,000	10,000	25,000	10,000	USD

Table A2. Cash flow input data and duration period.

Timeline of Alternative	Centralized	Decentralized
Cash flow year	2018	2018
Start of Construction	2025	2025
Start of Operation	2027	2026
End of Operation	2066	2045
Study Period	40	20
Planned Replacement Period	10	10
Construction Period	2	1
Plant Operation	40	20

Table A3. Economic data for performing cash flow study.

Economic Data	Centralized Methane	Centralized Electrolysis	Decentralized Methane	Decentralized Electrolysis
After-Tax Real IRR	10.0%	10.0%	10.0%	10.0%
Inflation Rate	3.9%	3.9%	3.9%	3.9%
Depreciation Length	40	40	20	20
Tax Rate	15%	15%	15%	15%
Working Capital (% Operating Cost)	15%	15%	15%	15%
Salvage Value (% Total Capital Investment)	10%	10%	10%	10%
% of Capital Investment During Construction year 1	40%	25%	100%	100%
% of Capital Investment During Construction year 2	60%	75%	0%	0%
% Fixed Operating Costs at Start up	100%	100%	100%	100%
% Revenue at Start up	50%	50%	100%	100%
% Variable operating Costs at start up	75%	75%	50%	50%
Decommissioning Costs	10%	10%	10%	10%

Table A4. Hydrogen Transportation Main Data *.

Data	Value	Unit
Capacity of vehicle	920	kg
Transport distance	300	km
Average speed of vehicle	80	km/h
Vehicle average consumption	0.094	L/km
Loading time	4	h

Based on available compressed hydrogen transport.

Table A5. Hydrogen Dispensing Main Data [45].

Hydrogen Dispensing	Value	Unit
Hydrogen Dispensing Pressure	40	Mpa
Hydrogen Dispensing Capacity	73,941	kg
Hydrogen Dispensing Compressor Power	66,145	kWe

Table A6. Fuel Cell Vehicle Data *.

FCV Data	Value
Fuel Tank	USD 975
Electric Motor	USD 1560
Inverter	USD 250
Battery	USD 3000
Fuel cell system	USD 15,985
Vehicle body	USD 2600
Other BOP materials Costs	USD 14,700
Average distance travelled	25,000 km/year
Mileage of FCV	300 km
Consumption for distance travelled	3 kg H_2

Based on available technical data and forecasted data for FCV.

References

1. Moliner, R.; Lázaro, M.J.; Suelves, I. Analysis of the strategies for bridging the gap towards the Hydrogen Economy. *Int. J. Hydrogn Energy* **2016**, *41*, 19500–19508. [CrossRef]
2. Milner-Elkharouf, L.; Khzouz, M.; Steinberger-Wilckens, R. Catalyst development for indirect internal reforming (IIR) of methane by partial oxidation. *Int. J. Hydrogn Energy* **2020**, *45*, 5285–5296. [CrossRef]
3. Lemus, R.G.; Martínez Duart, J.M. Updated hydrogen production costs and parities for conventional and renewable technologies. *Int. J. Hydrogn Energy* **2010**, *35*, 3929–3936. [CrossRef]
4. Balat, M. Potential importance of hydrogen as a future solution to environmental and transportation problems. *Int. J. Hydrogn Energy* **2008**, *33*, 4013–4029. [CrossRef]
5. *Prospects for Hydrogen and Fuel Cells*; Energy Technology Analysis; International Energy Agency: Paris, France, 2005.
6. FCH. *Hydrogen Roadmap Europe: A sustainable Pathway for the European Energy Transition*; Fuel Cells and Hydrogen: Bruxelles, Belgium, 2019.
7. IRENA. *Hydrogen from Renewable Power: Technology Outlook for the Energy Transition*; International Renewable Energy Agency: Abu Dhabi, UAE, 2018.
8. Brown, D.R. Worldwide Hydrogen Production Capacity at Refineries. Available online: https://h2tools.org/node/820 (accessed on 12 July 2020).
9. Moreno-Benito, M.; Agnolucci, P.; McDowall, W.; Papageorgiou, L.G.; Kravanja, Z.; Bogataj, M. Towards a sustainable hydrogen economy: Role of carbon price for achieving GHG emission targets. In *Computer Aided Chemical Engineering*; Elsevier: Amsterdam, The Netherlands, 2016; pp. 1015–1020.
10. Anandarajah, G.; McDowall, W.; Ekins, P. Decarbonising road transport with hydrogen and electricity: Long term global technology learning scenarios. *Int. J. Hydrogn Energy* **2013**, *38*, 3419–3432. [CrossRef]
11. Southall, G.D.; Khare, A. The feasibility of distributed hydrogen production from renewable energy sources and the financial contribution from UK motorists on environmental grounds. *Sustain. Cities Soc.* **2016**, *26*, 134–149. [CrossRef]
12. Khzouz, M.; Gkanas, E.I.; Girella, A.; Statheros, T.; Milanese, C. Sustainable hydrogen production via LiH hydrolysis for unmanned air vehicle (UAV) applications. *Int. J. Hydrogn Energy* **2020**, *45*, 5384–5394. [CrossRef]
13. Forsberg, C.W. Future hydrogen markets for large-scale hydrogen production systems. *Int. J. Hydrogn Energy* **2007**, *32*, 431–439. [CrossRef]
14. Khzouz, M.; Gkanas, E.I.; Du, S.; Wood, J. Catalytic performance of Ni-Cu/Al2O3 for effective syngas production by methanol steam reforming. *Fuel* **2018**, *232*, 672–683. [CrossRef]
15. Petrillo, A.; De Felice, F.; Jannelli, E.; Minutillo, M. Chapter 5—Life Cycle Cost Analysis of Hydrogen Energy Technologies. In *Hydrogen Economy*; Scipioni, A., Manzardo, A., Ren, J., Eds.; Academic Press: Cambridge, MA, USA, 2017; pp. 121–138. [CrossRef]

16. Khamis, I.; Malshe, U.D. HEEP: A new tool for the economic evaluation of hydrogen economy. *Int. J. Hydrogn Energy* **2010**, *35*, 8398–8406. [CrossRef]
17. Thengane, S.K.; Hoadley, A.; Bhattacharya, S.; Mitra, S.; Bandyopadhyay, S. Cost-benefit analysis of different hydrogen production technologies using AHP and Fuzzy AHP. *Int. J. Hydrogn Energy* **2014**, *39*, 15293–15306. [CrossRef]
18. Ally, J.; Pryor, T. Life cycle costing of diesel, natural gas, hybrid and hydrogen fuel cell bus systems: An Australian case study. *Energy Policy* **2016**, *94*, 285–294. [CrossRef]
19. Bicer, Y.; Dincer, I. Comparative life cycle assessment of hydrogen, methanol and electric vehicles from well to wheel. *Int. J. Hydrogn Energy* **2017**, *42*, 3767–3777. [CrossRef]
20. Li, Y.; Chen, D.W.; Liu, M.; Wang, R.Z. Life cycle cost and sensitivity analysis of a hydrogen system using low-price electricity in China. *Int. J. Hydrogn Energy* **2017**, *42*, 1899–1911. [CrossRef]
21. Sun, H.; He, C.; Wang, H.; Zhang, Y.; Lv, S.; Xu, Y. Hydrogen station siting optimization based on multi-source hydrogen supply and life cycle cost. *Int. J. Hydrogn Energy* **2017**, *42*, 23952–23965. [CrossRef]
22. Moreno-Benito, M.; Agnolucci, P.; Papageorgiou, L.G. Towards a sustainable hydrogen economy: Optimisation-based framework for hydrogen infrastructure development. *Comput. Chem. Eng.* **2017**, *102*, 110–127. [CrossRef]
23. Adami Mattioda, R.; Teixeira Fernandes, P.; Luiz Casela, J.; Canciglieri Junior, O. Chapter 7—Social Life Cycle Assessment of Hydrogen Energy Technologies A2—Scipioni, Antonio. In *Hydrogen Economy*; Manzardo, A., Ren, J., Eds.; Academic Press: Cambridge, MA, USA, 2017; pp. 171–188. [CrossRef]
24. Lucas, A.; Neto, R.C.; Silva, C.A. Energy supply infrastructure LCA model for electric and hydrogen transportation systems. *Energy* **2013**, *56*, 70–80. [CrossRef]
25. Sun, Y.; Ogden, J.; Delucchi, M. Societal lifetime cost of hydrogen fuel cell vehicles. *Int. J. Hydrogn Energy* **2010**, *35*, 11932–11946. [CrossRef]
26. Petrillo, A.; Mellino, S.; Petrillo, A.; Cigolotti, V.; Autorino, C.; Jannelli, E.; Ulgiati, S. A Life Cycle Assessment of lithium battery and hydrogen-FC powered electric bicycles: Searching for cleaner solutions to urban mobility. *Int. J. Hydrogn Energy* **2017**, *42*, 1830–1840. [CrossRef]
27. Cantuarias-Villessuzanne, C.; Weinberger, B.; Roses, L.; Vignes, A.; Brignon, J.-M. Social cost-benefit analysis of hydrogen mobility in Europe. *Int. J. Hydrogn Energy* **2016**, *41*, 19304–19311. [CrossRef]
28. Kuckshinrichs, W.; Ketelaer, T.; Koj, J.C. Economic Analysis of Improved Alkaline Water Electrolysis. *Front. Energy Res.* **2017**, *5*. [CrossRef]
29. Reuß, M.; Grube, T.; Robinius, M.; Stolten, D. A hydrogen supply chain with spatial resolution: Comparative analysis of infrastructure technologies in Germany. *Appl. Energy* **2019**, *247*, 438–453. [CrossRef]
30. Yang, C.; Ogden, J. Determining the lowest-cost hydrogen delivery mode. *Int. J. Hydrogn Energy* **2007**, *32*, 268–286. [CrossRef]
31. Reuß, M.; Grube, T.; Robinius, M.; Preuster, P.; Wasserscheid, P.; Stolten, D. Seasonal storage and alternative carriers: A flexible hydrogen supply chain model. *Appl. Energy* **2017**, *200*, 290–302. [CrossRef]
32. Zhao, G.; Nielsen, E.R.; Troncoso, E.; Hyde, K.; Romeo, J.s.S.n.; Diderich, M. Life cycle cost analysis: A case study of hydrogen energy application on the Orkney Islands. *Int. J. Hydrogn Energy* **2019**, *44*, 9517–9528. [CrossRef]
33. Zhang, Y.; Hua, Q.S.; Sun, L.; Liu, Q. Life Cycle Optimization of Renewable Energy Systems Configuration with Hybrid Battery/Hydrogen Storage: A Comparative Study. *J. Energy Storage* **2020**, *30*, 101470. [CrossRef]
34. Salkuyeh, Y.K.; Saville, B.A.; MacLean, H.L. Techno-economic analysis and life cycle assessment of hydrogen production from natural gas using current and emerging technologies. *Int. J. Hydrogn Energy* **2017**, *42*, 18894–18909. [CrossRef]
35. Pereira, S.R.; Coelho, M.C. Life cycle analysis of hydrogen—A well-to-wheels analysis for Portugal. *Int. J. Hydrogn Energy* **2013**, *38*, 2029–2038. [CrossRef]
36. Sabio, N.; Kostin, A.; Guillén-Gosálbez, G.; Jiménez, L. Holistic minimization of the life cycle environmental impact of hydrogen infrastructures using multi-objective optimization and principal component analysis. *Int. J. Hydrogn Energy* **2012**, *37*, 5385–5405. [CrossRef]
37. Yao, F.; Jia, Y.; Mao, Z. The cost analysis of hydrogen life cycle in China. *Int. J. Hydrogn Energy* **2010**, *35*, 2727–2731. [CrossRef]

38. Salkuyeh, Y.K.; Saville, B.A.; MacLean, H.L. Techno-economic analysis and life cycle assessment of hydrogen production from different biomass gasification processes. *Int. J. Hydrogn Energy* **2018**, *43*, 9514–9528. [CrossRef]
39. Barringer, H.P.; Weber, D.P. Life Cycle Cost Tutorial. In Proceedings of the Fifth International Conference on Process Plant Reliability, Marriott Houston Westside Houston, TX, USA, 2–4 October 1996.
40. Fabrycky, W.J.; Blanchard, B.S. *Life Cycle Cost and Economic Analysis*; Prentice Hall: Englewood Cliffs, NJ, USA, 1991.
41. Korpi, E.; Ala-Risku, T. Life cycle costing: A review of published case studies. *Manag. Audit. J.* **2008**, *23*, 240–261. [CrossRef]
42. Khzouz, M.; Gkanas, E.I. Experimental and numerical study of low temperature methane steam reforming for hydrogen production. *Catalysts* **2018**, *8*, 5. [CrossRef]
43. Serdaroglu, G.; Khzouz, M.; Gillott, M.; Shields, A.; Walker, G.S. The effect of environmental conditions on the operation of a hydrogen refuelling station. *Int. J. Hydrogn Energy* **2015**, *40*, 17153–17162. [CrossRef]
44. Mueller-Langer, F.; Tzimas, E.; Kaltschmitt, M.; Peteves, S. Techno-economic assessment of hydrogen production processes for the hydrogen economy for the short and medium term. *Int. J. Hydrogn Energy* **2007**, *32*, 3797–3810. [CrossRef]
45. Amos, W.A. *Costs of Storing and Transporting Hydrogen*; National Renewable Energy Laboratory: Golden, CO, USA, 1998.
46. Energy, D.O. Hydrogen Program. Available online: http://www.hydrogen.energy.gov/index.html (accessed on 12 July 2020).

Article

On the Role of Regulatory Policy on the Business Case for Energy Storage in Both EU and UK Energy Systems: Barriers and Enablers

Ahmed Gailani *, Tracey Crosbie, Maher Al-Greer, Michael Short and Nashwan Dawood

School of Computing, Engineering and Digital Technologies, Teesside University, Middleborough TS1 3BX, UK; T.Crosbie@tees.ac.uk (T.C.); M.Al-Greer@tees.ac.uk (M.A.-G.); M.Short@tees.ac.uk (M.S.); N.Dawood@tees.ac.uk (N.D.)

* Correspondence: a.fakhri@tees.ac.uk; Tel.: +44-(0)-16-4221-8121 (ext. 3660)

Received: 23 January 2020; Accepted: 25 February 2020; Published: 1 March 2020

Abstract: This paper presents a SWOT analysis of the impact of recent EU regulatory changes on the business case for energy storage (ES) using the UK as a case study. ES technologies (such as batteries) are key enablers for increasing the share of renewable energy generation and hence decarbonising the electricity system. As such, recent regulatory changes seek to improve the business case for ES technologies on national networks. These changes include removing double network charging for ES, defining and classifying ES in relevant legislations, and clarifying ES ownership along with facilitating its grid access. However, most of the current regulations treat storage in a similar way to bulk generators without paying attention to the different sizes and types of ES. As a result, storage with higher capacity receives significantly higher payment in the capacity market and can be exempt from paying renewable energy promotion taxes. Despite the recent regulatory changes, ES is defined as a generation device, which is a barrier to a wide range of revenue streams from demand side services. Also, regulators avoid disrupting the current energy market structure by creating an independent asset class for ES. Instead, they are encouraging changes that co-exist with the current market and regulatory structure. Therefore, although some of the reviewed market and regulatory changes for ES in this paper are positive, it can be concluded that these changes are not likely to allow a level playing field for ES that encourage its increase on energy networks.

Keywords: energy storage; regulatory barriers; storage policy; market regulations; SWOT analysis

1. Introduction

Climate change concerns are encouraging the international community to adopt policies to decarbonise the energy system by increasing the reliance on renewable energy sources (RES) such as wind and solar [1]. EU policies, for example, aim to increase the total consumed energy provided by RES to 20% in 2020 and 50% in 2050 [2,3]. Reflecting this, the total installed wind and solar energy capacity in the EU-28 countries increased by 104.1 GW and 100.4 GW in the last 10 years respectively [4]. By 2030, it is expected that the total wind and solar installed capacity in the EU will reach 327 GW and 270 GW respectively [5]. This increased penetration of RES in the electricity system poses network balance challenges to grid operators due to the intermittent nature of many clean energy sources such as PV and Wind turbines [6–8]. Also, it leads to increased reserve capacity [9], increased electricity system costs [10] and reduced system adequacy [11]. To ensure power system stability while allowing a high share of RES, several methods have been utilised such as demand side management [12], the introduction of capacity markets [13], and smart grid initiatives [14].

Energy storage (ES) is widely recognised as a key to resolve RES's intermittency by enabling electrical energy to be stored at the off-peak times and released during peak demand periods [15].

It is expected that the total global ES capacity needs to be tripled to reach 15.72 TWh if the share of RES is to be doubled in the energy system by 2030 [16]. Many studies evaluated the technical, economic and environmental performance for ES systems to identify their suitability to various grid applications [17–20].

Amongst the different types of ES systems, battery energy storage system (BESS) is interesting because of its suitability to many applications in grid-connected electricity networks such as peak shaving [21], energy arbitrage [22], reserve capacity [23], and frequency regulation [24] amongst many others. Furthermore, battery materials are constantly improving over the years to account for degradation [25], and its capital cost is expected to decrease over the next years. Bloomberg predicts that the price of battery energy storage system (BESS) for grid applications will be $70/kWh by 2030 in line with market growth for ES represented by a $620bn investment by 2040 [26]. As such, this work is concerned with BESS in particular when analyzing market and regulatory changes, and the different types of ES are beyond the scope of this paper.

BESS has the opportunity to provide different services across the electrical network if these services are technologically and operationally compatible [27]. However, despite these benefits, large-scale deployment of BESS on EU energy networks is hindered due to regulatory and market barriers that prevent storage from stacking multiple services across different markets [28,29]. Recent research [30] analysed the business model of two different BESSs and concluded that the current legislation in Europe hinders their value proposition in energy markets. Other researchers argued that the current regulatory framework forces storage developers to choose certain business models that may not be economically feasible [31]. The argument being that a reduction in capital cost of BESS technologies alone will not lead to an increase in their applications on energy networks [16]. As such, there is a need to mitigate market and regulatory barriers to make BESS commercially viable in different markets.

Energy regulators recognise the necessity of ES in modern energy networks and are exploring ways to enhance its business case. The European Commission (EC) recognises ES as a key component to accelerate clean energy transformation and is proposing a number of regulatory changes in [32]. Some of which have been adopted by UK's energy regulator (OFGEM) including: (i) defining 'Electricity Storage' in the main legislation; (ii) removing the double network and balancing charges for storage; (iii) co-locating storage with renewable generation sites that are supported through consumption levies policies; (iv) limiting storage operation by network owners; (v) facilitating ES planning permission and (vi) employing de-rating factors for storage in the capacity market (CM) [33].

The research presented here reviews the current proposals to amend the regulations governing ES and explores how these changes impact on business models for BESS taking the UK's regulator changes as a case study. This study applies the 'SWOT' analysis to examine the strengths (S), weaknesses (W), opportunities (O), and threats (T) of the future regulatory framework of ES in the UK and, by extension, in the EU since the regulatory changes are similar. Qualitative data from the EC, the UK government, energy regulators, journal articles, and reports are utilised to examine the internal (S/W) and external (O/T) factors concerning the proposed regulatory changes. Such analysis is vital to provide ES with a clearer insight into the regulatory framework surrounding future business cases for ES.

It should be noted that although we analyse the impact of the recent market and regulatory changes on the business case of BESS in particular, energy regulators are taking a neutral approach. Thus, the recent regulatory changes designed to support an increased use of ES on energy networks are equal for all types of ES including BESS.

The remainder of the paper is structured as follows: Section 2 provides an overview for ES classification; Section 3 reviews the main market/regulatory aspects that affect the business case of BESS and the regulators' proposed solutions; Section 4 presents SWOT analysis as a method; Section 5 presents and discusses SWOT analysis results; Section 6 provides a summary of the paper's findings and discusses the 'Brexit' issue; and Section 7 provides concluding remarks.

2. Energy Storage Classification

ES devices can convert electrical energy into several forms that can be stored and converted back to electrical energy again. The main types of ES systems can be categorised based on their storage form, storage size, and discharge time along with their applications. These are illustrated in Figure 1. The main types of ES depending on the stored energy are mechanical, thermal, electrochemical, hydrogen, and electrical. Depending on the application needed, the amount of energy/power required and the application suitable for discharge/charge time, a suitable storage type can be chosen. For instance, Pumped Hydro Storage (PHS) provides bulk power management normally associated with the electricity generation plants.

It can also be seen from Figure 1 that BESS represented by the different types of batteries can provide services across all the electricity network whether it is for short duration power supply, transmission/distribution or bulk power management. Recently, Lithium ion batteries have been used also to provide 100 MW bulk power management for the electricity grid in South Australia [34]. Many network operators use BESS to support grid reliability and initiate services to help BESS quantify its value. For example, a number of distribution network operators (DNOs) in the UK have successfully piloted several BESS technologies for different grid applications, such as peak shaving, voltage support, and renewable constraint management [35]. Therefore, the market and regulatory changes needed to increase the share of ES on energy networks should consider the different types, applications, and capability of ES.

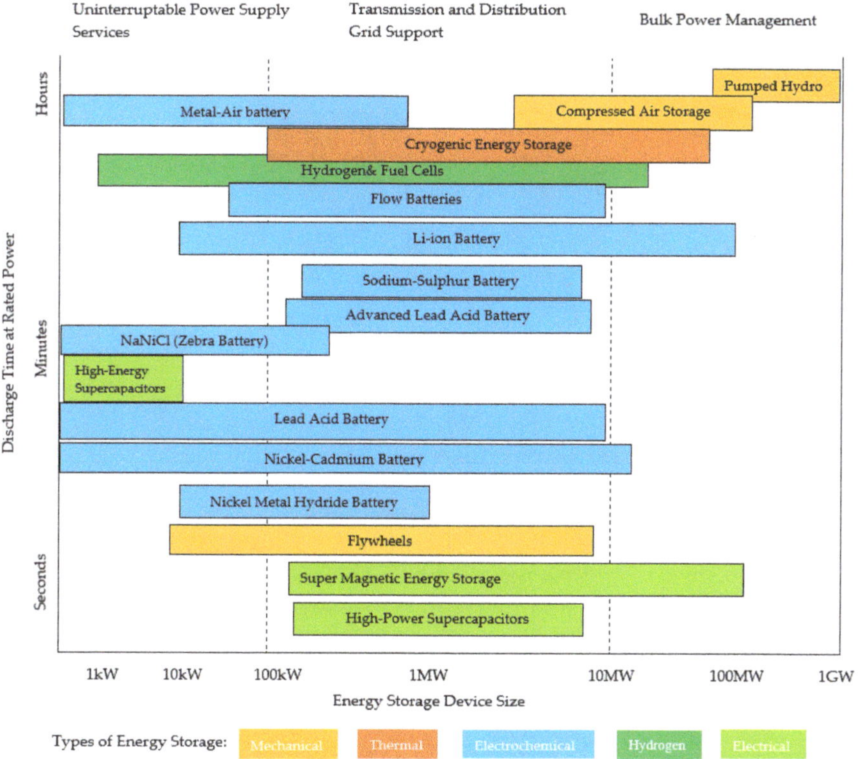

Figure 1. Classification of energy storage systems by the form of stored energy along with their power capacity, discharge time, and applications, adopted from [36].

3. Market and Regulatory Changes for Energy Storage

Market and regulatory barriers for ES in the EU and UK level are summarised in Table 1 as reported in earlier research. The following subsections analyse recent EU and UK regulatory changes that are being considered and/or implemented to mitigate some of these barriers.

Table 1. The main regulatory and market barriers for Energy Storage deployment in the EU/UK.

The Barrier to Energy Storage Deployment	References
The absence of ES definition and classification	[29,37–47]
Storage ownership by network operators	[37,38,40,41,43,45,46,48,49]
Payment of double network fees	[29,37,38,41,44,46,50,51]
Payment of final consumption levies (FCLs)	[37,38,43,46]
Lack of ancillary services remuneration	[29,37–41,43,52]
Reserve market requirements	[29,38,39,45]
Fixed electricity pricing	[37,43,53]
Absence of direct subsides	[29,38]

3.1. Definiation and Classification of Energy Storage

ES is not explicitly defined in most electricity markets as an activity or an asset. Therefore, it differs from other activities in the electricity market such as generation, transmission, distribution and supply. Historically, PHS is classified as a generation asset. This has resulted in all types of ES being classified as generation assets. According to EU Directive 2009/72/EC, ES is an "asset that produces electricity". Similarly, the UK Electricity Act 1989 provides a broad definition of the process of electricity generation as "generating at a relevant place". While this definition and classification may be adequate for large-scale ES such as PHS it poses investment risks for BESS because it limits the applications that ES can provide to those relating to generation.

BESS have a shorter lifecycle and lower energy capacity than PHS [54], making them suitable to help network operators effectively manage distribution and transmission networks in line with integrating distributed generators. Therefore, some network operators suggest creating an independent asset class to ES that identifies storage as a solution to integrate RES rather than competing with traditional generations [55]. Other network operators have proposed many solutions to the storage definition barrier for BESS by suggesting [56]:

- Defining storage as a discrete activity or asset class to ensure investment certainty.
- Introduce storage provisions within the distribution license so network operators can be free from some generation license rules.

EC proposed a definition for ES states that "Energy storage in the electricity system means the deferring of an amount of the energy that was generated to the moment of use, either as final energy or converted into another energy carrier" [32]. Similarly, OFGEM's proposed definition states that "Electricity Storage in the electricity system is the conversion of electrical energy into a form of energy which can be stored, the storing of that energy, and the subsequent reconversion of that energy back into electrical energy in a controllable manner" [57]. Both OFGEM and EC believe that storage should continue to be captured by the generation license in order not to distort competition.

3.2. The Interaction of Storage with Final Consumption Levies

The competitiveness of ES technologies are affected by policies that encourage RES deployment. For example, in Germany, there is no incentive for wind farm operators to store the energy generated because they are paid 95–100% of the relevant wind Feed in Tariff (FiT) for the curtailment of that energy [46,58]. In the UK, FiT, Renewables Obligation (RO), Contract for Difference (CfD), and Climate Change levy (CCL) are examples of FCLs policies introduced to encourage the deployment of RES. Consumers and storage are charged these levies for storage when importing electricity. However, it is argued that ES is not the final consumer of energy and should be exempted from such levies that increase the operational cost for storage owners [33]. It is found that the cost of RO and FiT levies accounts for 80% of all non-energy-related supply costs when charging a commercial grid-scale battery [59]. OFGEM recently proposed that any storage owner with the newly modified generation license (that define ES and its characteristics) could be exempted from paying those FCLs, if the main purpose of storage is to export electricity to the grid only (not for a self-consumption) [57]. If, however, an owner is exempt from obtaining a generation license, the payment of FCLs is still required (A generator and by extension a storage device could be exempt from applying for a license if it is generating at a rate below 100 MW in England and Wales).

3.2.1. Energy Storage Treatment in FiT/RO Sites

FiT is a scheme used to encourage customers and businesses to generate their own electricity from RES and be paid if there is energy surplus. RO works alongside FIT where electricity suppliers are obliged to buy certain amounts of electricity generated by large renewable generation plants. In both FiT and RO schemes, the technology type used, its installation, and capacity play a role in deciding the tariff rate, eligibility criteria and consequently the payment received by the owner [60,61]. Since the storage is essential for some technologies such as solar PV, it is the case of putting the accreditation at risk if storage is installed on FiT/RO sites. For example, if a battery is integrated with a PV solar site and a metering arrangement is installed in a way that leads the owner to receive payments from the electricity exported from the grid to the storage.

OFGEM'S recent regulations state that storage installation in FiT/RO sites should be permitted if the purpose of storage is to store electricity generated by renewable sources only and if the total contracted capacity is not changed [62,63]. To comply with these conditions, certain electricity meter arrangements need to be in place to distinguish between the imported and exported electricity.

3.2.2. Energy Storage Treatment under CfD

The basis of a CfD contract is to receive payment on the (clean) electricity generated by the CfD facility. Hence, if storage is defined within that CfD facility, there is a possibility for electricity to be imported from the grid, which cannot be necessarily generated by RES, and poured into it at another time (export time), which compromises the contract.

Two proposed options were introduced to mitigate this problem by the UK government [64]. First, any storage device in a CfD facility needs to be registered in a separate metering unit to ensure storage independency of the CfD facility. Second, storage can be registered in the same metering unit of a CfD facility only if certain metering arrangements are in place that prevent storing electricity other than that generated by the CfD facility generation equipment.

3.2.3. Energy Storage Interaction with the CCL

A climate change levy exempt certificate (LEC) can be issued if the electricity is generated by RES. Therefore, if the ES device imports electricity (from non-LEC generator), then a CCL is applicable. The worst scenario is if ES imports electricity (from non-LEC generator) and then exports that electricity to an industrial user, resulting in a double CCL being incurred by the end-user.

There is no regulatory clarification from the UK government about the above issue. However, based on the aforementioned regulatory changes, it is expected that storage will not pay the CCL when importing electricity but still pays the CCL as a generator.

3.3. Network Charges for Energy Storage

Generators, suppliers and consumers of electricity are required to pay network and balancing charges to cover the ongoing network costs. With the absence of clear EU legislation regarding the charging arrangement of ES, some countries (the UK, France, Germany, The Netherlands) impose double network charges for ES, once when charging and the other when discharging [65].

In the UK, if the storage capacity is above 100 MW (large-scale), two sets of Transmission Network Use of System (TNUoS) and Balancing Services Use of System (BSUoS) charges are incurred by the storage if it is connected on the transmission network. If the storage capacity is below 100 MW (small-scale) and connected on the distribution network, only Distribution Use of Services (DUoS) charges will be paid either based on the Common Distribution Charging Methodology (CDCM) or Extra High Voltage Distribution Charging Methodology (EDCM) rates. However, the current distribution charging methodology seems inconsistent. For instance, the CDCM charge for a recent storage project over 8 months was £54,149, while the EDCM charges over the same period were £10,668 [56]. Therefore, connecting storage to extra voltage lines is more cost-effective than high voltage lines.

OFGEM presented several solutions for the charges discusses above to support a level playing field for ES as equal to other generation technologies [66,67]. These changes are listed below and presented in Figure 2.

- Removing the TNUoS demand and generation residual charges;
- Removing DUoS demand residual charges;
- Removing BSUoS demand charges;
- Introducing new fixed charges to cover the increased implementation of 'behind the meter' storage.

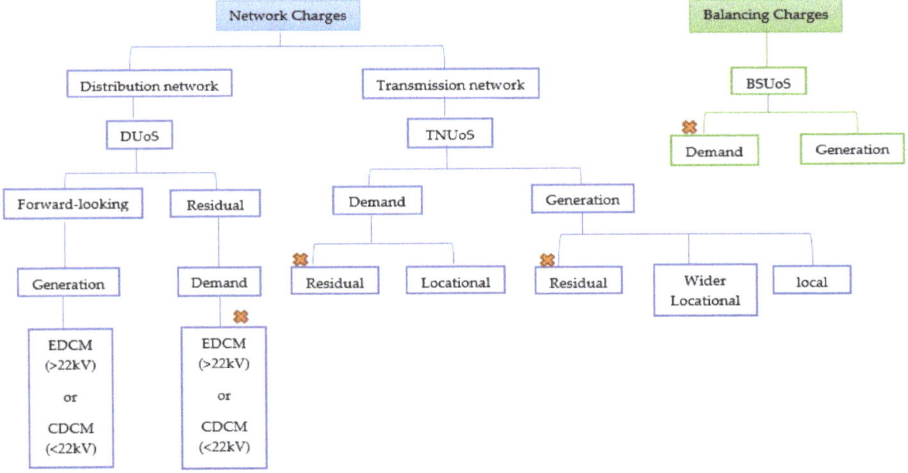

Figure 2. Network and Balancing system charges in the UK's electricity system, adopted from [68]. (x) means removed for ES per new legislations.

3.4. The Treatment of Energy Storage in the Capacity Market

The participation of ES in the CM is important to its business case [69]. Yet, the ability of ES to provide the necessary energy adequacy has been questioned due to its limited discharge capacity. As such, ES participation in this market is limited in some EU markets (for example, Ireland, Italy, Germany, France, Denmark, and the UK) [45]. For instance, ES was secured at a total of 3.2 GW in the UK's CM Tier-4 auction in 2016 in which 0.5 GW came from BESS [70]. In 2017, a study by the system operator found that a stress event could last for 2 h in the UK while the maximum duration of the current storage response is 30 min. This leads to linking the de-rating factor to the maximum discharge duration of storage [71], which means that maximum payment that a 0.5-h duration storage can receive is only 21.34% compared to 96.11% payment for a 4 h+ duration storage.

3.5. Energy Storage Ownership

The typical liberalised electricity market structure is illustrated in Figure 3. In this structure, transmission and distribution network operators (DNO, TNO) are legally required to separate network and non-network activities so as not to distort competition, which is referred to as 'Unbundling'. As storage is defined as a generation asset, this means that network owners cannot own, develop, manage, or operate storage assets for grid balancing or reinforcement under the EU rules, other than in the case of a few exceptions [72]. However, there is a lack of clarity regarding these exceptional circumstances leading to this rule being implemented differently in different EU's member states. For example, In the UK, if ES capacity is less than 100 MW, network operators can apply for a generation license exemption, thus owning storage [31]. Also, the Italian system operator (Terna) is granted permission to own and implement several ES projects to relieve the transmission network congestion [73]. This unclear ownership and operation status of ES creates uncertain investment environments, particularly for network operators.

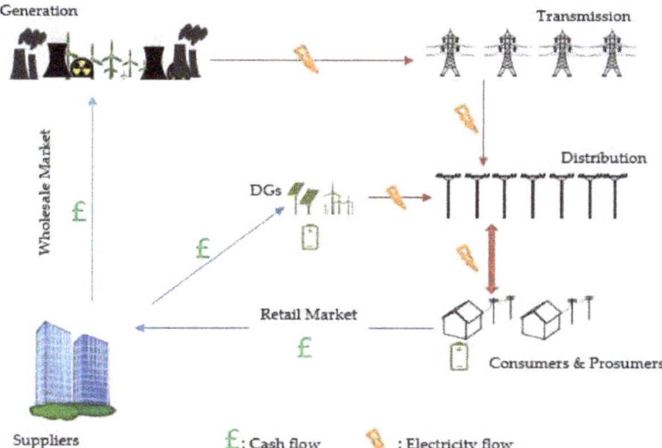

Figure 3. Example of a liberalised electricity market structure.

OFGEM aims, in line with the EC proposals, to strengthen the unbundling requirement by requiring separation of operation for storage assets owned by network operators even if it is below 100 MW (license exempted), while considering preventing DNO's ownership to storage in the near future [32,74]. Consequently, for now, storage can exceptionally be owned by network operators to provide services but not for trading in the electricity markets or provide balancing services.

3.6. Energy Storage Planning System

When network operators identify ES as a key solution to provide services to the network or prevents costly network expansions, ensuring access to infrastructure with appropriate time scale by the planning regimes is important. The EC is clear that storage should be granted access to grid infrastructure in a non-discriminatory way. Large-scale ES projects that are above 225 MW are included in the EU's cross-border infrastructure projects that link the energy system in at least two EU countries [75]. However, it remains unclear how small-scale ES such as BESS is to be treated in the planning system.

At the UK level, the government published a consultation to clarify the planning regimes for energy storage (small and large-scale) as follows [76]:

- If ES capacity is below 50 MW, the storage developer obtains planning consent from the local council.
- If ES capacity is above 50 MW, it is regarded as a nationally significant infrastructure project, and thus, storage owners obtain consent from the government.

- If storage is included in a composite project with other forms of generation, and the capacity of each of this installation is below 50 MW, this falls under the local planning regime.

4. SWOT Analysis Method

Earlier research published in [38] categorised 16 investment barriers that can be linked to four main regulatory and public attitudes barriers for ES. Since then, energy regulators in the EU and UK have acted to mitigate some of these barriers as detailed in the previous section. However, for the previous discussions, it is evident that the future of ES in national networks is not clear due to complex and interlinked market and regulatory changes. In an attempt to add further clarity to this issue, this study presents a SWOT analysis to highlight the future business potential of BESS in the future considering the recent market and regulatory changes.

A SWOT analysis, although criticised [77], is a widely used strategic planning tool that supports business to evaluate the strengths, weaknesses, threats, and opportunities of a proposed project [78]. Research into energy networks has been used to assess energy technologies [79], evaluate national policies [80] and assess international policies [81].

The SWOT analysis examines helpful and harmful aspects of the recent regulatory changes. The internal factors (Strengths/Weaknesses) are the direct regulatory and market changes proposed by energy regulators that affect storage's business avenue. External factors (Opportunities/Threats) are those affecting storage's business potential because of the indirect aspects beyond the stated market and regulatory changes. It should be noted, however, that SWOT analysis may provide incomplete qualitative examination such that the assessment may be subjective. As such, in this paper, all the SWOT analysis assessments will be supported by earlier research or quantitative reports.

5. Results

Table 2 provides the results and the general structure of SWOT analysis while more details and discussions of the results are provided below.

Table 2. SWOT analysis of the market/regulatory changes for ES in the UK.

Factors	Helpful to Storage Business Case	Harmful to Storage Business Case
	Strengths	Weaknesses
Internal	• Removing double network charges • ES co-location with FCLs sites • Facilitating grid access for ES	• ES definition and classification • Strengthen the unbundling requirements • Payment of FCLs for small-scale ES (<100 MW) • Employing de-rating factors in the capacity market • Introducing fixed charges for behind the meter ES
	Opportunities	Threats
External	• Encouraging private parties' investment in ES • Ancillary services aggregation	• High capital cost • Cannibalisation of revenue streams available for ES

5.1. *Strengths*

5.1.1. Removing Double Network Charges

The network and balancing system charges need to be included and paid by ES owners as part of operational costs for importing/exporting electricity to the grid. For example, the total cost of DUoS charges for one ES project owned by a DNO was found to be between £64,900 to £80,500 per annum, which is not cost-reflective [82]. Moreover, the TNUoS demand and generation residual charges represent approximately 80% of the total transmission network charges [68]. As such, removing these charges is a step forward towards reducing the operational cost for ES.

5.1.2. Energy Storage Co-Location with Final Consumption Levies Sites

Policies that support the deployment of low carbon technologies such as FiT and RO have significantly increased the uptake of RES across EU countries [83]. This makes ES co-location with RES essential to match the electricity supply and demand [84]. The new OFGEM regulations allowed ES to be co-located with FCLs sites without compromising the accreditation of these schemes. This triggered many positive responses from the industry and trade associations, because this means, for example, that home PV owners can install ES without the fear of compromising FiT payment scheme [85].

5.1.3. Facilitating Grid Access for Energy Storage

One of the factors affecting the investment decision in ES is the infrastructure access for the device and consequently the planning application. As discussed earlier, OFGEM has recently facilitated the planning regimes for ES technologies with a capacity below 50 MW by allowing ES developers to obtain planning permission from the local council. This is positive legislation for two reasons. First, the average capacity of the current and planned ES technologies projects is 27 MW, which is significantly below 50 MW [86]. Second, the composite projects that have total capacity above 50 MW (for example a generation unit with 40 MW and ES with 40 MW) are also applying for planning permission from the local council. This reduces the additional consent time (1–2 years) as well as the cost of application when compared to the application set at the national level [76].

5.2. Weaknesses

5.2.1. Energy Storage Definition and Classification

A unified and explicit definition for ES in the related legislations is a basic step to create certain investment environment. This clarifies the ownership and operation issues for ES. It also allows ES owners to have a clearer view over the available revenue streams during ES lifetime. However, ES definition proposed by both the EC and OFGEM limits its services related to generation. For instance, if a battery is used to curtail a wind turbine's surplus energy, it is exporting rather than generating electricity, which is a service to balance the system. The word 'generated' in the EU definition (see Section 3.1) implies that ES is a generation asset and therefore does not recognise other potential services and applications (please see Figure 1).

Classifying BESS as a generation asset puts it in direct competition with traditional bulk generators. This undermines its business case because BESS cannot trade in a large wholesale market as a generation asset due to its low capacity and low technical maturity [40]. Moreover, a unified ES definition is absent at the EU level due to national differences that prevent a fully integrated EU market design [37].

The current EU and UK ES definitions and classifications fall short behind California state legislation of ES services. The Assembly Bill No 2514 provides the following list of conditions applicable to ES that allow it to provide multiple services across the electricity network [87]:

- The ES system can be centralised or distributed.
- The ES system may be owned by a load serving entity, a customer, a third party, or local publicly-owned electric utility.
- The ES system can provide generation, transmission and distribution services.

5.2.2. Strengthen the Unbundling Requirements

If network operators are prevented from owning ES of any size, all ES devices connected to the distributing or transmission network will have to be operated by a legally separate party from the network operators. However, many DNOs have already installed ES devices in parts of their networks to defer conventional network upgrades and provide ancillary services. The latter have been shown to reduce the energy costs for customers and enhance network efficiency [35]. Some of the ES projects in the UK including their capacity, locations, ES type, and the type of the business model used are summarised in Table 3 [88–94]. Based on Table 3, two weaknesses can be noticed to this regulatory change. First, most DNOs are using smaller-size BESSs to increase network efficiency based on the 'DNO merchant' business model (this business model allows the DNO to procure and fully operate the storage device, and thus, use the storage services on its network and needs), which will not be legally valid if strengthening the unbundling requirements comes into force. Second, even though DNOs used 'DNO contracted' (the DNO procures, owns and operate the storage asset and use it in certain times only while a third party can have a contractual agreement with the DNO to commercially use the asset) and 'contracted services' (a third party procures, owns and operate the storage asset then sell the services to the DNO) business models for larger size BESS, they needed to enter into a complex contractual agreement with third parties to make revenue streams in the market because each party involved with the DNO needs to make a profit, which reduces the overall revenue.

Table 3. Some of BESS projects in the UK including their capacity, locations, type, DNO name, and the type of the business model.

Project Name	Capacity and Locations	ES	DNO	Type of BM
CLNR	RiseCarr: 5 MWh, H.Ngate: 0.2 MWh WoolerRamsay: 0.2 MWh Maltby: 0.1 MWh, Wooler St. Mary: 0.1 MWh Harrowgate Hill: 0.1 MWh	Li-ion	NPG	DNO Merchant
Short-Term Energy Storage	Hemsby: 200 kWh/200 kW	Li-ion	UKPN	DNO Merchant
SNS	Leighton Buzzard: 6 MW/10 MWh	Li-ion	UKPN	DNO Contracted
FALCON	Milton Keynes: 100 kWh	$NaNiCl_2$	WPD	DNO Merchant
Orkney Park	Kirkwall: 500 kWh, 2 MW	Li-ion	SSEPD	Contracted Services
NINES	Shetland Iceland 3 MWH, 1 MW	Lead-Acid	SSEPD	DNO Merchant
Low Voltage Connected Energy Storage	Chalvey, Berkshire 2 kWh, 25 kW	Li-ion	SSEPD	DNO Merchant

5.2.3. Payment of FCLs for Small-Scale ES

Large-scale ES are exempted from paying FCLs, which is seen as positive in determining its commercial availability [95]. However, this is not in favor of BESS given that the power is 50 MW for the largest battery energy storage project in Europe [96], and the average capacity of the current and planned ES technologies projects is 27 MW [86]. It has been demonstrated that the payment of these FCLs by ES can cost up to £20k–£50k per annum for the SNS project in Table 3 and makes the project's business model unprofitable outside the peak demand months [89].

5.2.4. Employing De-Rating Factors in the Capacity Market

Before introducing the de-rating factors to ES in this market, its participation was not penalised according to its discharge capacity, which meant that all storage devices received full payment of the clearing price. However, due to de-rating factor changes, discussed in Section 3.4, BESS participation in this market has decreased. For instance, in the T-4 auction of 2018, only 158 MW of BESS have secured a contract compared to 500 MW in the previous auction [97]. As depicted in Figure 4, the number of 0.5 h and 1 h duration batteries decreased from 40 and 91 in T-1 2018 auction to 17 and 32 in the T-1 2019 auction respectively. Moreover, the number of 1.5 h duration batteries increased from 8 in the T-4 2021 to 18 in the T-4 2022. This is in line with first time appearance of 11 and 2 batteries providing 3 h and 4 h discharge duration, respectively. The decline in the shorter duration storage (0.5 h and 1 h) and the increase in longer duration storage (1.5 h, 3 h, 4 h) may be a result of other services commitments. However, it is an indication of a regulatory change that encourages the battery industry to increase its energy and power density by introducing higher payment for longer duration storage.

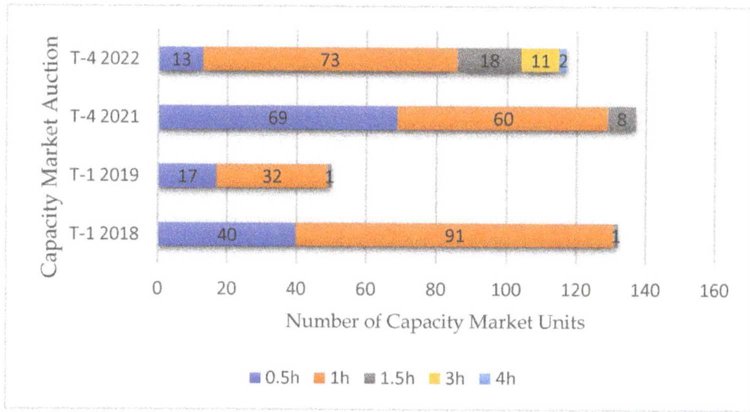

Figure 4. The number of batteries participating in the UK's capacity market from T-1 2018 to T-4 2022 as reported in the CM register (adopted from [98]).

A basic rule of the CM requires participants to remain ready in case of a system stress event. This means that storage must be fully charged at all-times, thus increasing its rate of degradation. This is one of the main barriers to BESS in the CM [38]. However, the degradation economic losses are neither remunerated nor studied in the de-rating factor study by the system operator [71].

5.2.5. Introducing Fixed Charges for behind the Meter Energy Storage

Behind-the-meter storage is normally used to reduce the electricity bill for commercial users. They are normally charged based on their consumption during the peak demand periods. However, because more users and businesses are able to predict these periods, they reduce their exposure to these charges when reducing their consumption due to using ES or demand side management techniques. Thus, OFGEM introduced fixed consumption charges although users can reduce the network charges for themselves; other electricity users need to compensate and cover the network fixed cost. This would seem to suggest that storage is regarded by energy regulators as a disruptive technology that can displace the existing energy regime [99].

5.3. Opportunities

5.3.1. Encouraging Private Parties' Investment in Energy Storage

Energy regulators in most liberalised countries are seeking to encourage competition between energy sector entities and regulate the revenue for market monopolies. The ES applications that can be provided by each electricity market entity are summarised in Table 4. It can be seen that private operators under the current market arrangements can provide all ES applications compared to other actors. As discussed in Section 3.5, energy regulators are considering strict unbundling requirements. Therefore, TSOs and DNOs will need to procure these services from the private sector that is increasingly interested in investing in ES [99].

Table 4. ES applications mapped to electricity industry actors [35].

Application	DNO	TSO	Energy Supplier	Generators	Private Operators
Energy Arbitrage	x	x	√	√	√
Peak shaving	√	x	x	x	√
Voltage support	√	√	x	x	√
Constraint management	√	x	x	√	√
System balancing	x	x	√	√	√
Portfolio balancing services	x	x	√	√	√

5.3.2. Ancillary Services Aggregation

The act of aggregation refers to the grouping of several units (consumers, or prosumers) in the power system to act as a single entity when trading in the electricity market [100]. Several studies identified the role of aggregators for ES in providing different services in the energy markets. However, the business case for aggregators is seen as hampered by the regulatory frameworks that prevent wider market access [41,101,102]. In 2016, the EC required member states to facilitate direct market access to the retail market [103]. Similarly, OFGEM identified some of the barriers to service aggregation and acted to amend some of the balancing market codes to allow aggregators to stack multiple revenue streams from different services [104].

5.4. Threats

5.4.1. High Capital Cost

Although the capital cost of BESS continues to fall, a recent study found that a cumulative investment of US$175–510 billion is needed in order for the capital cost of battery packs to reach US$175 ± 25/kWh, which is expected between 2027–2040 [105]. In another study, battery technology experts found that even with the recent advances in battery pack manufacturing capabilities and chemistry changes, the battery pack cost will not significantly decrease by 2020 [106]. As a result, a capital cost barrier is still a threat to the potential market growth of many BESSs.

5.4.2. Cannibalisation of Revenue Streams Available for Energy Storage

With the increased deployment of ES technologies combined with the necessity for a sustainable business case, there is a risk of revenue streams cannibalisation as a result of market competition. This risk is considered in the UK's system operator studies when the forecast of ES deployment falls from 18.3 GW by 2040 to 10.7 GW by 2050 [107].

6. Summary and Brexit Discussion

This section offers a summary to the obtained results in Table 2. It also discusses the 'Brexit' issue on the policy implication for both the EU and UK with regard to ES. Table 2 shows that the current ES policy has both strengths and weaknesses. Also, it shows an indirect implication to this policy represented by some threats and opportunities.

One of the main strength points of the UK's ES policy is the reduction of the operational cost for ES devices due to the removal of double network charges in the electricity markets. This is backed up by facilitating the planning permission for ES devices on a national and local scale. Another ES enabler factor is the co-location of ES devices with FiT and RO supporting schemes, which is a step forward in recognizing ES as a key enabler to greener electricity systems.

In terms of the weaknesses, the energy storage definition limits ES's role of generation only while this generation classification prevents a wide range of other revenue streams. Other barriers are preventing network owners from owning and operating ES from all sizes, payment of FCLs for small-scale ES introducing fixed charged for behind the meter storage, and de-rating the ES's capacity in the CM. These barriers can reduce the overall profit for ES assets.

Despite the aforementioned weaknesses, a number of opportunities have arisen. For instance, strengthening the 'unbundling' requirements for network owners can encourage private parties to invest in ES and stack multiple revenues from different services. Another opportunity is the shift towards allowing multiple smaller-scale ES assets to be aggregated together to provide vital ancillary services. Although stacking multiple revenues may be attractive for all ES assets, a threat of cannibalisation of the available revenue streams by all ES assets may occur. Another threat is the current high capital cost of ES.

The above analysed SWOT factors and their effects on ES's business case can be hugely changed due to the UK's exit from the EU (Brexit). The UK's electricity markets and the regulatory landscape are currently compliant with the EU regulation. Indeed, some studies such as in [108] analysed the implications of Brexit on the electricity sector and how the UK could lose the economic benefits of the interconnectors of some EU countries. These interconnectors usually provide valuable services to the UK's electricity network by managing intermittent RESs. In the case of a Brexit agreement that does not involve electricity market integration or interconnector share, there is a significant need for ES and other flexibility options to mitigate RESs intermittency [109]. Therefore, it seems that Brexit can raise the deployment of ES devices in the UK. A recent study also confirmed this finding by stating that the economic investment in large-scale ES to increase the UK's peaking capacity may be boosted without the EU's interconnectors sharing [110].

In case of a Brexit scenario that limits the UK's access to the EU's single market or a trade arrangement that sees tariffs imposed between the UK and the bloc, importing batteries from the EU may not be cost-effective to meet the local UK's demand. Therefore, the UK government should adopt policies to build battery gigafactories in the UK to cover the local demand whether it is for electric vehicles or the grid. The Faraday institution estimates that at least eight gigafactories are needed in the UK by 2040 [111].

7. Conclusions

ES is recognised as a key technology to mitigate the intermittency of many RES. Earlier research found many barriers to the rollout of ES in current energy networks. Since 2017, energy regulators in the EU and the UK proposed changes to enable a level playing field for ES and remove these barriers. The changes include (i) defining 'Electricity Storage' in the main legislation; (ii) removing double network and balancing charges for storage; (iii) co-locating storage with renewable generation sites that are supported through consumption levies policies; (iv) limiting the storage operation by network owners; (v) facilitating ES planning permission; and (vi) employing de-rating factors for storage in the capacity market (CM). Since the proposed regulatory changes at the UK and EU level are similar, the UK is taken as a case study considering its market design and relevant regulations.

This paper presented a SWOT analysis to explore the impact of the recent changes on the business potential of BESS and examined whether these regulatory changes have been supportive to BESS's business case in the UK.

Three main benefits of the recent regulatory changes were found. First, the removal of the double network charging for ES by eliminating the demand residual charges (when importing electricity from the grid) can reduce its operational cost. Second, ES co-location with RES sites that receive government subsidies can not only boost its business case but also recognise ES as a key player in integrating these intermittent resources. Third, facilitating ES access to the grid by allowing cost-effective infrastructure access and planning permission can positively affect investment decision in ES. The recent regulation supported faster implementation of composite projects that have a total capacity above 50 MW (for example, a generation unit with 40 MW and ES with 40 MW).

However, a number of drawbacks were found in the recent regulatory changes that may outweigh these benefits. First, the new definition for ES recognised it as a generation device, thus it has to compete with traditional generator assets in many instances, which it cannot do cost effectively. Second, the introduced regulations do not consider the different types and sizes of ES and tend to support large-scale ES with a capacity of 100 MW or above (similar to generators). However, there is a key role for smaller-scale ES in current energy systems in the form of BESS as discussed in the paper. From the perspective of the regulators, this makes sense as they do not want to disrupt the current energy market structure by creating an independent storage asset class. Instead, they are looking for a technological advance for different types of ES to increase its energy and power density to place it in the traditional generator's category. Third, only large-scale ES assets (above 100 MW) are exempt from paying FCLs, however, most of the current ES projects especially BESS are far below 100 MW. As such, in the case of BESS, this might be a longtime coming, limiting its value in current energy systems under current regulatory and market regimes. In the CM, for example, higher-duration BESS receives more payment than smaller-duration.

Despite the previous drawbacks and a high current capital cost for BESS, many opportunities have been found for private parties who are encouraged to own and operate ES devices to stack multiple revenues from different services. The recent regulation meant that private parties are in the best place to provide ES services. This is along with the suggestion that allows several ES units to be aggregated, and thus, provides services in the wholesale market.

Finally, we suggest a number of policy implications from the above SWOT analysis results. First, an independent asset class should be created for ES because the current energy markets are designed without electricity storage in mind. Second, a unique definition for ES that reflect its features is needed. For instance, the current definition for both the EU and the UK does not recognise the ES service when charging the device to help store the exceeded power from a wind farm. Third, in the CM where the capacity of the BESS asset is de-rated, the economic assessment of the degradation cost should be taken into account. This is because degradation can affect the availability of these assets, which in turn can affect the reliability and the energy security of the electricity network should the implementation of ES be increased.

Author Contributions: Conceptualization, A.G. and T.C.; Methodology, A.G. and T.C.; Formal analysis, A.G., M.A.-G., M.S., N.D. and T.C.; Investigation, T.C., N.D. and M.S.; Writing—Original draft preparation, A.G. and T.C.; Writing—Review and editing, T.C., M.A.-G., N.D. and M.S.; Visualization, T.C.; Supervision, T.C., N.D., M.A.-G. and M.S. All authors have read and agreed to the published version of the manuscript.

Funding: This research received no external funding.

Acknowledgments: Teesside University is gratefully acknowledged for fully supporting Ahmed's PhD scholarship. We are grateful also to the two anonymous reviewers for their suggestions to improve the paper.

Conflicts of Interest: The authors declare no conflict of interest.

References

1. Narayanan, A.; Mets, K.; Strobbe, M.; Develder, C. Feasibility of 100% renewable energy-based electricity production for cities with storage and flexibility. *Renew. Energy* **2019**, *134*, 698–709. [CrossRef]
2. Rogge, N. EU countries' progress towards 'Europe 2020 strategy targets'. *J. Policy Model.* **2019**, *41*, 255–272. [CrossRef]
3. Hübler, M.; Löschel, A. The EU Decarbonisation Roadmap 2050—What way to walk? *Energy Policy* **2013**, *55*, 190–207. [CrossRef]
4. IRENA. *Renwable Capacity Statistics*; International Renewable Energy Agency: Abu Dhabi, UAE, 2019; Available online: https://www.irena.org/publications/2019/Mar/Capacity-Statistics-2019 (accessed on 30 April 2019).
5. IRENA; European Commission. *Renewable Energy Prospects for the European Union*; International Renewable Energy Agency: Abu Dhabi, UAE, 2018; Available online: https://www.irena.org/-/media/Files/IRENA/Agency/Publication/2018/Feb/IRENA_REmap-EU_2018_summary.pdf?la=en&hash=818E3BDBFC16B90E1D0317C5AA5B07C8ED27F9EF (accessed on 26 June 2019).
6. Notton, G.; Nivet, M.-L.; Voyant, C.; Paoli, C.; Darras, C.; Motte, F.; Fouilloy, A. Intermittent and stochastic character of renewable energy sources: Consequences, cost of intermittence and benefit of forecasting. *Renew. Sustain. Energy Rev.* **2018**, *87*, 96–105. [CrossRef]
7. McCormick, P.G.; Suehrcke, H. The effect of intermittent solar radiation on the performance of PV systems. *Sol. Energy* **2018**, *171*, 667–674. [CrossRef]
8. Ren, G.; Liu, J.; Wan, J.; Guo, Y.; Yu, D. Overview of wind power intermittency: Impacts, measurements, and mitigation solutions. *Appl. Energy* **2017**, *204*, 47–65. [CrossRef]
9. Brouwer, A.S.; van den Broek, M.; Seebregts, A.; Faaij, A. Impacts of large-scale Intermittent Renewable Energy Sources on electricity systems, and how these can be modeled. *Renew. Sustain. Energy Rev.* **2014**, *33*, 443–466. [CrossRef]
10. Brouwer, A.S.; van den Broek, M.; Zappa, W.; Turkenburg, W.C.; Faaij, A. Least-cost options for integrating intermittent renewables in low-carbon power systems. *Appl. Energy* **2016**, *161*, 48–74. [CrossRef]
11. Brouwer, A.S.; van den Broek, M.; Seebregts, A.; Faaij, A. Operational flexibility and economics of power plants in future low-carbon power systems. *Appl. Energy* **2015**, *156*, 107–128. [CrossRef]
12. Williams, S.; Short, M.; Crosbie, T. On the use of thermal inertia in building stock to leverage decentralised demand side frequency regulation services. *Appl. Therm. Eng.* **2018**, *133*, 97–106. [CrossRef]
13. Chattopadhyay, D.; Alpcan, T. Capacity and Energy-Only Markets Under High Renewable Penetration. *IEEE Trans. Power Syst.* **2016**, *31*, 1692–1702. [CrossRef]
14. Hossain, M.S.; Madlool, N.A.; Rahim, N.A.; Selvaraj, J.; Pandey, A.K.; Khan, A.F. Role of smart grid in renewable energy: An overview. *Renew. Sustain. Energy Rev.* **2016**, *60*, 1168–1184. [CrossRef]
15. Peker, M.; Kocaman, A.S.; Kara, B.Y. Benefits of transmission switching and energy storage in power systems with high renewable energy penetration. *Appl. Energy* **2018**, *228*, 1182–1197. [CrossRef]
16. IRENA. *Electricity Storage and Renewables: Costs and Markets to 2030*; International Renewable Energy Agency: Abu Dhabi, UAE, 2017; Available online: https://www.irena.org/publications/2017/Oct/Electricity-storage-and-renewables-costs-and-markets (accessed on 10 November 2019).
17. Lorenzi, G.; da Silva Vieira, R.; Santos Silva, C.A.; Martin, A. Techno-economic analysis of utility-scale energy storage in island settings. *J. Energy Storage* **2019**, *21*, 691–705. [CrossRef]
18. Ding, J.; Xu, Y.; Chen, H.; Sun, W.; Hu, S.; Sun, S. Value and economic estimation model for grid-scale energy storage in monopoly power markets. *Appl. Energy* **2019**, *240*, 986–1002. [CrossRef]
19. Obi, M.; Jensen, S.M.; Ferris, J.B.; Bass, R.B. Calculation of levelized costs of electricity for various electrical energy storage systems. *Renew. Sustain. Energy Rev.* **2017**, *67*, 908–920. [CrossRef]
20. Zakeri, B.; Syri, S. Electrical energy storage systems: A comparative life cycle cost analysis. *Renew. Sustain. Energy Rev.* **2015**, *42*, 569–596. [CrossRef]
21. Martins, R.; Hesse, C.H.; Jungbauer, J.; Vorbuchner, T.; Musilek, P. Optimal Component Sizing for Peak Shaving in Battery Energy Storage System for Industrial Applications. *Energies* **2018**, *11*, 2048. [CrossRef]
22. Kocer, C.M.; Cengiz, C.; Gezer, M.; Gunes, D.; Cinar, A.M.; Alboyaci, B.; Onen, A. Assessment of Battery Storage Technologies for a Turkish Power Network. *Sustainability* **2019**, *11*, 3669. [CrossRef]

23. Gailani, A.; Al-Greer, M.; Short, M.; Crosbie, T. Degradation Cost Analysis of Li-Ion Batteries in the Capacity Market with Different Degradation Models. *Electronics* **2020**, *9*, 90. [CrossRef]
24. Rancilio, G.; Lucas, A.; Kotsakis, E.; Fulli, G.; Merlo, M.; Delfanti, M.; Masera, M. Modeling a Large-Scale Battery Energy Storage System for Power Grid Application Analysis. *Energies* **2019**, *12*, 3312. [CrossRef]
25. Harlow, J.E.; Ma, X.; Li, J.; Logan, E.; Liu, Y.; Zhang, N.; Ma, L.; Glazier, S.L.; Cormier, M.M.E.; Genovese, M.; et al. A Wide Range of Testing Results on an Excellent Lithium-Ion Cell Chemistry to be used as Benchmarks for New Battery Technologies. *J. Electrochem. Soc.* **2019**, *166*, A3031–A3044. [CrossRef]
26. BloombergNEF. *New Energy Outlook*; Bloomberg: New York, NY, USA, 2018; Available online: https://bnef.turtl.co/story/neo2018?teaser=true (accessed on 3 April 2019).
27. Sandia National Laboratories. *Energy Storage for the Electricity Grid: Benefits and Market Potential Assessment Guide*; Sandia National Laboratories: Sandia, CA, USA, 2010. Available online: https://www.sandia.gov/ess-ssl/publications/SAND2010-0815.pdf (accessed on 4 April 2019).
28. Forrester, S.P.; Zaman, A.; Mathieu, J.L.; Johnson, J.X. Policy and market barriers to energy storage providing multiple services. *Electr. J.* **2017**, *30*, 50–56. [CrossRef]
29. Anuta, O.H.; Taylor, P.; Jones, D.; McEntee, T.; Wade, N. An international review of the implications of regulatory and electricity market structures on the emergence of grid scale electricity storage. *Renew. Sustain. Energy Rev.* **2014**, *38*, 489–508. [CrossRef]
30. Hamelink, M.; Opdenakker, R. How business model innovation affects firm performance in the energy storage market. *Renew. Energy* **2019**, *131*, 120–127. [CrossRef]
31. Poyry. *Storage Business Models in the GB Market*; Pöyry Management Consulting: London, UK, 2014; Available online: http://www.poyry.com/sites/default/files/374_elexon_storagebusinessmodelsandgbmarket_v2_0.pdf (accessed on 10 April 2019).
32. EU Commission. *Energy Storage—The Role of Electricity*; EU Commission: Brussels, Belgium, 2017. Available online: https://ec.europa.eu/energy/sites/ener/files/documents/swd2017_61_document_travail_service_part1_v6.pdf (accessed on 11 April 2019).
33. OFGEM. *Upgrading Our Energy System: Smart Systems and Flexibility Plan*; OFGEM: London, UK, 2017. Available online: https://www.gov.uk/government/publications/upgrading-our-energy-system-smart-systems-and-flexibility-plan (accessed on 11 April 2017).
34. Faunce, T.A.; Prest, J.; Su, D.; Hearne, S.J.; Iacopi, F. On-grid batteries for large-scale energy storage: Challenges and opportunities for policy and technology. *MRS Energy Sustain.* **2018**, *5*, E11. [CrossRef]
35. EA Technology. *A Good Practice Guide on Electrical Energy Storage*; EA Technology: Chester, UK, 2014; Available online: https://www.eatechnology.com/engineering-projects/electrical-energy-storage/ (accessed on 4 April 2019).
36. Argyrou, M.C.; Christodoulides, P.; Kalogirou, S.A. Energy storage for electricity generation and related processes: Technologies appraisal and grid scale applications. *Renew. Sustain. Energy Rev.* **2018**, *94*, 804–821. [CrossRef]
37. EuroBat. *Battery Energy Storage in the EU—Barriers, Opportunities, Services & Benefits*; Eurobat: Düsseldorf, Germany, 2016; Available online: https://www.eurobat.org/news-publications/publications (accessed on 10 April 2019).
38. Castagneto Gissey, G.; Dodds, P.E.; Radcliffe, J. Market and regulatory barriers to electrical energy storage innovation. *Renew. Sustain. Energy Rev.* **2018**, *82*, 781–790. [CrossRef]
39. Staffell, I.; Rustomji, M. Maximising the value of electricity storage. *J. Energy Storage* **2016**, *8*, 212–225. [CrossRef]
40. Ruz, F.C.; Pollitt, M.G. Overcoming Barriers to Electrical Energy Storage: Comparing California and Europe. *Compet. Regul. Netw. Ind.* **2016**, *17*, 123–149. [CrossRef]
41. Rappaport, R.D.; Miles, J. Cloud energy storage for grid scale applications in the UK. *Energy Policy* **2017**, *109*, 609–622. [CrossRef]
42. He, X.; Delarue, E.; D'Haeseleer, W.; Glachant, J.-M. A novel business model for aggregating the values of electricity storage. *Energy Policy* **2011**, *39*, 1575–1585. [CrossRef]
43. Dusonchet, L.; Favuzza, S.; Massaro, F.; Telaretti, E.; Zizzo, G. Technological and legislative status point of stationary energy storages in the EU. *Renew. Sustain. Energy Rev.* **2019**, *101*, 158–167. [CrossRef]
44. Wasowicz, B.; Koopmann, S.; Dederichs, T.; Schnettler, A.; Spaetling, U. Evaluating Regulatory and Market Frameworks for Energy Storage Deployment in Electricity Grids with High Renewable Energy Penetration.

In Proceedings of the 2012 9th International Conference on the European Energy Market, Florence, Italy, 10–12 May 2012; pp. 1–8.

45. Usera, I.; Rodilla, P.; Burger, S.; Herrero, I.; Batlle, C. The Regulatory Debate About Energy Storage Systems: State of the Art and Open Issues. *IEEE Power Energy Mag.* **2017**, *15*, 42–50. [CrossRef]
46. The Fuel Cells and Hydrogen Joint Undertaking. *Commercialisation of Energy Storage in Europe*; FCH Europe: Brussels, Belgium, 2015; Available online: https://www.fch.europa.eu/sites/default/files/CommercializationofEnergyStorageFinal_3.pdf (accessed on 9 April 2019).
47. Zame, K.K.; Brehm, C.A.; Nitica, A.T.; Richard, C.L.; Schweitzer, G.D., III. Smart grid and energy storage: Policy recommendations. *Renew. Sustain. Energy Rev.* **2018**, *82*, 1646–1654. [CrossRef]
48. Taylor, P.G.; Bolton, R.; Stone, D.; Upham, P. Developing pathways for energy storage in the UK using a coevolutionary framework. *Energy Policy* **2013**, *63*, 230–243. [CrossRef]
49. Ferreira, R.; Matos, M.; Lopes, J.P. Regulatory Issues in the Deployment of Distributed Storage Devices in Distribution Networks. In Proceedings of the 2016 13th International Conference on the European Energy Market (EEM), Porto, Portugal, 6–9 June 2016; pp. 1–6.
50. Yan, X.; Gu, C.; Li, F.; Xiang, Y. Network pricing for customer-operated energy storage in distribution networks. *Appl. Energy* **2018**, *212*, 283–292. [CrossRef]
51. Carbon Trust; Imperial College London. Can Storage Help Reduce the Cost of a Future UK Electricity System? 2016. Available online: https://www.carbontrust.com/media/672486/energy-storage-report.pdf (accessed on 10 April 2019).
52. Strbac, G.; Aunedi, M.; Konstantelos, I.; Moreira, R.; Teng, F.; Moreno, R.; Pudjianto, D.; Laguna, A.; Papadopoulos, P. Opportunities for Energy Storage: Assessing Whole-System Economic Benefits of Energy Storage in Future Electricity Systems. *IEEE Power Energy Mag.* **2017**, *15*, 32–41. [CrossRef]
53. Hu, J.; Harmsen, R.; Crijns-Graus, W.; Worrell, E.; van den Broek, M. Identifying barriers to large-scale integration of variable renewable electricity into the electricity market: A literature review of market design. *Renew. Sustain. Energy Rev.* **2018**, *81*, 2181–2195. [CrossRef]
54. Abdon, A.; Zhang, X.; Parra, D.; Patel, M.K.; Bauer, C.; Worlitschek, J. Techno-economic and environmental assessment of stationary electricity storage technologies for different time scales. *Energy* **2017**, *139*, 1173–1187. [CrossRef]
55. National Grid. *Future Energy Scenarios*; National Grid: London, UK, 2015; Available online: http://fes.nationalgrid.com/fes-document/ (accessed on 16 April 2019).
56. Bradbury, S.; Hayling, J.; Papadopoulos, P.; Heyward, N. Smarter Network Storage Electricity Storage in GB: SNS 4.7 Recommendations for Regulatory and Legal Framework (SDRC 9.5). 2015. Available online: http://innovation.ukpowernetworks.co.uk/innovation/en/Projects/tier-2-projects/Smarter-Network-Storage-(SNS)/ (accessed on 16 April 2019).
57. OFGEM. *Clarifying the Regulatory Framework for Electricity Storage: Licensing*; Office of Gas and Electricity Markets: London, UK, 2017. Available online: https://www.ofgem.gov.uk/publications-and-updates/clarifying-regulatory-framework-electricity-storage-licensing (accessed on 17 April 2019).
58. Schermeyer, H.; Vergara, C.; Fichtner, W. Renewable energy curtailment: A case study on today's and tomorrow's congestion management. *Energy Policy* **2018**, *112*, 427–436. [CrossRef]
59. Bradbury, S.; Laguna, A.; Papadopoulos, P. Electricity Storage in GB: Final Evaluation of the Smarter Network Storage Solution (SDRC 9.8). 2016. Available online: http://innovation.ukpowernetworks.co.uk/innovation/en/Projects/tier-2-projects/Smarter-Network-Storage-(SNS)/ (accessed on 20 April 2019).
60. Pearce, P.; Slade, R. Feed-in tariffs for solar microgeneration: Policy evaluation and capacity projections using a realistic agent-based model. *Energy Policy* **2018**, *116*, 95–111. [CrossRef]
61. Woodman, B.; Mitchell, C. Learning from experience? The development of the Renewables Obligation in England and Wales 2002–2010. *Energy Policy* **2011**, *39*, 3914–3921. [CrossRef]
62. OFGEM. *Guidance for Generators: Co-Location of Electricity Storage Facilities with Renewable Generation Supported under the Renewables Obligation or Feed-in Tariff Schemes*; OFGEM: London, UK, 2018. Available online: https://www.ofgem.gov.uk/publications-and-updates/guidance-generators-co-location-electricity-storage-facilities-renewable-generation-supported-under-renewables-obligation-or-feed-tariff-schemes-version-1 (accessed on 17 April 2019).

63. OFGEM. *Feed-in Tariffs: Guidance for Licensed Electricity Suppliers (Version 10)*; OFGEM: London, UK, 2018. Available online: https://www.ofgem.gov.uk/publications-and-updates/feed-tariffs-guidance-licensed-electricity-suppliers-version-10 (accessed on 18 April 2019).
64. BEIS. *Contracts For Difference Government Response to the Consultation on Changes to the CFD Contract and CFD Regulations*; UK Government: London, UK, 2017. Available online: https://assets.publishing.service.gov.uk/government/uploads/system/uploads/attachment_data/file/589996/FINAL_-_Government_Response_to_the_CFD_Contract_Changes_Consultation.pdf (accessed on 17 April 2019).
65. Norton Rose Fulbright. Regulatory Progress for Energy Storage in Europe. Available online: https://www.nortonrosefulbright.com/en/knowledge/publications/8b5285f4/regulatory-progress-for-energy-storage-in-europe (accessed on 18 April 2019).
66. OFGEM. *Open Letter on Implications of Charging Reform on Electricity Storage*; OFGEM: London, UK, 2019. Available online: https://www.ofgem.gov.uk/publications-and-updates/open-letter-implications-charging-reform-electricity-storage (accessed on 18 April 2019).
67. OFGEM. *Targeted Charging Review: Minded to Decision and Draft Impact Assessment*; OFGEM: London, UK, 2018. Available online: https://www.ofgem.gov.uk/publications-and-updates/targeted-charging-review-minded-decision-and-draft-impact-assessment (accessed on 19 April 2019).
68. OFGEM. *Targeted Charging Review: A Consultation*; OFGEM: London, UK, 2017. Available online: https://www.ofgem.gov.uk/publications-and-updates/targeted-charging-review-consultation (accessed on 19 April 2019).
69. Khan, A.S.M.; Verzijlbergh, R.A.; Sakinci, O.C.; De Vries, L.J. How do demand response and electrical energy storage affect (the need for) a capacity market? *Appl. Energy* **2018**, *214*, 39–62. [CrossRef]
70. National Grid. *T-4 Capacity Market Auction for 2020/21*; National Grid: London, UK, 2016; Available online: https://www.emrdeliverybody.com/CM/T-4-Auction-2016.aspx (accessed on 18 June 2019).
71. National Grid. *Duration-Limited Storage De-Rating Factor Assessment—Final Report*; National Grid: London, UK, 2017; Available online: https://www.emrdeliverybody.com/Lists/Latest%20News/Attachments/150/Duration%20Limited%20Storage%20De-Rating%20Factor%20Assessment%20-%20Final.pdf (accessed on 12 February 2018).
72. EU Commission. *Directive of the European Parliament and of the Council on Common Rules for the Internal Market in Electricity (Recast)*; EU Commission: Brussels, Belgium, 2016. Available online: https://eur-lex.europa.eu/legal-content/EN/TXT/?uri=CELEX%3A52016PC0864R%2801%29 (accessed on 13 November 2019).
73. Graditi, G.; Ippolito, M.G.; Telaretti, E.; Zizzo, G. Technical and economical assessment of distributed electrochemical storages for load shifting applications: An Italian case study. *Renew. Sustain. Energy Rev.* **2016**, *57*, 515–523. [CrossRef]
74. OFGEM. *Enabling the Competitive Deployment of Storage in a Flexible Energy System: Decision on Changes to the Electricity Distribution Licence*; OFGEM: London, UK, 2018. Available online: https://www.ofgem.gov.uk/publications-and-updates/enabling-competitive-deployment-storage-flexible-energy-system-statutory-consultation-changes-electricity-distribution-licence (accessed on 18 September 2019).
75. EU Commission. *Technical Information on Projects of Common Interest*; EU Commission: London, UK, 2018. Available online: http://ec.europa.eu/energy/sites/ener/files/technical_document_3rd_list_with_subheadings.pdf (accessed on 3 November 2019).
76. BEIS. *Consultation on Proposals Regarding the Planning System for Electricity Storage*; UK Government: London, UK, 2019. Available online: https://www.gov.uk/government/consultations/the-treatment-of-electricity-storage-within-the-planning-system (accessed on 23 December 2019).
77. Chang, H.-H.; Huang, W.-C. Application of a quantification SWOT analytical method. *Math. Comput. Model.* **2006**, *43*, 158–169. [CrossRef]
78. Huang, J.; Fan, J.; Furbo, S. Feasibility study on solar district heating in China. *Renew. Sustain. Energy Rev.* **2019**, *108*, 53–64. [CrossRef]
79. Njoh, A.J. The SWOT model's utility in evaluating energy technology: Illustrative application of a modified version to assess the sawdust cookstove's sustainability in Sub-Saharan Africa. *Renew. Sustain. Energy Rev.* **2017**, *69*, 313–323. [CrossRef]

80. Igliński, B.; Iglińska, A.; Koziński, G.; Skrzatek, M.; Buczkowski, R. Wind energy in Poland—History, current state, surveys, Renewable Energy Sources Act, SWOT analysis. *Renew. Sustain. Energy Rev.* **2016**, *64*, 19–33. [CrossRef]
81. Wang, Q.; Li, R. Impact of cheaper oil on economic system and climate change: A SWOT analysis. *Renew. Sustain. Energy Rev.* **2016**, *54*, 925–931. [CrossRef]
82. Scottish and Southern Electricity Power Distribution. *Trial of Orkney Energy Storage Park SSET1009*; SSEPD: Edinburgh, UK, 2013; Available online: https://www.ssen.co.uk/WorkArea/DownloadAsset.aspx?id=7305 (accessed on 23 November 2017).
83. Krajačić, G.; Duić, N.; Tsikalakis, A.; Zoulias, M.; Caralis, G.; Panteri, E.; Carvalho, M.d.G. Feed-in tariffs for promotion of energy storage technologies. *Energy Policy* **2011**, *39*, 1410–1425. [CrossRef]
84. Grantham, A.; Pudney, P.; Ward, L.A.; Whaley, D.; Boland, J. The viability of electrical energy storage for low-energy households. *Sol. Energy* **2017**, *155*, 1216–1224. [CrossRef]
85. Solar Trade Association. *Ofgem Clarification Removes Barrier to Battery Storage for 900,000 Solar Homes*; STA: London, UK, 2018; Available online: https://www.solar-trade.org.uk/press-release-ofgem-clarification-removes-barrier-to-battery-storage-for-900000-solar-homes/ (accessed on 4 June 2016).
86. Renewables UK. *Energy Storage Capacity Set to Soar, 300 UK-Based Companies Involved in New Sector*; Renewables UK: London, UK, 2018; Available online: https://www.renewableuk.com/news/425522/Energy-storage-capacity-set-to-soar-300-UK-based-companies-involved-in-new-sector.htm (accessed on 3 April 2019).
87. Zillman, D.; Godden, L.; Paddock, L.R.; Roggenkamp, M. *Innovation in Energy Law and Technology: Dynamic Solutions for Energy Transitions*; OUP: Oxford, UK, 2018.
88. Northern Powergrid. *Customer-Led Network Revolution Project Closedown Report*; Northern Powergrid: Newcastle Upon Tyne, UK, 2015; Available online: http://www.networkrevolution.co.uk/project-library/project-closedown-report-2/ (accessed on 2 May 2018).
89. UK Power Networks. *Smarter Network Storage Close-Down Report*; UKPN: London, UK, 2017; Available online: http://innovation.ukpowernetworks.co.uk/innovation/en/Projects/tier-2-projects/Smarter-Network-Storage-(SNS)/ (accessed on 2 May 2018).
90. Western Power Distrbituion. *Developing Future Power Networks Project FALCON Close Down Report*; WPD: Bristol, UK, 2016. Available online: https://www.ofgem.gov.uk/publications-and-updates/wpd-s-falcon-project-closedown-report (accessed on 2 May 2018).
91. Scottish and Southern Energy Power Distribution. *LCNF Tier 1 Close-Down Report Orkney Energy Storage Park*; Scottish and Southern Energy Power Distribution: Edinburgh, UK, 2012; Available online: https://www.ssen.co.uk/WorkArea/DownloadAsset.aspx?id=7305 (accessed on 2 May 2018).
92. Scottish and Southern Energy Power Distribution. *NINES Project Closedown Report*; Scottish and Southern Energy Power Distribution: Edinburgh, UK, 2017; Available online: https://www.ninessmartgrid.co.uk/library/nines-closedown-report/ (accessed on 3 May 2018).
93. Scottish and Southern Energy Power Distribution. *LCNF Tier 1 Closedown Report Low Voltage Connected Energy Storage*; SSEPD: Scotland, UK, 2017. Available online: https://www.ofgem.gov.uk/sites/default/files/docs/2014/08/sset1008_lv_connected_batteries_closedown_2nd_submission_0.pdf (accessed on 3 May 2018).
94. UK Power Networks. *Smarter Network Storage—Business Model Consultation*; UKPN: London, UK, 2017; Available online: https://www.ukpowernetworks.co.uk/internet/en/community/documents/Smarter-Network-Storage-Business-model-consultation.pdf (accessed on 3 January 2018).
95. National Grid. *Future Energy Scenarios*; National Grid: London, UK, 2017; Available online: http://fes.nationalgrid.com/fes-document/fes-2017/ (accessed on 3 April 2019).
96. SMA System Technology. SMA System Technology Deployed in Europe's Largest Battery Storage Project. Available online: https://www.sma.de/en/newsroom/current-news/news-details/news/3411-sma-system-technology-deployed-in-europes-largest-battery-storage-project.html (accessed on 17 April 2019).
97. National Grid. *T-4 Capacity Market Auction for 2021/22*; National Grid: London, UK, 2018; Available online: https://www.emrdeliverybody.com/Lists/Latest%20News/DispForm.aspx?ID=175&IsDlg=1 (accessed on 18 April 2019).
98. National Grid. *Capacity Market Registers*; National Grid: London, UK, 2019; Available online: https://www.emrdeliverybody.com/CM/Registers.aspx (accessed on 25 April 2019).
99. Winfield, M.; Shokrzadeh, S.; Jones, A. Energy policy regime change and advanced energy storage: A comparative analysis. *Energy Policy* **2018**, *115*, 572–583. [CrossRef]

100. Burger, S.; Chaves-Ávila, J.P.; Batlle, C.; Pérez-Arriaga, I.J. A review of the value of aggregators in electricity systems. *Renew. Sustain. Energy Rev.* **2017**, *77*, 395–405. [CrossRef]
101. Contreras-Ocana, J.; Ortega-Vazquez, M.; Zhang, B. Participation of an Energy Storage Aggregator in Electricity Markets. In Proceedings of the 2018 IEEE Power & Energy Society General Meeting (PESGM), Portland, OR, USA, 5–10 August 2018; pp. 1171–1183.
102. Castagneto Gissey, G.; Subkhankulova, D.; Dodds, P.E.; Barrett, M. Value of energy storage aggregation to the electricity system. *Energy Policy* **2019**, *128*, 685–696. [CrossRef]
103. EU Commission. *Common Rules for the Internal Electricity Market*; EU Commission: Brussels, Belgium, 2019. Available online: http://www.europarl.europa.eu/RegData/etudes/BRIE/2017/595924/EPRS_BRI(2017)595924_EN.pdf (accessed on 3 May 2019).
104. Bray, R.; Woodman, B. *Barriers to Independent Aggregators in Europe*; University of Exeter: Exeter, UK, 2019; Available online: https://bit.ly/2FH6sr5 (accessed on 3 September 2019).
105. Schmidt, O.; Hawkes, A.; Gambhir, A.; Staffell, I. The future cost of electrical energy storage based on experience rates. *Nat. Energy* **2017**, *2*, 17110. [CrossRef]
106. Few, S.; Schmidt, O.; Offer, G.J.; Brandon, N.; Nelson, J.; Gambhir, A. Prospective improvements in cost and cycle life of off-grid lithium-ion battery packs: An analysis informed by expert elicitations. *Energy Policy* **2018**, *114*, 578–590. [CrossRef]
107. The Energyst. A Business Case for Battery Storage. 2017. Available online: http://www.electricitystorage.co.uk/files/7615/0900/7557/Battery_Storage_Report_2017_28pp_cropped0.1.pdf (accessed on 5 October 2019).
108. Lockwood, M.; Froggatt, A.; Wright, G.; Dutton, J. The implications of Brexit for the electricity sector in Great Britain: Trade-offs between market integration and policy influence. *Energy Policy* **2017**, *110*, 137–143. [CrossRef]
109. Mayer, P.; Ball, S.C.; Vögele, S.; Kuckshinrichs, W.; Rübbelke, D. Analyzing Brexit: Implications for the Electricity System of Great Britain. *Energies* **2019**, *12*, 3212. [CrossRef]
110. Geske, J.; Green, R.; Staffell, I. Elecxit: The cost of bilaterally uncoupling British-EU electricity trade. *Energy Econ.* **2020**, *85*, 104599. [CrossRef]
111. The Faraday Institution. *The Gigafactory Boom: The Demand for Battery Manufacturing in the UK*; The Faraday Institution: Oxford, UK, 2019; Available online: https://faraday.ac.uk/wp-content/uploads/2019/08/Faraday_Insights-2_FINAL.pdf (accessed on 21 February 2020).

Article

Time-Frequency Image Analysis and Transfer Learning for Capacity Prediction of Lithium-Ion Batteries

Ma'd El-Dalahmeh *, Maher Al-Greer, Mo'ath El-Dalahmeh and Michael Short

School of Computing, Engineering and Digital Technologies, Teesside University, Middlesbrough TS1 3BX, UK; M.Al-Greer@tees.ac.uk (M.A.-G.); Mo'ath.El-Dalahmeh@tees.ac.uk (M.E.-D.); M.Short@tees.ac.uk (M.S.)

* Correspondence: Ma'd.El-Dalahmeh@tees.ac.uk; Tel.: +44-(0)-16-4221-8121

Received: 18 September 2020; Accepted: 16 October 2020; Published: 19 October 2020

Abstract: Energy storage is recognized as a key technology for enabling the transition to a low-carbon, sustainable future. Energy storage requires careful management, and capacity prediction of a lithium-ion battery (LIB) is an essential indicator in a battery management system for Electric Vehicles and Electricity Grid Management. However, present techniques for capacity prediction rely mainly on the quality of the features extracted from measured signals under strict operating conditions. To improve flexibility and accuracy, this paper introduces a new paradigm based on a multi-domain features time-frequency image (TFI) analysis and transfer deep learning algorithm, in order to extract diagnostic characteristics on the degradation inside the LIB. Continuous wavelet transform (CWT) is used to transfer the one-dimensional (1D) terminal voltage signals of the battery into 2D images (i.e., wavelet energy concentration). The generated TFIs are fed into the 2D deep learning algorithms to extract the features from the battery voltage images. The extracted features are then used to predict the capacity of the LIB. To validate the proposed technique, experimental data on LIB cells from the experimental datasets published by the Prognostics Center of Excellence (PCoE) NASA were used. The results show that the TFI analysis clearly visualised the degradation process of the battery due to its capability to extract different information on electrochemical features from the non-stationary and non-linear nature of the battery signal in both the time and frequency domains. AlexNet and VGG-16 transfer deep learning neural networks combined with stochastic gradient descent with momentum (SGDM) and adaptive data momentum (ADAM) optimization algorithms are examined to classify the obtained TFIs at different capacity values. The results reveal that the proposed scheme achieves 95.60% prediction accuracy, indicating good potential for the design of improved battery management systems.

Keywords: lithium-ion battery; capacity prediction; state of health estimation; time–frequency image analysis; continuous wavelet transform (CWT)

---

1. Introduction

1.1. Motivation

Lithium-ion batteries (LIBs) are recognised as a key future form of technology for renewable energy and electric vehicles (EVs) due to their high power and energy densities, low maintenance cost, long lifetime, and low self-discharge rate [1]. However, to optimise the energy performance of the LIBs, prolong their life cycle, and reduce their cost, it is thus critical to monitor the internal state of the battery, such as state-of-charge (SoC), state-of-health (SoH), and remaining useful life (RUL) [2,3]. During operation, the continuous determination of the LIB's internal state is achieved through the battery management system (BMS), which guarantees the reliability and efficiency of LIBs.

Over the battery's lifetime, however, the capability for LIBs to provide a certain amount of power and store energy is reduced due to internal ageing phenomena [4,5]. Therefore, monitoring SoH is an important and difficult task and one of many functions performed by the BMS [6]. SoH is typically the ability of LIBs to store energy compared to its initial value. The battery's SoH is often quantified by determining its capacity or resistance parameters [7]. When the capacity parameter is utilised as an indicator of the battery's SoH, the SoH is known as the ratio of the battery capacity at current cycle to its rated capacity at the beginning of life or initial capacity provided by the battery manufacturer [8]. An accurate evaluation of the battery's capacity provides reliable battery performance and forecasts failure conditions, to avoid the risks posed by the battery. Therefore, the useful life of the cell can be used completely without affecting LIB safety. Nonetheless, LIBs are complicated electrochemical devices which have unique non-linear functioning that is reliant on different internal and external conditions [9]. This makes the prediction of the capacity a relatively challenging task [10].

1.2. Previous Work

Many researchers have reported interesting methods for the online evaluation of the capacity of an LIB [11,12]. These capacity estimation techniques can be categorised into two types: (1) model-based approaches, including equivalent circuit models (ECMs) [13–15] and electrochemical models (EMs) [16,17]; (2) the data-driven approaches using neural network (NN) methods [18] and kernel regression methods [19]. In the case of ECM-based models, which are a combination of lumped elements such as resistors, capacitors, and inductors, their main advantage is ease of modelling and implementation in BMS. Plett [13] presented an online estimation of LIB cell capacity by combining an extended Kalman filter (EKF) and an enhanced self-correcting equivalent circuit model. However, ECMs are unable to capture the dynamic behaviour of the LIB and incapable of describing its internal reactions because of the lack of a physical-chemical representation [20].

Thus, special attention has been given to EMs for the capacity estimation of LIBs. The EMs are based on a set of coupled partial differential equations (PDEs) to explain the actual electrochemical reaction process within LIBs and describe its internal reaction process [21]. In doing so, it can capture the cell's dynamic behaviour with a greater degree of accuracy than ECMs. The authors in [22] proposed two sliding mode observers to determine the SoC and SoH of an LIB combined with a reduced order EM. However, the number of partial differential equations in EMs is large and, thus, their solution requires a significant amount of computational time. Creating precise LIB models is not easy since, usually, detailed physical understanding and in depth experimental data are needed in a controlled situation, and these tend to be either unfeasible or too costly [11]. On the other hand, learning algorithms or data-driven estimation techniques have recently been applied to estimate LIB capacity [23]. Data-driven approaches have gained more popularity in recent years for capacity estimation, since these methods do not require an understanding of LIB working principles and are only dependent on the collected/measured experimental data [24]. Examples of such methods are neural networks (NN) [18], support vector machines (SVM) [25], and k-nearest neighbour (kNN) regression [19]. These methods have been used to estimate capacity by learning the dependence of the cell's features as extracted from the measured signal of the voltage, current, and temperature [26]. The capacity estimation framework generally encompasses three key steps: (1) data acquisition; (2) exploration of historical data such as voltage, current, and voltage to extract and construct promising health indicators or features; and (3) using the selected health indicators to build a machine learning model to learn the correlation between the capacity and chosen indicators [12]. Of these three steps, the main challenge is to extract useful indicators from the measured signals to describe more precisely the degradation phenomena of the LIB over its lifetime [27]. For instance, the authors in [28] manually selected five diagnostic features from a charging curve to indicate LIB capacity. These diagnostic features were the initial charge voltage, the constant current (CC) charge capacity, the constant voltage (CV) charge capacity, the final charge voltage, and the final charge current. The selected diagnostic features were then input to a relevance vector machine (RVM) model to estimate the capacity of the

LIB. The above study investigated the diagnostic indicators under predetermined operation conditions, such as certain current or voltage ranges, which involve CC charge and cycling discharge. Nonetheless, in real-world applications like EV applications, the LIB never operates in a static scenario.

Few studies have evaluated a battery's capacity estimation under a dynamic current profile. Venugopal et al. [29] extracted 18 time-domain diagnostic features which affect the battery capacity from the measured voltage, current, and temperature of an LIB operated under a dynamic charging/discharging current profile. The selected features were fed into an independently recurrent NN (IndRNN) for capacity evaluation. The authors in [30] developed a data-driven LIB health diagnosis method based on time and frequency domain features such as mean, covariance, and kurtosis from the time domain features; from the frequency domain features, the authors selected median frequency, band-power, signal-to-noise ratio (SNR), and total harmonic distortion (THD) as the features to determine the battery's capacity. The extracted time and frequency features are then evaluated and ordered based on trendability and monotonicity metrics to select the most features that provide high accuracy and lower computational burden to estimate the battery's capacity. A Gaussian process regression (GPR) model was then designed to capture the correlation between the selected features and the capacity of the LIB.

To enhance the quality of the extracted diagnostic characteristics under a variable current profile, Jonghoon Kim [31] proposed an advanced signal processing method known as discrete wavelet transform (DWT) with multiresolution analysis (MRA) to analyse the non-linearity and non-stationarity behaviour of the battery terminal voltage under dynamic load profile. More useful diagnostic features were extracted for capacity evaluation in the time and frequency domains, respectively. Similarly, the authors in [32] proposed a fast wavelet transform to extract the dynamic features of voltage and current. Then, an xD-Markov machine learning model was established to obtain the battery's capacity from the extracted features. All of the above mentioned studies considered the time or frequency domain for the extraction of the most useful diagnostic features to describe the degradation process inside the LIB. Analyses are usually used to extract indicators from the measured signal and the time and frequency domains, known as conventional feature extraction techniques. However, conventional feature extraction and selection are time-consuming and labour-intensive processes requiring detailed knowledge of the relevant features of the system. This selection process can introduce uncertainty and biased results [33]. Moreover, the performance of machine learning techniques is based on the quality of the features extracted from the measured data. That is, if the extracted characteristic features are weak, this can negatively impact the performance of these approaches [34,35]. Another study [33] reported the difficulty of identifying suitable features with the relevant information required for capacity evaluation.

1.3. Contributions

To overcome the aforementioned issues, a time–frequency image (TFI) analysis-based approach has been proposed to analyse the non-stationary and non-linear behaviour of the LIB terminal voltage in both time and frequency domains and for online prediction of LIB capacity. Moreover, in this paper, we concentrate on the operation of the battery under the variable current load profiles of different battery cycles. The key benefit of the proposed solution is that the battery's terminal voltage is presented in a two-dimensional (2D) representation (time–frequency domain), which provides more helpful characteristics information related to the battery degradation than a one-dimensional (1D) representation (time or frequency domain). Wavelet transforms combined with a deep learning method have been developed for the analysis of the characteristics of LIBs under different operating conditions. The proposed method utilises a deep learning featureless methodology to learn the features of the data automatically, unlike traditional machine learning methods. This avoids manual feature extraction, which relies heavily on human knowledge and experience. Finally, the proposed scheme is a non-parametric estimation method, and, thus, offline testing and modelling are not required. To validate the proposed technique, experimental data on LIB cells from the experimental datasets published by the Prognostics Center of Excellence (PCoE) NASA were used. The results reveal that the proposed scheme improves over competing baseline schemes and achieves 95.60% prediction accuracy. This indicates good potential for the design of improved battery management systems based upon this method.

1.4. Structure

The remainder of the paper is organised as follows: Section 2 introduces the randomised battery dataset analysis. Section 3 presents the proposed capacity imaging analysis scheme. Section 4 presents the results and discussion of the obtained results. Conclusions and future work directions are presented in Section 5.

2. Randomised Battery Dataset

The practical operating environment of a real-world EV battery includes a dynamic and partial driving pattern. Most of the literature utilises a battery dataset with limited assumptions; for example, the battery is cycled using only a constant current profile and specific voltage limit. Nevertheless, these assumptions do not cover the real operating situation of EV batteries. In this study, a randomised battery usage dataset was adopted from the NASA Ames Prognostics Center of Excellence [36,37], to ascertain the impact of actual, dynamic EV driving cycles. This dataset contains the ageing results of four LIBs named RW9, RW10, RW11, and RW12, acquired at room temperature. The general properties of the battery are summarised in Table 1. These four LIBs were cycled using two cycling protocols known as random walk cycling mode and reference charge and discharge cycling mode. The cycle process is shown in Figure 1 [36].

Table 1. The general characteristics of the tested cells.

Battery Properties	18650 LIBs
Manufacture	LG Chem
Chemistry	18650 lithium cobalt oxide vs. graphite
Nominal capacity	2.10 Ah
Capacity range	2.10 Ah–0.80 Ah
Voltage range	4.2–3.2 V

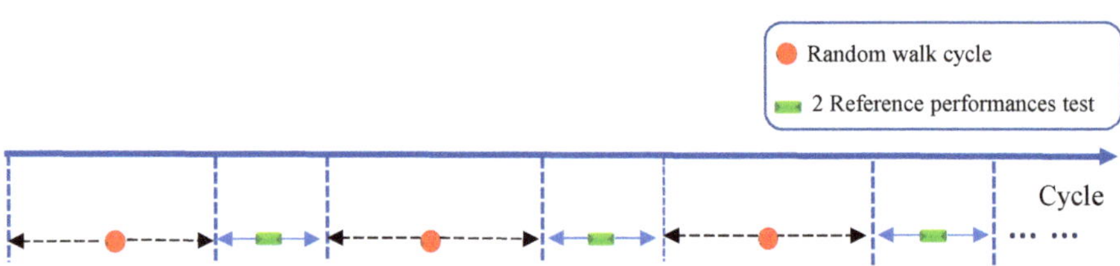

Figure 1. The cycle processes.

2.1. Random Walk Cycling Mode

A sequence of currents set varying between (−4.5 A, −3.75 A, −3 A, −2.25 A, −1.5 A, −0.75 A, 0.75 A, 1.5 A, 2.25 A, 3 A, 3.75 A, 4.5 A) is used for charging and discharging; the four LIBs were continuously cycled. Hence, negative currents are related to the charging operation, while positive currents denote the discharge operation of the batteries. This mode of charging and discharging protocol is known as random walk (RW) operation mode. The aforementioned current sequence is randomly applied to an LIB for five minutes, which is identified as a step in the dataset. It is important to mention that a single random walk (RW) cycle consist of 1500 RW steps and 1500 rests; each RW profile consists of numerous RW cycles. After every RW profile, the battery undergoes several reference charging and discharging tests to measure its capacity and calculate its *SoH* value. Figures 2 and 3 show the measured voltage, current, and temperature of the first and last 100 RW charging and discharging steps of battery RW9, respectively [29].

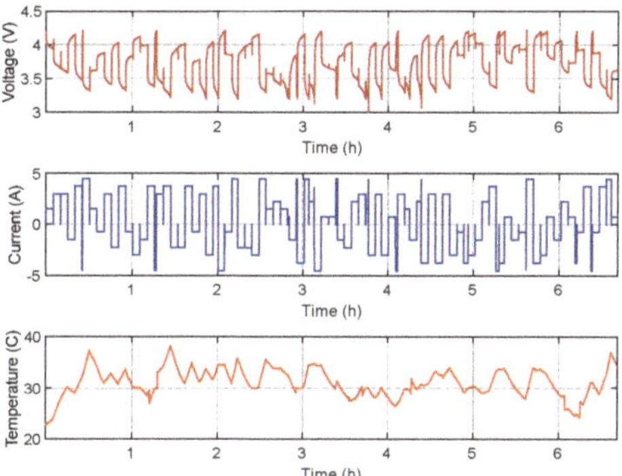

Figure 2. Voltage, current, and temperature of the first 100 steps measured for battery RW9.

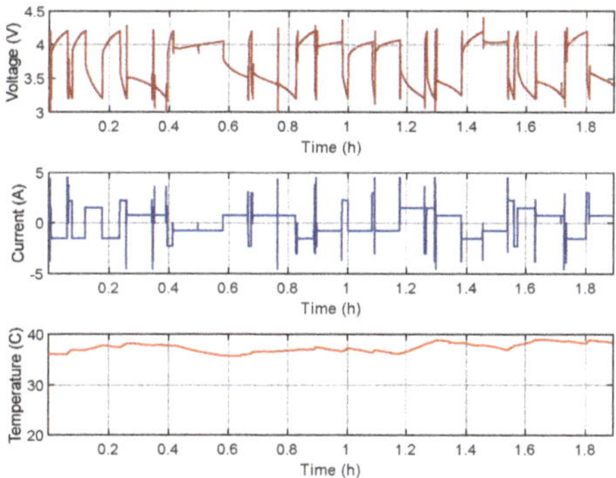

Figure 3. Voltage, current, and temperature of the last 100 steps measured for battery RW9.

2.2. Reference Charge and Discharge Cycle Mode

After the start of each RW cycle, a sequence of reference charging and discharge current profiles is implemented to set a standard benchmark for the battery's SoH. Initially, by applying a 2 A constant current profile, the LIB is charged to its maximum voltage and then a 4.2 V constant voltage is sustained until its current falls to 0.01 A, then it rests for 20 min. Afterwards, 2 A discharge constant current profile is applied to the LIB until its voltage reduces to its minimum voltage, 3.2 V. This procedure is known as reference charging and discharging cycle mode [38]. To measure the LIB's capacity and calculate its *SoH* value, after every RW cycle, the current cell capacity ($Q_{current}$) is calculated using the Coulomb counting method as follows:

$$Q_{current}(\mathrm{t}) = \int_0^t I_d(t)\mathrm{dt} \tag{1}$$

where I_d is the total discharge current of the battery cell. Once the battery capacity is calculated, the *SoH* of the battery can be calculated, as shown in (2).

$$SoH = \frac{Q_{Current}}{Q_{Fresh}} \cdot 100\% \tag{2}$$

where $Q_{Current}$ is the measured capacity after every RW cycle and Q_{Fresh} is the measured capacity of the battery at the beginning of its life. The capacities of the measured cells of the four batteries are shown in Figure 4.

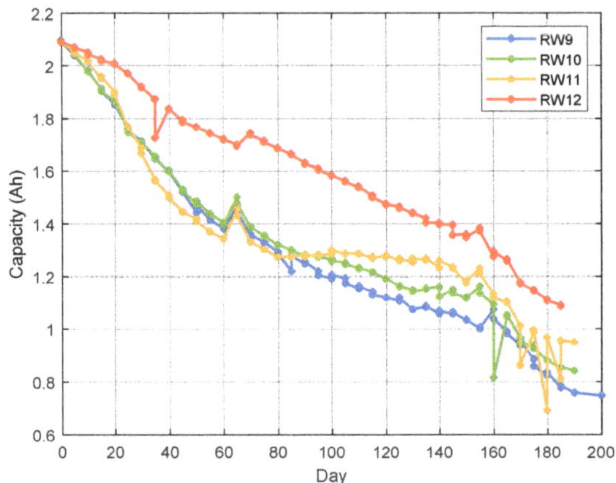

Figure 4. Capacities of the measured cells for four batteries, RW9, RW10, RW11, and RW12.

As illustrated in Figure 4, the capacity fade is a non-linear and non-homogeneous process since each cell degrades in different ways, even under the same test conditions. Therefore, the main aim of this study is to investigate if the time–frequency image analysis method can capture the non-stationary behaviour of the battery during cycling.

3. The Proposed Capacity Imaging Analysis Scheme

Figure 5 shows the block diagram of the proposed TFI paradigm. Here, the terminal voltage of the cycled battery using randomised current profile data is utilised to predict the capacity of the battery. Initially, the terminal voltage of the battery is measured at different capacities, and then the measured voltage is transformed from 1D raw data to a 2D time-frequency image (TFI) by applying a CWT algorithm. The raw data of the measured voltage for the LIB at different capacities contains only the time-domain information, but the converted 2D TFI features of the raw data clearly represent the time and frequency domain information at once. Finally, the generated TFIs for different battery capacities are fed into a deep learning convolutional neural network for feature extraction and image classification. The following subsections explain the capacity prediction steps in detail.

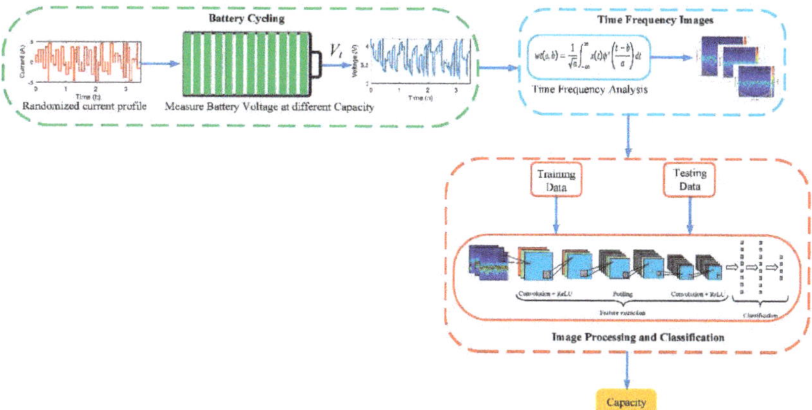

Figure 5. General flow chart of the proposed method for TFI capacity estimation.

3.1. Time–Frequency Image (TFI) Analysis

The main aim of this research was the analysis and classification of the measured voltage of LIBs at different capacities using TFI analysis. As stated in Section 1, most studies have used traditional methods of time or frequency domain to analyse the LIB's measured signals and extract diagnostic characteristics to describe the degradation inside the battery cell. Like most real-life signals, the measured LIB terminal voltage can exhibit non-stationary characteristics [31,32]. Thus, analysing the LIB's measured voltage using traditional time or frequency domains is insufficient to extract comprehensive features about the degradation process inside the battery [39]. Nevertheless, time–frequency approaches can extract several domain features in order to assess the measured signal for the time and frequency domains, precisely extracting better features from the measured signal [40]. The main goal of a time-frequency signal processing analysis is to extract useful information features from the measured battery terminal voltage by converting the time series signal into the time–frequency domain [41,42].

Various methods can be used for time-frequency representation, such as CWT, Wigner-Ville distribution, and short-time Fourier transform (STFT). Of these, CWT is employed to transform the measured signal from the time domain to a TFI, whose energy concentrations visualise changes in frequency components over time as the battery degrades [43]. In the proposed prediction scheme, shown in Figure 5, a CWT was applied to the measured terminal voltage, to transform it from a time domain signal into a time-frequency domain scalogram image. The aim of this conversion process was to extract more useful information about the degradation process inside the battery during a randomised charging/discharging current profile. Through a set of wavelet basis functions, the wavelet transforms (WT) degraded the LIB's terminal voltage in the time-frequency domain. This transform employed a wavelet function of a finite bandwidth in terms of the time and frequency domains. Via the scaling and translation of the wavelet basis function, the signal was degraded with various resolutions at differing scales of time and frequency [44]. Equation (3) describes the mathematical scaling and translation of basic wavelet function:

$$\psi_{a,b}(t) = \frac{1}{\sqrt{a}}\psi\left(\frac{t-b}{a}\right) \tag{3}$$

where $\psi_{a,\,b}(t)$ is an analysis wavelet and is called a child wavelet, $a \in (0, +\infty)$ is the scale parameter of the wavelet transform, and $b \in (-\infty, +\infty)$ is the translation or shift parameter in time. The function $\psi(t)$ is known as the mother wavelet with finite length and fast depletion. Two limit conditions must be satisfied in the mother wavelet signal, which are ($\int_{-\infty}^{+\infty} \psi(t)dt = 0$ and $\int_{-\infty}^{+\infty} |\psi(t)|^2 dt = 1$) [32]. For the given time series signal $x(t)$, the wavelet coefficients $wt(a,b)$ are obtained by the convolution integral of the mother wavelet $\psi(t)$ and the given signal $x(t)$, as presented in (4):

$$wt(a,b) = \frac{1}{\sqrt{a}} \int_{-\infty}^{\infty} x(t)\psi^*\left(\frac{t-b}{a}\right)dt \tag{4}$$

where ψ^* means the complex conjugate of function ψ. Throughout this procedure, the signal $x(t)$ is divided into a sequence of scaled and shifted wavelet coefficients, in which the wavelet set is the basis function. After this, the signal $x(t)$ is altered by CWT and sent to the 2D time-frequency domains [45]. With this approach, the 1D time domain signals are transformed into TFI. Figure 6 shows an example schematic of the battery measured voltage at three different capacities and the transformed time-frequency domain representation features for each capacity. It can be observed that there are clear discriminative variations in the CWT coefficients at each capacity in the TFI. Therefore, battery degradation is well reflected by the multi-domain TFI features, which can thus be used as input for the deep learning convolutional neural network (CNN) to classify the battery's capacity during the degradation.

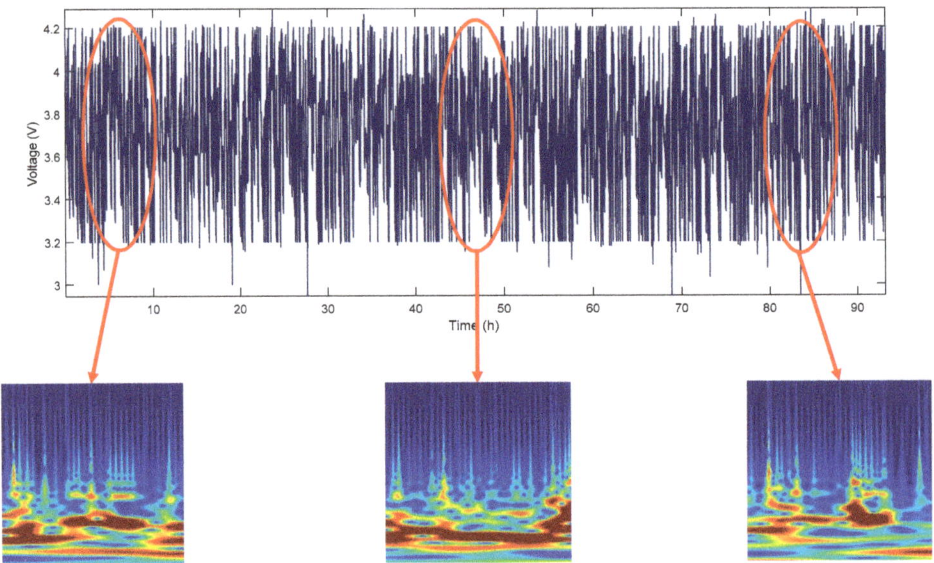

Figure 6. Measured voltage of lithium-ion battery and transformed TFI energy concentration spectrum information of the measured voltage at three different capacity values.

3.2. *Time-Frequency Image Analysis and Classification Using Deep Learning Algorithm*

The obtained TFIs from the battery measured voltage at different capacities are fed into the deep learning predictive model to extract the relationship between the TFIs and corresponding capacity. In this research, a data-driven method based on transfer deep learning model is adopted to perform capacity classification in LIB.

Generally, a DL-CNN is built from three types of layers, namely (1) convolution layers, (2) pooling layers, and (3) fully connected layers [44]. The convolution and pooling layers are linked to form convolution blocks, and many of these blocks are stacked to create a deep architecture. The fully connected layer has generally been used as the final layer in the classification or regression. The general architecture of the DL-CNN is shown in Figure 7 [46].

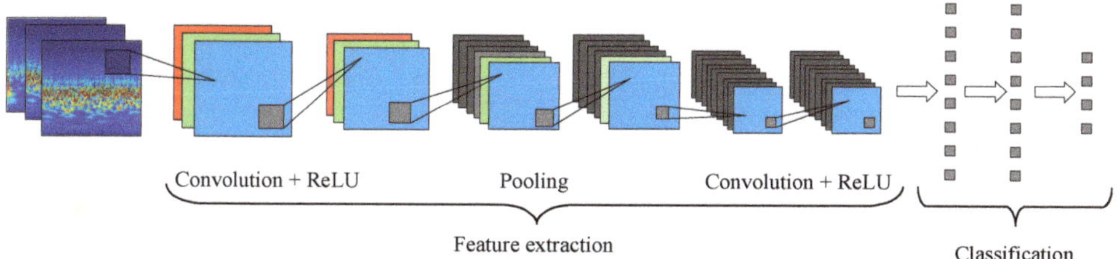

Figure 7. Deep learning convolutional neural network architecture.

The convolutional layer combines the input image from the initial layer with several filters, known as kernels, and these are then fed to the activation function to create a features map [47]. The output features map is the convolutional result of the input maps and can be calculated as in (5) [48]:

$$x_j{}' = f\left(\sum_{i \in M_j} x_i{}^{l-1} * k_{ij}{}^{l} + b_j{}^{l}\right) \tag{5}$$

Here, $*$ is the convolutional operation, x_i is ith input map, k is a F × F convolutional filter, b_j is the additive bias, M_j is the feature map of the convolutional layer, and l is the lth layers in the neural network. Lastly, the obtained results from the convolutional layers are fed to the activation function. The common used activation function is the rectified linear unit (ReLU), defined as in (6) [47–49].

$$ReLU(x) = \max(0, x) \tag{6}$$

For the pooling layer, which follows the convolutional layer, low-resolution maps are created from the most significant local information. The maximum value is derived from each region by using the max pooling layer, as shown in Figure 8 [48]. In the fully connected layer, a 1D vector is utilized to represent all the feature maps, which is fully connected to the output layer. The output of the fully connected layer is described as given in (7):

$$O_i = f\left(\sum_{j=1}^{d} x_j{}^{F} w_{ij} + b_i\right) \tag{7}$$

where O_i is the output layer, $x_j{}^F$ is the jth neuron in the fully connected layer, w_{ij} is the weight related to O_i and $x_j{}^F$, b_i is the bias corresponding to O_i, and f is the activation function [48,50].

4	0	4	4
1	2	5	8
3	6	3	2
2	4	5	0

Max pool with 2 × 2 filters

4	8
6	5

Figure 8. Max pooling layer.

In this study, two deep transfer learning algorithms were applied to classify the generated results from TFI. The classification results are presented in the Results and Discussion section.

4. Results and Discussion

The proposed scheme (see Figure 5) was implemented using the MATLAB environment. First, the results of TFI estimation using complex Morlet mother wavelet transform (CMMW) will be presented in this section, and then the results of image analysis and classification using deep learning algorithms will be demonstrated.

4.1. Time–Frequency Image (TFI) Results

Figure 9a shows the measured voltage signal of the first cycle of RW9 at the beginning of the life of the LIB, indicating that no degradation had occurred at this point and the battery's capacity was full (here, C = 2.1 Ah). Figure 9c presents the measured voltage at the end of the LIB's life and shows that the battery degraded, and its capacity decreased to its minimum value (C = 0.8 Ah). From both results, it can be clearly observed that the battery had aged and that the time period of the measured voltage (31.07 h) at the end of life was less than that for the first cycle of the battery (93.18 h). This is because the battery has reached its threshold voltage level very frequently due to an increase in the internal resistance of the LIB, and this inherently reduces the LIB capacity.

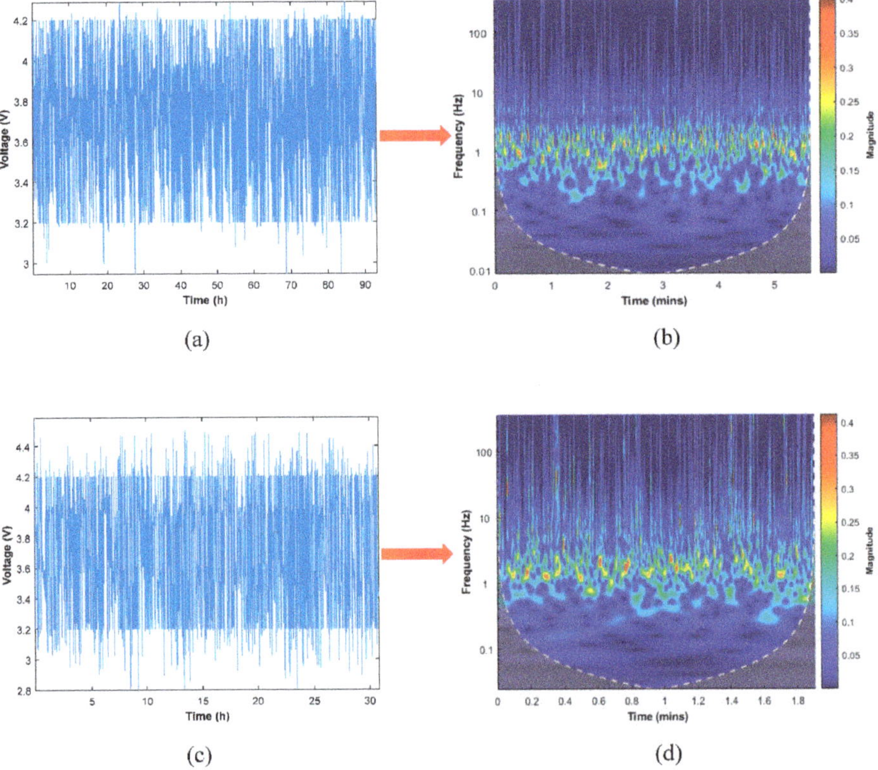

Figure 9. Time-domain measured voltage and plot of TFI energy concentration spectrum of battery cell (RW9): (**a**) measured voltage signal (beginning of life), (**b**) wavelet energy concentration spectrum (beginning of life), (**c**) measured voltage signal (end of life), and (**d**) wavelet energy concentration spectrum (end of life).

To extract appropriate features capable of describing the degradation phenomena inside the battery from the measured voltage, CMMW was applied using Equations (3) and (4) to the LIB terminal voltage, as illustrated in Figure 9b,d. In the TFI, the energy concentration of the measured voltage of the LIB is shown at different capacities, with the horizontal and vertical axes representing time and frequency, respectively. The amplitudes of the wavelet coefficients are represented in different colours on the TFI [47]; red colour indicates that the level of energy density is high. In Figure 9b, when the LIB is at the beginning of its life, the energy distribution is concentrated in the middle frequency range of the image and varies from 1 to 2 Hz [51]. In contrast, Figure 9d shows the energy distributed in the middle and high frequency bands (up to 100 Hz) of the image, meaning that the amplitude of the wavelet coefficients increased to cover the rapid changes between the upper and lower voltage range. A comparison of the TFIs for the battery at the beginning and end of its life clearly shows a significant difference in the energy density distribution. Therefore, the capacity degradation of the battery is well reflected by the TFI information and multiple domain features can be calculated using TFI.

The datasets for the four test batteries were processed by the proposed CMMW. Figure 10 presents a sample of the results of the time-frequency images of the measured LIB terminal voltage at different capacities. They show the energy concentration distribution of the battery terminal voltage according to the level of degradation in the TFI. As shown in Figure 10, as the battery's capacity decreases, the distribution of the energy in the TFI changes, clearly illustrating the degradation process inside the battery cell. From the extracted wavelet coefficients using CMMW, the energy concentrations for wavelet coefficients are calculated in the form of TFIs, which are then fed to the pre-trained CNN for image pattern recognition and classification.

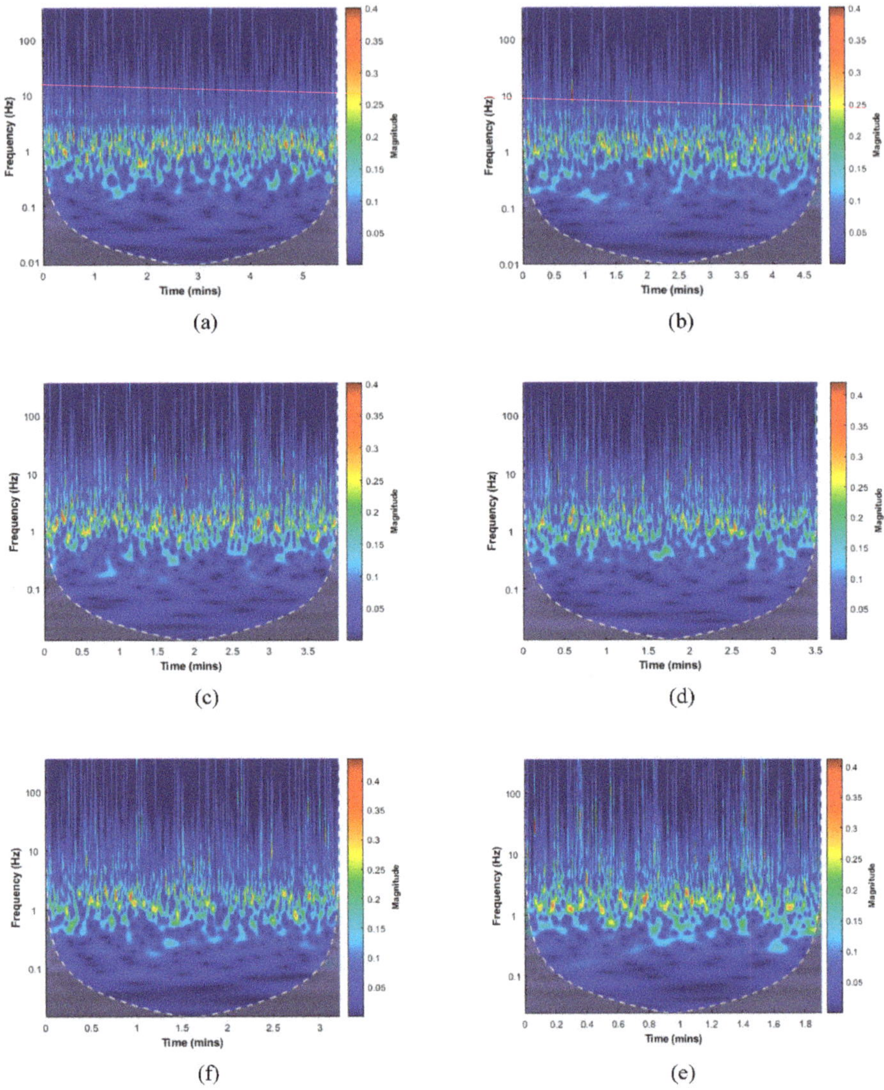

Figure 10. TFI energy concentration spectrum results of the battery RW9 measured voltage at different capacities: (**a**) 2.1, (**b**) 1.8, (**c**) 1.6, (**d**) 1.4, (**f**) 1.2, (**e**) 0.8 Ah.

To further validate the proposed technique, the experimental datasets of the LIBs RW10, RW11, and RW12 at different capacity ranges from 2.1 to 1.2 Ah have been tested. Table 2 demonstrates the effectiveness of the proposed TFI method to predict the capacity of these cells. Ongoing research is mainly attempting to apply the proposed technique to quantify which type of degradation mode most affects the battery's capacity.

Table 2. TFIs energy concentration spectrum results of RW10, RW11, and RW12 LIBs measured voltage at different capacities.

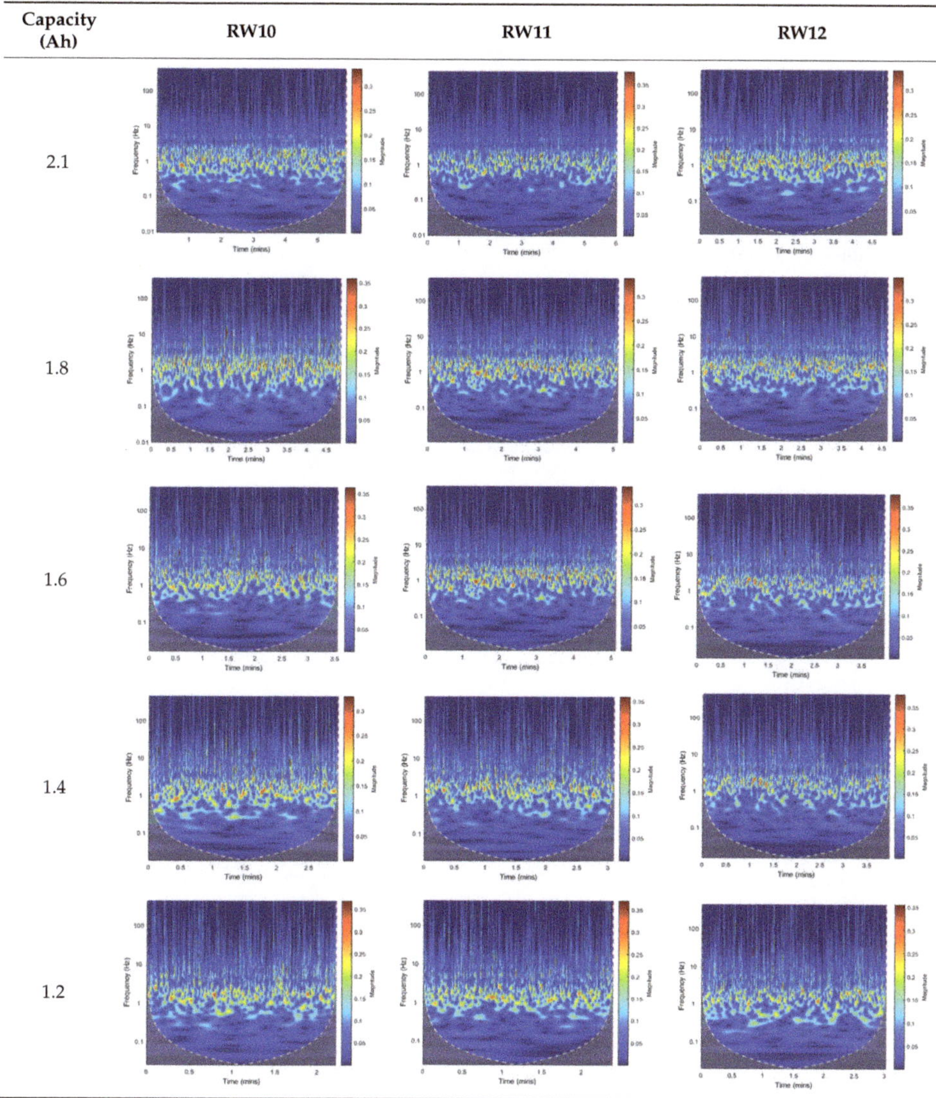

Capacity (Ah)	RW10	RW11	RW12
2.1			
1.8			
1.6			
1.4			
1.2			

4.2. DL-CNN Results

In this subsection, we demonstrate how the generated TFIs from the measured terminal voltage at different capacities through CWT were fed into AlexNet and VGG-16 DL-CNNs for capacity classification, after they had been trained for TFI classification. The Stochastic gradient descent (SGDM) and adaptive moment estimation (ADAM) optimisation algorithms were used to train the generated TFIs.

4.2.1. AlexNet Neural Network

The AlexNet neural network was created in 2012 by Krizhevsky et al. [52] and is built from eight layers (five convolutional and three fully connected layers). In this study, we assumed that the battery's capacity is classified into five classes from the first measured capacity value, which is 2.1 Ah, to the measured capacity value of 1.20 Ah in the experimental dataset—similar to the work presented in [53,54]. The measured voltage of each battery per RW cycle was labelled to show its corresponding capacity. The datasets of LIBs RW9–12 were split into training data (70% of all TFI in the four folds) and 30% for testing data. Therefore, the LIB capacity estimation can be considered a 5-class classification. The size of the generated TFIs is 227 × 227 × 3, which is suitable for the pretrained AlexNet network. The training settings for modelling the network and the AlexNet architecture are given in Tables 3 and 4. The AlexNet output layer was replaced with a new output layer with five neurons corresponding to five classes of capacity values. Since the AlexNet network is not trained to classify TFIs, only the weights of the last convolution layer and fully connected layer were trained, while the first four convolution layers were not. Training only the last layers of the AlexNet network reduced the training time of the model [44].

Table 3. Hyperparameter settings for the trained AlexNet model.

Hyperparameters	Values
Momentum	0.9
Initial learning rate	0.0001
Learning rate drop factor	0.1
Learning rate drop period	10
Number of epochs	50
Batch size	15
Optimiser	SGDM, ADAM

Table 4. AlexNet architecture.

Name	Type	Activations	Learnable
Data 227 × 227 × 3 images	Image input	227 × 227 × 3	-
Conv 1	Convolution	55 × 55 × 96	Weights 11 × 11 × 3 × 96 Bias 1 × 1 × 96
Pool 1	Max Pooling	27 × 27 × 96	-
Conv 2	Convolution	27 × 27 × 256	Weights 5 × 5 × 48 × 256 Bias 1 × 1x256
Pool 2	Max Pooling	13 × 13 × 256	-
Conv 3	Convolution	13 × 13 × 384	Weights 3 × 3 × 256 × 384 Bias 1 × 1 × 384
Conv 4	Convolution	13 × 13 × 384	Weights 3 × 3 × 192 × 384 Bias 1 × 1 × 384
Conv 5	Convolution	13 × 13 × 256	Weights 3 × 3 × 192 × 256 Bias 1 × 1x256
Pool 5	Max Pooling	6 × 6 × 256	-
Fc6	Fully Connected	1 × 1 × 4096	Weights 4096 × 9216 Bias 4096 × 1
Fc7	Fully Connected	1 × 1 × 4096	Weights 4096 × 4096 Bias 4096 × 1
Fc8	Fully Connected	1 × 1 × 1000	Weights 1000 × 4096 Bias 1000 × 1
Prob Softmax layer	Softmax	1 × 1 × 1000	-
Output	Classification	-	-

During the training phase, SGDM and ADAM optimisers were utilised with an initial learning rate of 0.0001, and the batch size was set at 50. Once the training process was finished, the accuracy of the classification by the model was evaluated using the test dataset. A deep learning toolbox from MATLAB was employed to train and test the model. Classification accuracy was calculated to evaluate the model's accuracy, and the test dataset results are given in Table 5. The results show that the trained model correctly classified the test data with an accuracy rate of 95.69%. Moreover, the SDGM and ADAM optimisers achieved good performance regarding updating the weights of our model. Thus, these classification results demonstrate that the proposed method achieved accurate predictions of battery capacity for each battery cell.

Table 5. Capacity prediction accuracy for each battery cell using AlexNet.

	RW9		RW10		RW11		RW12	
Optimiser	SGDM	Adam	SGDM	ADAM	SGDM	ADAM	SGDM	ADAM
Accuracy	95.0673%	95.69%	91.96%	94.20%	93.39%	94.27%	90.25%	91.5%

4.2.2. VGG-16 Neural Network

The second pretrained model tested in this study is a convolutional neural network from the Oxford Visual Geometry Group (VGG) which is a 16-layer network [55]. VGG-16 has achieved classification accuracy performance on the ImageNet dataset and Table 6 shows the layer structure of VGG-16. With the same hyperparameter settings, the VGG-16 model is compared with the AlexNet model and the prediction accuracy results are presented in Table 7. The comparison classification results from Tables 5 and 7 for AlexNet and the VGG-16 models show that as the DL-CNN becomes deeper, the accuracy of the model will increase. However, as the DL-CNN becomes deeper, the training time will increase, and more computational complexity will be added to the model.

Table 6. VGG-16 architecture hyperparameters.

Name	Type	Activations	Learnable
Data 224 × 224 × 3 images	Image input	224 × 224 × 3	-
Block 1-Conv 1	Convolution	224 × 224 × 64	Weights 3 × 3 × 3 × 64, Bias 1 × 1x 64
Block 1-Conv 2	Convolution	224 × 224 × 64	Weights 3 × 3 × 3 × 64, Bias 1 × 1 × 64
Block 1-Pool	Max Pooling	112 × 112 × 64	-
Block 2-Conv 1	Convolution	112 × 112 × 128	Weights 3 × 3 × 64 × 128, Bias 1 × 1 × 128
Block 2-Conv 2	Convolution	112 × 112 × 128	Weights 3 × 3 × 128 × 128, Bias 1 × 1 × 128
Block 2-Pool	Max Pooling	56 × 56 × 128	-
Block 3-Conv 1	Convolution	56 × 56 × 256	Weights 3 × 3 × 128 × 256, Bias 1 × 1 × 256
Block 3-Conv 2	Convolution	56 × 56 × 256	Weights 3 × 3 × 128 × 256, Bias 1 × 1 × 256
Block 3-Pool	Max Pooling	28 × 28 × 256	-
Block 4-Conv 1	Convolution	28 × 28 × 512	Weights 3 × 3 × 256 × 512, Bias 1 × 1 × 512
Block 4-Conv 2	Convolution	28 × 28 × 512	Weights 3 × 3 × 256 × 512, Bias 1 × 1 × 512
Block 4-Pool	Max Pooling	14 × 14 × 512	-
Block 5-Conv 1	Convolution	14 × 14 × 512	Weights 3 × 3 × 512 × 512, Bias 1 × 1 × 512
Block 5-Conv 2	Convolution	14 × 14 × 512	Weights 3 × 3 × 512 × 512, Bias 1 × 1 × 512
Block 5-Pool	Max Pooling	7 × 7 × 512	-
Fc1	Fully Connected	1 × 1 × 4096	Weights 4096 × 4096, Bias 4096 × 1
Fc2	Fully Connected	1 × 1 × 4096	Weights 4096 × 4096, Bias 4096 × 1
Prob Softmax layer	Softmax	1 × 1 × 1000	-
Output	Classification	-	-

Table 7. Capacity prediction accuracy for each battery cell using the VGG-16 model.

	RW9		RW10		RW11		RW12	
Optimiser	SGDM	ADAM	SGDM	ADAM	SGDM	ADAM	SGDM	ADAM
Accuracy	95.52%	95.52%	95.09%	95.60%	94.29%	94.92%	92.25%	95.5%

5. Conclusions

This paper proposed a new capacity evaluation method for LIBs using multi-domain features obtained from a TFI algorithm. The terminal voltage of the battery was transformed into a 2D image feature using CWT instead of 1D raw data of terminal voltage or the extraction of multiple statistical features. The proposed method was applied CWT to produce a 2D multi-domain features time–frequency representation of the battery terminal voltage, known as a TFI. To this end, experimental data on four LIBs cells published by the Prognostics Center of Excellence (PCoE) NASA were adopted to validate the effectiveness of the proposed method. The generated TF images clearly demonstrated the degradation process of the battery throughout the distribution of the energy concentration of the measured voltage for the battery at different capacity values. Two pre-trained DL-CNN were utilised to classify the generated TFIs at various capacities into five classes. The classification results achieved 95.69% accuracy using the AlexNet network and 95.52% accuracy from the VGG-16 network. The accuracy of the proposed method indicates that the proposed technique can be an effective health prognostic tool for managing LIBs for various applications such as electric vehicles and grid applications. However, the proposed method should be tested under different operating conditions (i.e., different temperatures and high-discharge current profile) to increase the accuracy of prediction and to develop a general prediction model. It is important to emphasise that the TFIs generated can be improved since the TFIs were generated by the use of a classical CWT. The resolution of the obtained energy concentration spectra affects the performance of the deep learning algorithm for capacity estimation. Our future work will therefore explore the applicability of the multi-domain features using energy concentration in time–frequency image analysis to provide robust results under different operating and temperature conditions. In addition, other time–frequency techniques such as STFT and Winger–Ville distribution will be investigated and compared to each other in terms of time–frequency resolution and computational complexity for online capacity prediction. Finally, future work will also compare computation load for the proposed algorithms.

Author Contributions: Conceptualization, M.E.-D. (Ma'd El-Dalahmeh) and M.A.-G.; methodology, M.E.-D. (Ma'd El-Dalahmeh) and M.A.-G.; formal analysis, M.E.-D. (Mo'ath El-Dalahmeh), M.A.-G., and M.E.-D. (Ma'd El-Dalahmeh); investigation, M.E.-D. (Ma'd El-Dalahmeh) and M.A.-G.; writing—original draft preparation, M.E.-D. (Mo'ath El-Dalahmeh); writing—review and editing, M.A.-G., M.S.; supervision, M.A.-G., M.S. All authors have read and agreed to the published version of the manuscript.

Funding: This research received no external funding.

Conflicts of Interest: The authors declare no conflict of interest.

References

1. Ren, G.; Ma, G.; Cong, N. Review of electrical energy storage system for vehicular applications. *Renew. Sustain. Energy Rev.* **2015**, *41*, 225–236. [CrossRef]
2. Waag, W.; Fleischer, C.; Sauer, D.U. Critical review of the methods for monitoring of lithium-ion batteries in electric and hybrid vehicles. *J. Power Sources* **2014**, *258*, 321–339. [CrossRef]
3. Zhang, R.; Xia, B.; Li, B.; Cao, L.; Lai, Y.; Zheng, W.; Wang, H.; Wang, W. State of the Art of Lithium-Ion Battery SOC Estimation for Electrical Vehicles. *Energies* **2018**, *11*, 1820. [CrossRef]
4. Cadini, F.; Sbarufatti, C.; Cancelliere, F.; Giglio, M. State-of-life prognosis and diagnosis of lithium-ion batteries by data-driven particle filters. *Appl. Energy* **2019**, *235*, 661–672. [CrossRef]
5. Vetter, J.; Novák, P.; Wagner, M.R.; Veit, C.; Möller, K.C.; Besenhard, J.O.; Winter, M.; Wohlfahrt-Mehrens, M.; Vogler, C.; Hammouche, A. Ageing mechanisms in lithium-ion batteries. *J. Power Sources* **2005**, *147*, 269–281. [CrossRef]
6. Lu, L.; Han, X.; Li, J.; Hua, J.; Ouyang, M. A review on the key issues for lithium-ion battery management in electric vehicles. *J. Power Sources* **2013**, *226*, 272–288. [CrossRef]
7. Farmann, A.; Waag, W.; Marongiu, A.; Sauer, D.U. Critical review of on-board capacity estimation techniques for lithium-ion batteries in electric and hybrid electric vehicles. *J. Power Sources* **2015**, *281*, 114–130. [CrossRef]
8. Tian, H.; Qin, P.; Li, K.; Zhao, Z. A review of the state of health for lithium-ion batteries: Research status and suggestions. *J. Clean. Prod.* **2020**, *261*, 120813. [CrossRef]
9. Berecibar, M.; Gandiaga, I.; Villarreal, I.; Omar, N.; Van Mierlo, J.; Van den Bossche, P. Critical review of state of health estimation methods of Li-ion batteries for real applications. *Renew. Sustain. Energy Rev.* **2016**, *56*, 572–587. [CrossRef]
10. Lipu, M.S.H.; Hannan, M.A.; Hussain, A.; Hoque, M.M.; Ker, P.J.; Saad, M.H.M.; Ayob, A. A review of state of health and remaining useful life estimation methods for lithium-ion battery in electric vehicles: Challenges and recommendations. *J. Clean. Prod.* **2018**, *205*, 115–133. [CrossRef]

11. Xiong, R.; Li, L.; Tian, J. Towards a smarter battery management system: A critical review on battery state of health monitoring methods. *J. Power Sources* **2018**, *405*, 18–29. [CrossRef]
12. Meng, H.; Li, Y.-F. A review on prognostics and health management (PHM) methods of lithium-ion batteries. *Renew. Sustain. Energy Rev.* **2019**, *116*, 109405. [CrossRef]
13. Plett, G.L. Extended Kalman filtering for battery management systems of LiPB-based HEV battery packs: Part 3. State and parameter estimation. *J. Power Sources* **2004**, *134*, 277–292. [CrossRef]
14. Bian, X.; Liu, L.; Yan, J. A model for state-of-health estimation of lithium ion batteries based on charging profiles. *Energy* **2019**, *177*, 57–65. [CrossRef]
15. Zou, Y.; Hu, X.; Ma, H.; Li, S.E. Combined State of Charge and State of Health estimation over lithium-ion battery cell cycle lifespan for electric vehicles. *J. Power Sources* **2015**, *273*, 793–803. [CrossRef]
16. Bartlett, A.; Marcicki, J.; Onori, S.; Rizzoni, G.; Yang, X.G.; Miller, T. Electrochemical Model-Based State of Charge and Capacity Estimation for a Composite Electrode Lithium-Ion Battery. *IEEE Trans. Control Syst. Technol.* **2016**, *24*, 384–399. [CrossRef]
17. Lotfi, N.; Li, J.; Landers, R.G.; Park, J. Li-ion Battery State of Health Estimation based on an improved Single Particle model. In Proceedings of the 2017 American Control Conference (ACC), Seattle, WA, USA, 24–26 May 2017; pp. 86–91.
18. Wu, B.; Han, S.; Shin, K.G.; Lu, W. Application of artificial neural networks in design of lithium-ion batteries. *J. Power Sources* **2018**, *395*, 128–136. [CrossRef]
19. Hu, C.; Jain, G.; Zhang, P.; Schmidt, C.; Gomadam, P.; Gorka, T. Data-driven method based on particle swarm optimization and k-nearest neighbor regression for estimating capacity of lithium-ion battery. *Appl. Energy* **2014**, *129*, 49–55. [CrossRef]
20. Hu, X.; Li, S.; Peng, H. A comparative study of equivalent circuit models for Li-ion batteries. *J. Power Sources* **2012**, *198*, 359–367. [CrossRef]
21. Li, J.; Adewuyi, K.; Lotfi, N.; Landers, R.G.; Park, J. A single particle model with chemical/mechanical degradation physics for lithium ion battery State of Health (SOH) estimation. *Appl. Energy* **2018**, *212*, 1178–1190. [CrossRef]
22. Lin, C.; Xing, J.; Tang, A. Lithium-ion Battery State of Charge/State of Health Estimation Using SMO for EVs. *Energy Procedia* **2017**, *105*, 4383–4388. [CrossRef]
23. Li, Y.; Liu, K.; Foley, A.M.; Zülke, A.; Berecibar, M.; Nanini-Maury, E.; Van Mierlo, J.; Hoster, H.E. Data-driven health estimation and lifetime prediction of lithium-ion batteries: A review. *Renew. Sustain. Energy Rev.* **2019**, *113*, 109254. [CrossRef]
24. Ng, M.-F.; Zhao, J.; Yan, Q.; Conduit, G.J.; Seh, Z.W. Predicting the state of charge and health of batteries using data-driven machine learning. *Nat. Mach. Intell.* **2020**, *2*, 161–170. [CrossRef]
25. Nuhic, A.; Terzimehic, T.; Soczka-Guth, T.; Buchholz, M.; Dietmayer, K. Health diagnosis and remaining useful life prognostics of lithium-ion batteries using data-driven methods. *J. Power Sources* **2013**, *239*, 680–688. [CrossRef]
26. Severson, K.A.; Attia, P.M.; Jin, N.; Perkins, N.; Jiang, B.; Yang, Z.; Chen, M.H.; Aykol, M.; Herring, P.K.; Fraggedakis, D.; et al. Data-driven prediction of battery cycle life before capacity degradation. *Nat. Energy* **2019**, *4*, 383–391. [CrossRef]
27. Pan, H.; Lü, Z.; Wang, H.; Wei, H.; Chen, L. Novel battery state-of-health online estimation method using multiple health indicators and an extreme learning machine. *Energy* **2018**, *160*, 466–477. [CrossRef]
28. Hu, C.; Jain, G.; Schmidt, C.; Strief, C.; Sullivan, M. Online estimation of lithium-ion battery capacity using sparse Bayesian learning. *J. Power Sources* **2015**, *289*, 105–113. [CrossRef]
29. Venugopal, P.; Vigneswaran, T. State-of-Health Estimation of Li-ion Batteries in Electric Vehicle Using IndRNN under Variable Load Condition. *Energies* **2019**, *12*, 4338. [CrossRef]
30. Khaleghi, S.; Firouz, Y.; Van Mierlo, J.; Van den Bossche, P. Developing a real-time data-driven battery health diagnosis method, using time and frequency domain condition indicators. *Appl. Energy* **2019**, *255*, 113813. [CrossRef]

31. Kim, J. Discrete Wavelet Transform-Based Feature Extraction of Experimental Voltage Signal for Li-Ion Cell Consistency. *IEEE Trans. Veh. Technol.* **2016**, *65*, 1150–1161. [CrossRef]
32. Cai, Y.; Yang, L.; Deng, Z.; Zhao, X.; Deng, H. Online identification of lithium-ion battery state-of-health based on fast wavelet transform and cross D-Markov machine. *Energy* **2018**, *147*, 621–635. [CrossRef]
33. You, G.-W.; Park, S.; Oh, D. Real-time state-of-health estimation for electric vehicle batteries: A data-driven approach. *Appl. Energy* **2016**, *176*, 92–103. [CrossRef]
34. Shen, S.; Sadoughi, M.; Chen, X.; Hong, M.; Hu, C. A deep learning method for online capacity estimation of lithium-ion batteries. *J. Energy Storage* **2019**, *25*, 100817. [CrossRef]
35. Shen, S.; Sadoughi, M.; Li, M.; Wang, Z.; Hu, C. Deep convolutional neural networks with ensemble learning and transfer learning for capacity estimation of lithium-ion batteries. *Appl. Energy* **2020**, *260*, 114296. [CrossRef]
36. Bole, B.; Kulkarni, C.S.; Daigle, M. Randomized battery usage data set. *NASA AMES Progn. Data Repos.* **2014**, *70*. Available online: https://ti.arc.nasa.gov/tech/dash/groups/pcoe/prognostic-data-repository/ (accessed on 10 October 2020).
37. Bole, B.; Kulkarni, C.S.; Daigle, M. Adaptation of an electrochemistry-based Li-ion battery model to account for deterioration observed under randomized use. *Proc. Annu. Conf. Progn. Health Manag. Soc.* **2014**, *2*, 1–9.
38. Yu, J.; Mo, B.; Tang, D.; Yang, J.; Wan, J.; Liu, J. Indirect State-of-Health Estimation for Lithium-Ion Batteries under Randomized Use. *Energies* **2017**, *10*, 2012. [CrossRef]
39. Feltane, A. Time-Frequency Based Methods for Non-Stationary Signal Analysis with Application To EEG Signals. Ph.D. Thesis, The University of Rhode Island, Kingston, RL, USA, 2016.
40. Sejdić, E.; Orović, I.; Stanković, S. Compressive sensing meets time–frequency: An overview of recent advances in time–frequency processing of sparse signals. *Digit. Signal Process.* **2018**, *77*, 22–35. [CrossRef]
41. Li, X.; Ding, Q.; Sun, J.-Q. Remaining useful life estimation in prognostics using deep convolution neural networks. *Reliab. Eng. Syst. Saf.* **2018**, *172*, 1–11. [CrossRef]
42. Boashash, B.; Ouelha, S. Automatic signal abnormality detection using time-frequency features and machine learning: A newborn EEG seizure case study. *Knowl. Based Syst.* **2016**, *106*, 38–50. [CrossRef]
43. Yan, R.; Gao, R.X.; Chen, X. Wavelets for fault diagnosis of rotary machines: A review with applications. *Signal Process.* **2014**, *96*, 1–15. [CrossRef]
44. Shao, S.; McAleer, S.; Yan, R.; Baldi, P. Highly Accurate Machine Fault Diagnosis Using Deep Transfer Learning. *IEEE Trans. Ind. Inform.* **2019**, *15*, 2446–2455. [CrossRef]
45. Boashash, B.; Ouelha, S. Designing high-resolution time–frequency and time–scale distributions for the analysis and classification of non-stationary signals: A tutorial review with a comparison of features performance. *Digit. Signal Process.* **2018**, *77*, 120–152. [CrossRef]
46. Nielsen, M.A. *Neural Networks and Deep Learning*; Determination press: San Francisco, CA, USA, 2015; Volume 2018.
47. Yoo, Y.; Baek, J.-G. A Novel Image Feature for the Remaining Useful Lifetime Prediction of Bearings Based on Continuous Wavelet Transform and Convolutional Neural Network. *Appl. Sci.* **2018**, *8*, 1102. [CrossRef]
48. Schmidhuber, J. Deep learning in neural networks: An overview. *Neural Netw.* **2015**, *61*, 85–117. [CrossRef]
49. Aggarwal, C.C. *Neural Networks and Deep Learning*; Springer: Cham, Switzerland, 2018.
50. Liao, Y.; Zeng, X.; Li, W. Wavelet transform based convolutional neural network for gearbox fault classification. In Proceedings of the 2017 Prognostics and System Health Management Conference (PHM-Harbin), Harbin, China, 9–12 July 2017; pp. 1–6.
51. Zhang, Y.; Tang, Q.; Zhang, Y.; Wang, J.; Stimming, U.; Lee, A.A. Identifying degradation patterns of lithium ion batteries from impedance spectroscopy using machine learning. *Nat. Commun.* **2020**, *11*, 1706. [CrossRef] [PubMed]
52. Krizhevsky, A.; Sutskever, I.; Hinton, G.E. Imagenet classification with deep convolutional neural networks. In Proceedings of the Advances in Neural Information Processing Systems 25 (NIPS 2012), Lake Tahoe, NV, USA, 3–8 December 2012; pp. 1097–1105.
53. Patil, M.A.; Tagade, P.; Hariharan, K.S.; Kolake, S.M.; Song, T.; Yeo, T.; Doo, S. A novel multistage Support Vector Machine based approach for Li ion battery remaining useful life estimation. *Appl. Energy* **2015**, *159*, 285–297. [CrossRef]

54. Ali, M.U.; Zafar, A.; Nengroo, S.H.; Hussain, S.; Park, G.-S.; Kim, H.-J. Online Remaining Useful Life Prediction for Lithium-Ion Batteries Using Partial Discharge Data Features. *Energies* **2019**, *12*, 4366. [CrossRef]
55. Simonyan, K.; Zisserman, A. Very deep convolutional networks for large-scale image recognition. *arXiv* **2014**, arXiv:1409.1556.

Article

Stacked Boosters Network Architecture for Short-Term Load Forecasting in Buildings

Tuukka Salmi *, Jussi Kiljander and Daniel Pakkala

VTT Technical Research Centre of Finland, FI-02044 Espoo, Finland; jussi.kiljander@vtt.fi (J.K.); daniel.pakkala@vtt.fi (D.P.)

* Correspondence: tuukka.salmi@vtt.fi

Received: 19 March 2020; Accepted: 3 May 2020; Published: 9 May 2020

Abstract: This paper presents a novel deep learning architecture for short-term load forecasting of building energy loads. The architecture is based on a simple base learner and multiple boosting systems that are modelled as a single deep neural network. The architecture transforms the original multivariate time series into multiple cascading univariate time series. Together with sparse interactions, parameter sharing and equivariant representations, this approach makes it possible to combat against overfitting while still achieving good presentation power with a deep network architecture. The architecture is evaluated in several short-term load forecasting tasks with energy data from an office building in Finland. The proposed architecture outperforms state-of-the-art load forecasting model in all the tasks.

Keywords: deep neural networks; short-term load forecasting

---

1. Introduction

Due to increasing utilization of variable renewable energy (VAE), demand flexibility is becoming crucial part of the stabilization of smart grids. Buildings constitute 32% of global final energy consumption [1] and possess natural thermal storage capacity making them important resource for demand flexibility. Accurate short-term load forecasting is crucial part of demand-side flexibility management.

Short-term energy load forecasting has been studied extensively, but there are some limitations in the existing state-of-the-art solutions. According to a recent study by Amasyali et al. [2], Artificial Neural Networks (ANNs) are the most commonly used machine learning model in short-term forecasting of building energy loads. The challenge with artificial neural networks is that they require a lot of data to prevent overfitting. This is usually tackled by reducing the number of layers, which in turn reduces the presentation power of the model. For instance, a very typical neural network architecture in building load forecasting is a single hidden layer multilayer perceptron (MLP) model [3–6]. Besides neural network methods, support vector regression (SVR) is a common machine learning model for building load forecasting [7–9]. Although SVR is not a neural network model, it can be though as a single hidden layer ANN from the point view of the model capacity.

In addition to the shallow MLP and SVR models, deeper model architectures for building load forecasting have been proposed. Marino et al. [10] compare standard long short-term memory (LSTM) and LSTM sequence-to-sequence architectures (also known as encode-decoder) for building energy load forecasting. LSTM based architecture is also studied by Kong et al. in two papers [11,12]. Although the LSTM architectures studied in all three papers consist of only two stacked LSTM layers, the recurrent neural networks (RNN) such as LSTM are deep models due to the recursive call made at every time step. Amarasinghe et al. [13] present study on a convolutional neural networks (CNN) model for building load forecasting. The model consist of three CNN layers and two fully connected layers. Yan et al. [14] propose a CNN-LSTM network consisting of two CNN layers and a single LSTM layer. The model is evaluated in five different building data sets. Mocanu et al. [15] study the performance of conditional restricted Boltzmann machine (CRBM) and factored conditional restricted Boltzmann machine (FCRBM) neural network architectures in short-term load forecasting.

A very deep ANN architecture for energy load forecasting is presented by Chen et al. [16]. The authors adopt residual connections [17], a successful approach for building deep CNNs for image processing, and form a 60 layer deep network for energy load forecasting. The proposed ResNet borrows some ideas from RNN type network since it contains connections from previous hour forecast to next hour forecasts, but the connection is not an actual feedback since there are no parameter sharing among different hour forecasters. The proposed ResNet combined with an ensemble approach achieves state of the art results in three public energy forecasting benchmarks. Although none of these data sets focuses on individual building loads, the proposed work is general energy load forecasting model and thus good state-of-the-art benchmark also for building level forecasting tasks.

The aforementioned deep learning models have been evaluated using data sets with large amount of relatively static training data, which allows the models to avoid overfitting and outperform shallower models. However, in many situations (e.g., with new buildings or locations with extreme weather conditions) it is useful to be able to avoid overfitting also in situations with limited amount of training data. Moreover, a big limitation of the machine learning methods in general is that they assume that the data distribution does not change. This assumption can cause problems when machine learning methods such as neural networks or SVR are used for modelling real-life systems that change over time. In the context of buildings these changes in the data distribution can be caused, for example, by changes in the people (e.g., habits of people change over time or new people move in) or in the building infrastructure (e.g., new appliances, or configuration of existing systems such as HVAC change). A good example on how the data distributions can change dramatically is demonstrated by the Covid-19 pandemic, which has reduced usage rate of office building dramatically and therefore affected to energy consumption of these buildings.

We propose a novel hierarchical neural network architecture for short-term load forecasting that has been designed to address the abovementioned limitations of state-of-the-art load forecasting models. The architecture, called Stacked Booster Network (SBN), tries to achieve the good properties of deep models while still avoiding overfitting in small sample size situations. The core idea is to reduce the model parameter space with following principles: (1) sparse interactions, (2) parameter sharing, and (3) equivariant representations. Another key idea of the architecture is a novel boosting technique, which makes it possible to transform the original multivariate time series problem into univariate one. This further reduces the model parameters while keeping the network capacity high enough. Additionally, the proposed boosting technique enables the model to correct systematic mistakes by utilizing residual information on historical forecasts. With these ideas we can build a deep learning framework for short-term load forecasting with following properties:

- Minimal number of parameters leading to robust training even with small amount of training data
- Sufficiently large number of layers leading to enough presentation power for modelling real phenomenon of the modelled load
- Boosting technique that allows the model to adapt to changing data distributions, which are typical in real-life data

The paper is organized as follows. In Section 2, a general network architecture of the SBN is presented. Section 3 introduces the case studies and presents the evaluation where the SBN is compared to the state-of-the-art ResNet model [16]. Section 4 concludes the presentation and presents directions for future work.

2. SBN Architecture

The SBN is a general neural network architecture style, which can be instantiated in different ways. This section presents the general SBN architecture style and its main ideas and design principles. A concrete instance of the SBN for a particular case study with associated implementation details is presented in Section 3.3.

General model architecture is described in Figure 1. The architecture is composed of forecasting submodels at four levels:

- *Instant forecaster*
- *Hourly boosting forecaster*
- *Daily boosting forecaster*
- *Weekly boosting forecaster*

The architecture and configuration of the Instant forecaster depends on the energy system to be modelled. Typically, it takes as input a multivariate time series consisting of some combination of weather, temporal and energy features. A key idea in the SBN architecture, is that the *Instant forecaster* is the only part of the model that needs to manipulate the multivariate timeseries. The boosting forecasters are more general and designed to be stacked over previous forecaster in a way that the next forecaster boosts the previous. To this end, each booster gets a timeseries of past residuals as input and forecast the error made by the previous layer of models.

The architecture of the whole boosting system is shown in Figure 1 while the architecture of an individual booster is shown Figure 2. Architecture of the *Instant forecaster* is shown in Figure 3.

The next three subsections cover each of the forecasters with more details with their limitations. Then, we present a more detailed demonstration on how the SBN works with illustrative figures. Final subsection covers some general implementation notes of the SBN.

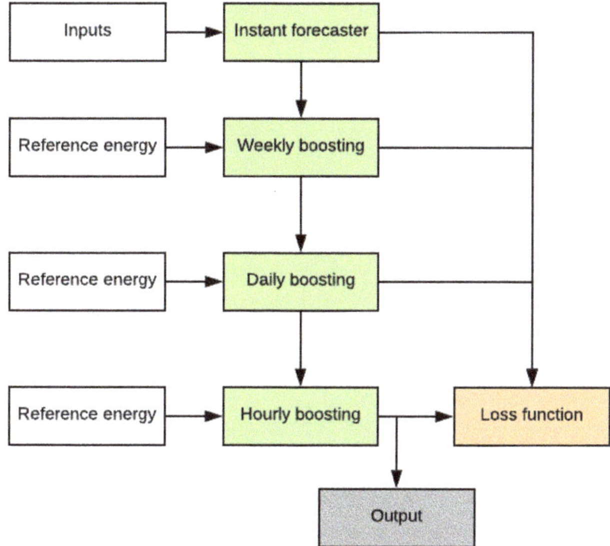

Figure 1. SBN architecture.

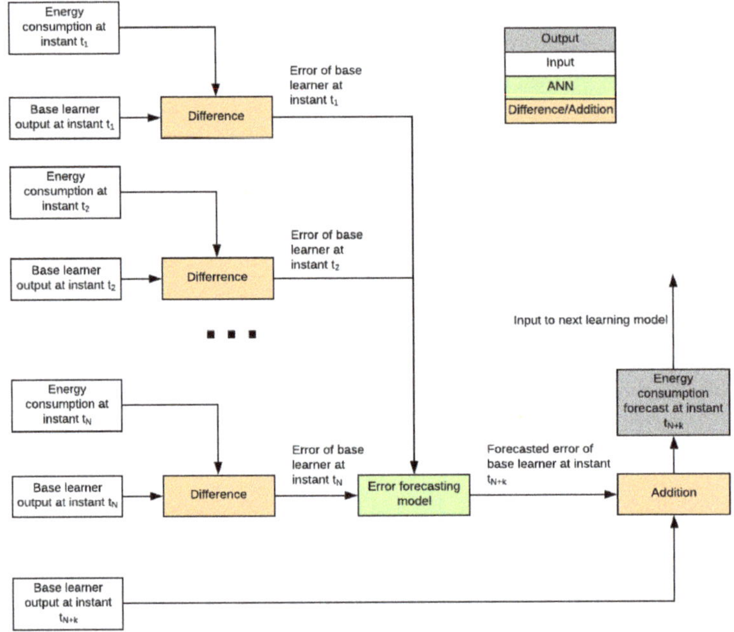

Figure 2. Architecture of a single boosting forecasters. Each boosting forecaster has the same general architecture. Base learner in the figure is previous booster or the instant forecaster in case of the first booster. In the figure, value N is the amount of previous errors used in forecasting future error and value k is the amount of time steps to be forecasted.

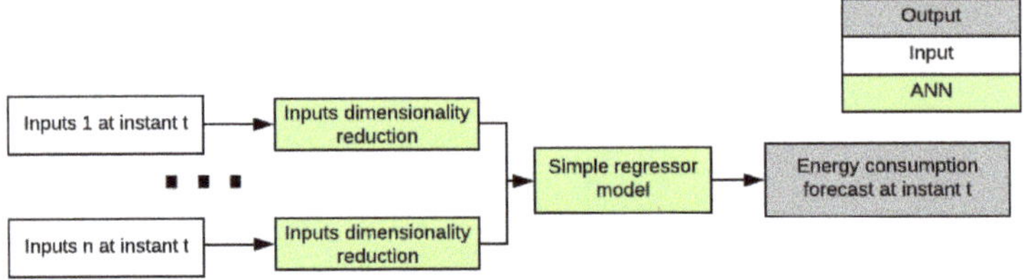

Figure 3. Architecture of the *Instant forecaster*.

2.1. Instant Forecaster

The purpose of the *Instant forecaster* is to forecast energy consumption based on independent variables that are assumed to be fully observable without any information on past energy consumptions. In our case study, described with details in Section 3, the only independent time series are temperature, hour of the day, and day of the week. Here also temperature can be considered fully observable by replacing future temperature values by their forecasts. These independent variables are first feed with dimensionality reduction submodels to reduce time dimension of these univariate time series to one. Considering our case study, hour of the day and day of the week are deterministic time series and do not therefore require dimensionality reduction submodel.

It should be noted that we make a reasonable assumption that different temperature profiles behave exactly same way at every weekday and every hour. Therefore, with this assumption reducing temperature sequence dimension to one before merging to other inputs does not lose any usable information. If some other information sources would be used, it is crucial to analyse whether the approach loses information and whether extra complexity needs to be addressed in the submodel.

2.2. *Weekly Boosting Forecaster*

Energy time series have typically periodical loads that occur same time on specific days of the week. If the loads would be always at the same level at the same time, the *Instant forecaster* would estimate them perfectly. However, it is quite typical that the timing and volume of these loads varies over time. This is the phenomena that the *Weekly boosting forecaster* should compensate. Consider, for example, a load occurring at some specific weekday rapidly increases by fixed amount. Then the *Weekly boosting forecaster* gets time series having this constant value as an input, and forecasts this constant value for the future errors. Therefore, in this case, the *Weekly boosting forecaster* is capable for estimating this dynamic change perfectly after short reaction time.

2.3. *Daily Boosting Forecaster*

The *Daily boosting forecaster* works exactly same way as the *Weekly boosting forecaster* but it focuses on changes on loads that occurs daily basis rather than weekly. Consider, for example that daily occurring load rapidly increases by fixed amount. The *Weekly boosting forecaster* would need weeks of data to be able to estimate and correct it. Therefore, it remains uncorrected at the beginning. The role of the *Daily boosting forecaster* is address this delay.

2.4. *Hourly Boosting Forecaster*

Some energy time series have periodic fluctuations in the energy consumption that cannot be explained purely by the input data. This can be explained by phenomena where some devices may be turned on and off periodically and this periodicity can be changing by some unknown conditions. Therefore, this phenomena can be seen as error signal of the *Instant forecaster*. Moreover, if the periodicity is not divisible by 24 h, the *Weekly boosting forecaster* and the *Daily boosting forecaster* cannot compensate it. This is the phenomena that the *Hourly boosting forecaster* should compensate. Another more minor phenomena that hourly boosting forecaster should compensate is a situation where a constantly occurring load changes somehow. In this case weekly and daily boosting forecasters react with days delays but the *Hourly boosting forecaster* can react and estimate the errors more rapidly.

2.5. *Demonstration of Operation*

This section demonstrates the SBN network operation with figures. To simplify demonstration, the *Hourly boosting forecaster* is omitted. In this demonstration, the *Weekly Booster forecaster* uses three weeks of data and the *Daily booster forecaster* uses seven days of data.

In this setup, we consider forecasting some fixed hour on Day 1 on Week 4. First, we run the *Instant forecaster* for all the historical values (Figure 4). Then, we calculate the errors for these historical values using the real historical knowledge of energy consumption (Figure 5). Next step is to use the *Weekly boosting forecaster* to forecast error of Week 3 for each day by using errors of *Instant forecaster* on Weeks 1 and 2 (Figure 6). We correct errors of the *Instant forecaster* for Week 3 by estimated errors of the *Weekly booster forecasters* for Week 3. This way we get seven weekly boosted errors for Week 3. By using these values we forecast the error of our final forecasted time on Day 1 on Week 4.

Now, we are ready for forecasting energy consumption for our target time on Day 1 on Week 4. This step is demonstrated in Figure 7. Here we first run the *Instant forecaster* on the target day. Then we subtract the estimated weekly boosted error from this instant forecast. Finally, we subtract daily boosted forecast error from the weekly boosted forecast to get the final forecast.

From this illustration, one should note that the *Weekly boosting forecaster* uses two data points for forecasting the third and the *Daily boosting forecaster* uses seven data points for forecasting the eight. In more general, the last boosting forecaster uses the whole input length for making forecast while the previous boosting forecasters must use one data point less.

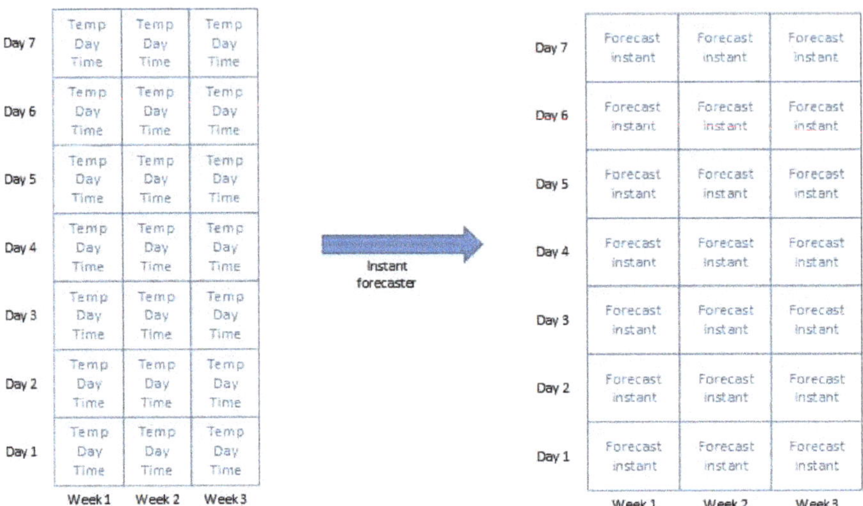

Figure 4. Step 1: Run historical values for given window through the *Instant forecaster*.

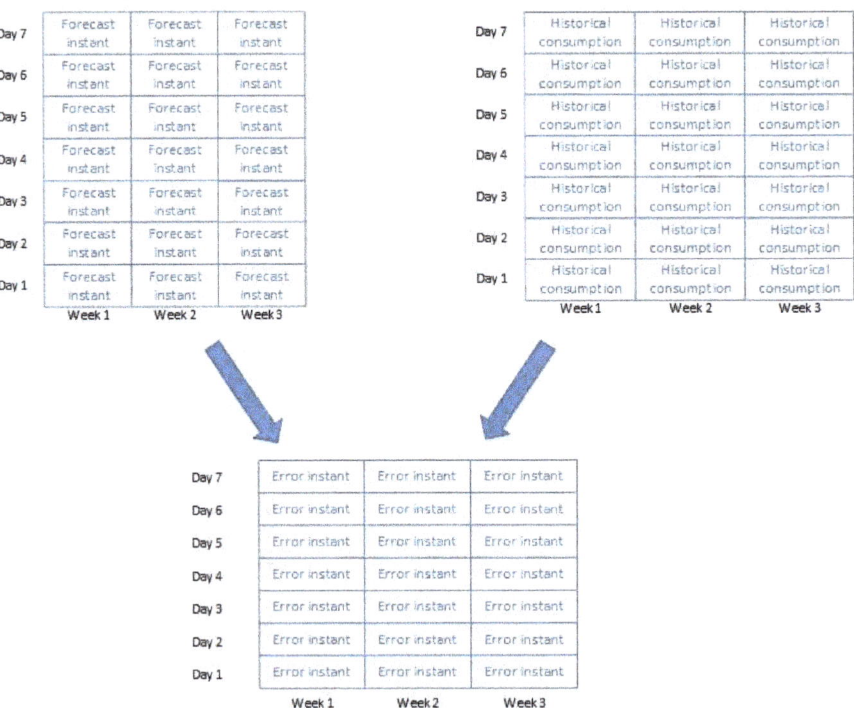

Figure 5. Step 2: Calculate errors of the *Instant forecasters* for historical data.

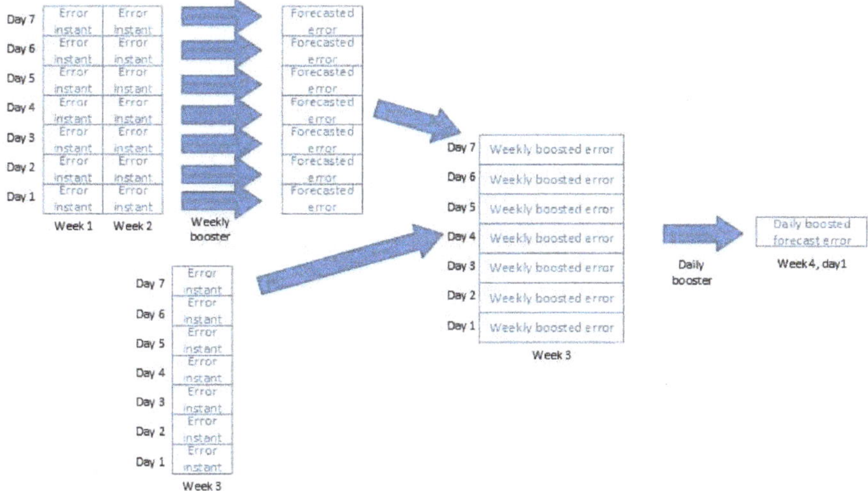

Figure 6. Step 3: Run historical values through the *Weekly boosting forecaster* and the *Daily boosting forecaster*.

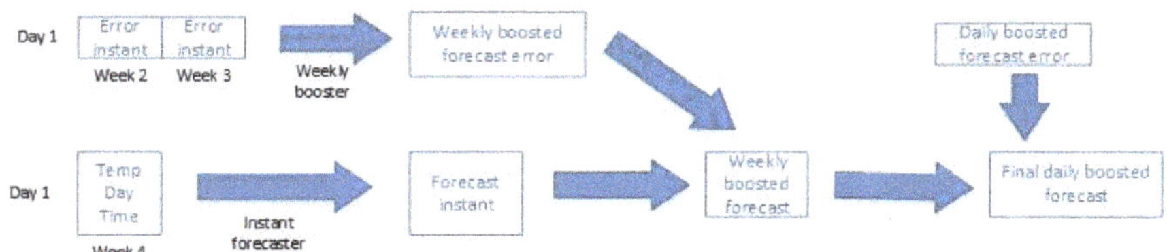

Figure 7. Make the final forecast using the *Instant forecaster* and compensate the forecast with the *Weekly boosting forecaster* and the *Daily boosting forecaster*.

2.6. Implementation Notes

It should be noted that stacking of the *Hourly boosting forecaster*, the *Daily boosting forecaster* and the *Weekly boosting forecaster* can be done in any order. Moreover, some of the forecasters may not be needed at all. In fact, higher time resolution boosting forecaster can correct exactly the same issues than the lower time resolution forecaster when the time window it gets as input is big enough. However, due to fact that some energy load occurs typically in daily and weekly basis, these energy loads are more easily estimated and compensated by the *Daily boosting forecaster* and the *Weekly boosting forecaster*. We have not experimented different ordering of the boosters and our experiments presented in Section 3 uses only one fixed stacking order. This order is weekly booster, daily booster and lastly hourly booster.

A nice feature of the architecture is that it splits the big problem into small problems that can be optimized separately when stacking the forecasters. First, the network performing the *Instant forecaster* should be implemented and optimized. This network architecture can be optimized without the boosting forecasters. Then, the first boosting forecaster is implemented and stacked on top of the *Instant forecaster*. The network architecture of this boosting forecaster can now be optimized separately. Similarly all the network architectures of the other booster forecasters can be optimized one by one. It is still an open issue whether this iterative approach will provide an optimal total network architecture. However, it seems to offer a reasonable good architecture with an easy design flow.

3. Case Study

The SBN architecture is applied for forecasting thermal energy consumption of a Finnish office building located in Espoo. The building is a typical office building constructed in 1960s and having four floors.

The main problem is to forecast hourly thermal energy consumption of the building for 24 h in advance. In addition to this main problem, we evaluate the performance in 48 h and 96 h forecasts also. Moreover, the effect of the training data size is evaluated by altering the training data size from 6 months to 6 years. The target metric is Normalized Root Mean Square Error (NRMSE) that is calculated using formula

$$NRMSE = MSE/(\max - \min),$$

where MSE is standard mean square error and maximum and minimum values are calculated over evaluation period.

Available data are the past hourly energy consumption and temperature measurements. Temperature measurements are measured from a weather station being a few kilometres away from the building. For the future timestamps, temperature forecasts are not used in this experiment but real temperature readings from the future timestamps.

3.1. Data Set

Thermal energy data set contains hourly energy measurements and hourly temperature measurements from 1.2.2012 to 31.12.2018. The data set is visualized in Figure 8 with an example forecast.

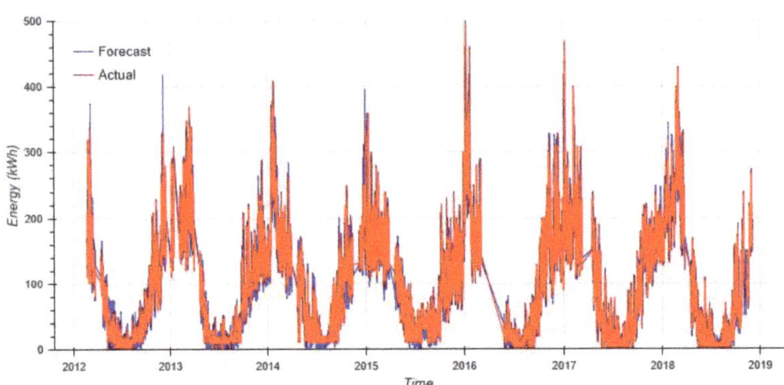

Figure 8. The whole data set and 24 h forecasts. Here one can see that there are some missing data that causes direct lines in the figure.

3.2. Methodology

The model architecture was optimized by using years 2012–2016 for training and year 2017 for the validation using manual architecture search. When reasonable model architecture was found, the final performance metric was run from year 2018. All the runs that was executed for year 2018 are presented in this paper. After running the metrics for year 2018 no modifications to model parameters was made.

3.3. Model

3.3.1. Instant Forecaster

The *Instant forecaster* has three inputs:

1. Twelve consecutive hours temperature readings before each forecast hour are used.

2. Dummy encoded vector of length 2 determining whether the day is Saturday, Sunday or regular weekday.
3. Predicted hour. Hour is encoded onto unit circle to get continuous variable. Hour is therefore two length vector.

It should be noted that there are many instances of the *Instant forecaster* submodel, and all the instances share the same parameters. Twelve temperature values are first fed through dimensionality reduction submodel that reduces the temperature dimension to one and then all the inputs are processed by two layer densely connected network to get simple forecasts where the hidden layer contains 32 units with dropout regularization.

3.3.2. Boosting Forecasters

The problem shall be solved by stacking all the three boosting forecasters in the following order: weekly, daily, hourly. For simplicity all the boosting forecasters shall use the same structure having two fully connected layers where the hidden layer contains 32 units with dropout regularization. The *Weekly boosting forecaster* contains three weeks of data (i.e., two weeks as input and one week to be forecasted). Therefore, we have values $N = 2$ and $k = 1$ in Figure 2. The *Daily boosting forecaster* contains seven days of data indicating values $N = 7$ and $k = 1$ for 24 h forecast in Figure 2. Finally, the *Hourly boosting forecaster* contains 24 h data, indicating values $N = 24$ and $k = 24$ for 24 h forecast. It should be noted that there are multiple instances of each boosting forecaster submodels and the submodels share the same weights. However, for different booster levels, submodel weights are naturally different.

Some statistics of network architectures for each of the used booster setups is shown in Table 1.

Table 1. Architectures of different booster combinations.

Method	Num Layers	Num Parameters
Instant forecaster	3	238
Daily booster	5	527
Weekly booster	5	399
Weekly and daily boosters	7	624
Weekly, daily and hourly boosters	9	1457

3.4. Training Details

The proposed architecture offers multiple different ways of training the network:

- Train the whole network at once using the final prediction value only in loss function
- Train the whole network at once using all the forecasters outputs in loss function but possible with different weights.
- Train first the first prediction submodel and freeze it. Then, train add one boosting forecasters one by one freezing all the earlier forecasters.
- Train iteratively as in previous bullet but do not freeze the earlier forecasters.

We have evaluated all the approaches. The second and the fourth provided roughly equal good results in terms of NRMSE. However, the second one is naturally faster since it needs only one training round. It seems that providing some weights to earlier forecaster loss functions fights well against overfitting and it can be considered as a kind of regularization technique. This technique can also be considered as a kind of a short-cut connection used in ResNets [16,17].

It seems that the training is not very sensitive to the weight given to the *Instant forecaster* and the earlier boosting forecasters. In our case study, we used weight 0.1 for the *Instant forecaster* and earlier boosters and weight 0.9 for the final booster.

We used Adam optimizer [18] in training the network. Besides of learning rate decay of the built-in Adam optimizer, we used an additional exponential learning rate decay. The learning rate decay is dependent of training data size such that for 5 year training data learning rate is reduced by 2% after each epoch. For other training data sizes the decay is altered such that the total decay is same per batch. Initial learning rate is set to 0.0025. Batch size was 256 in our experiments.

Due to learning rate decay and other regularization techniques, early stopping is not needed and the model can be trained with fixed number of epochs.

The network was implemented with Keras library ([19]) using Tensorflow backend ([20]).

3.5. Forecast Analysis

This section provides some views of the target data sets and forecasts. First, the whole data set is shown in Figure 8 with an example forecast.

Figure 9 shows the forecasts for our evaluation year 2018. To get better understanding of the forecast and the actual energy consumption, one zoomed snapshot is shown in Figure 10.

For an ideal forecaster the difference between the actual energy consumption and the forecast is a pure white noise. This difference is shown in Figures 11 and 12. From the closer view one can see that the error is correlated but not very much.

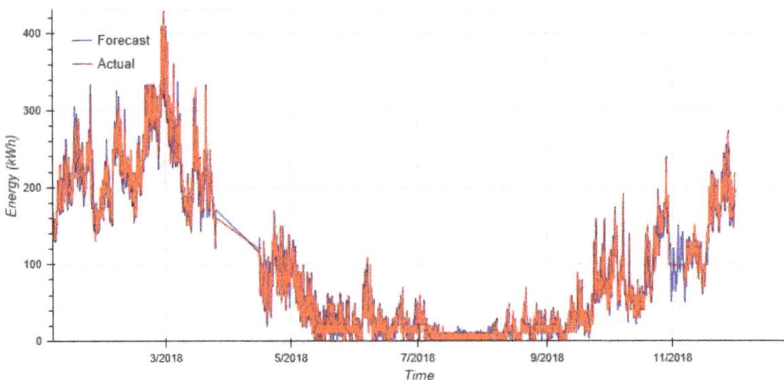

Figure 9. 24 h forecasts of the test year 2018.

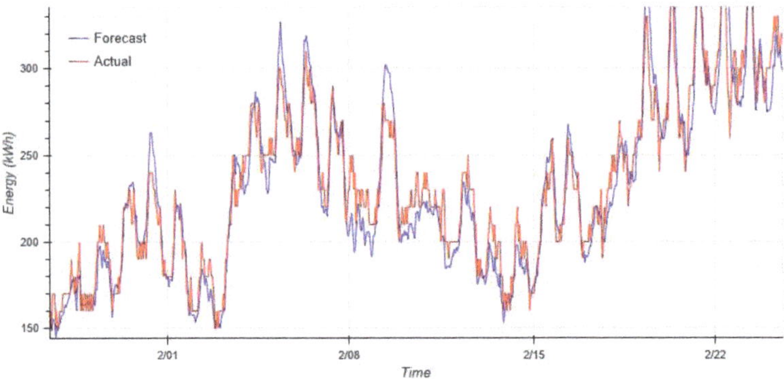

Figure 10. 24 h forecasts of a part of the test year 2018.

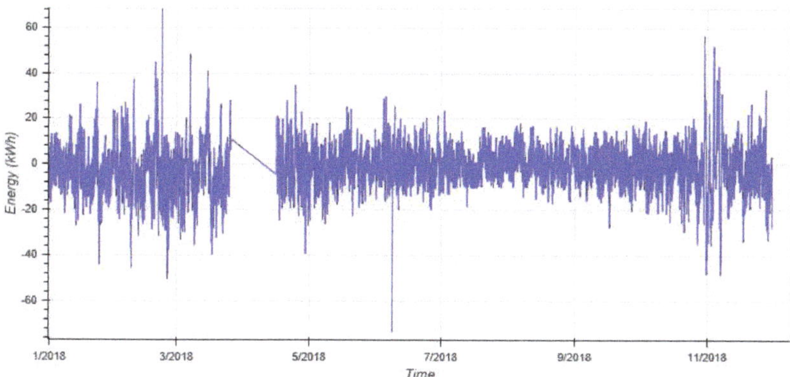

Figure 11. Difference of real energy consumption and 24 h forecasted one in the test year 2018.

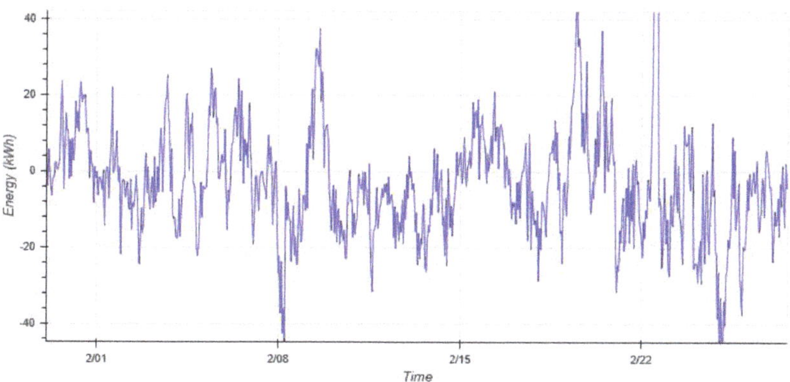

Figure 12. Difference of real energy consumption and 24 h forecasted one in a part of year 2018.

3.6. Performance Comparison

Performance comparisons of the usage of the different booster setups are presented in Tables 2 and 3, Figures 13 and 14. First, Table 2 and Figure 13 compare the setups by altering the training data size. Here one should note that the basic *Instant forecaster* is the most sensitive to the training data size. When the training data size is only 6 months, and there are no training data available from all temperature conditions, the performance of the *Instant forecaster* decreases a lot. Nevertheless, boosting forecasts are still able to correct these errors to some extent. Moreover, when the training data size increases to 6 years, aging of the data (the building energy consumption has changed over the years so that it does not reflect the situation captured by the older data) starts to decrease the performance of the *Instant forecaster*. Again the boosting forecasters can compensate these errors to some extent. From Table 3 and Figure 14 one should note that usage of the *Hourly boosting forecaster* gets useless when the forecasting period increases since the old hourly energy consumptions do not provide any correlation to future data of different hours. It eventually decreases performance due to overfitting.

Table 2. Boosters NRMSE values in percentages in 24 h forecast.

Method/Training Data Size	6 Months	1 Year	6 Years
Instant forecaster	6.28	3.64	4.46
Daily booster	4.27	2.46	2.49
Weekly booster	5.43	2.62	2.79
Weekly and daily boosters	4.07	2.37	2.42
Weekly, daily and hourly boosters	5.83	2.34	2.35

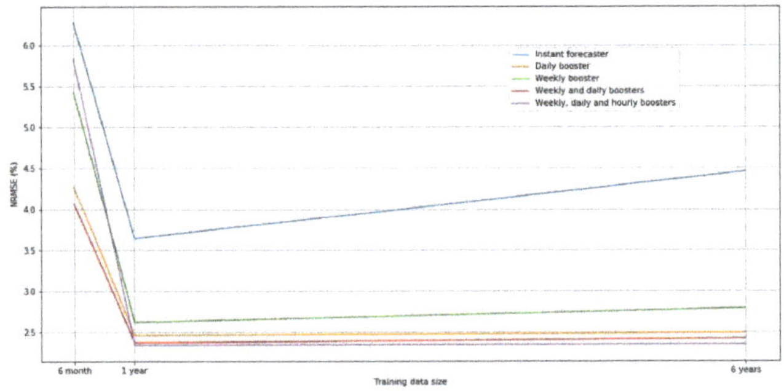

Figure 13. NRMSE values of different booster setups in percentages in 24 h forecast.

Table 3. Boosters NRMSE values in percentages.

Method	24 h Forecast	48 h Forecast	96 h Forecast
Instant forecaster	3.64	3.64	3.64
Daily booster	2.46	2.62	2.64
Weekly booster	2.62	2.62	2.62
Weekly and daily boosters	2.37	2.51	2.57
Weekly, daily and hourly boosters	2.34	2.51	2.66

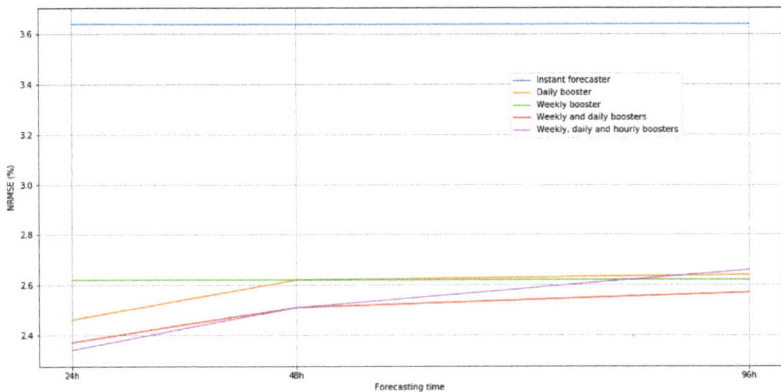

Figure 14. NRMSE values of different booster setups for different forecast lengths.

Performance Comparison to the State-of-Art

We used Residual network approach [16] as a presentation of state-of-the-art since it has implementation available (https://github.com/yalickj/load-forecasting-resnet) and it has done good job in public data sets. Let us call this solution simply ResNet. However, the comparison is not completely fair due to following reasons:

- ResNet does not in fact solve the same problem we have stated. The model is suitable for a little bit easier problem where one forecasts energy consumption of each of the hours on the next day given all the energy consumptions of the current day. It utilizes therefore 0–23 more recent observations that we do not utilise depending on the forecast hour.
- The architecture of the SBN is tuned for this particular data set while ResNet is used in plug and play style. However, our network architecture is not very sensitive to changes of each submodel architecture.

The target metrics for comparison is shown in Table 4 and in Figure 15. ResNet solutions seems to perform quite well for plug and play solution when there are a lot of training data available. However, when reducing the amount of training data, it starts to overfit and performance decreases dramatically. However, our SBN solution behaves very well on even a relative small training data size.

Table 4. Performance comparison between the SBN and the ResNet. Values are in percentages.

Training Data	ResNet ([16]) NRMSE	SBN (Ours) NRMSE
6 years	2.57	2.35
5 years	2.75	2.37
4 years	3.11	2.34
3 years	3.51	2.40
2 years	3.95	2.29
1 years	13.40	2.34
6 months	41.91	5.83

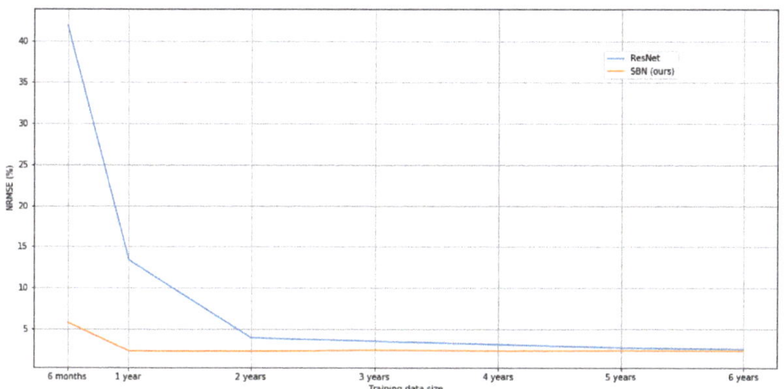

Figure 15. NRMSE values of the ResNet ([16]) NRMSE and the SBN(ours) in different training data lengths.

4. Conclusions and the Future Work

This paper presented a novel neural network architecture style, called SBN, for short-term energy load forecasting. The SBN aims to achieve the representation capacity of deep learning while at the same time (1) reduce overfitting and (2) provide the capability to adapt to changing data distributions caused by changes in the modelled energy system. This is achieved with following key ideas, which have driven the design of the SBN: (1) sparse interactions, (2) parameter sharing, (3) equivariant representations, and (4) feedback of residuals from past forecasts. In practice, the SBN architecture consists of a base learner and multiple boosting systems that are modelled as a single deep neural network.

A specific instance of the SBN was implemented on top of Keras and Tersorflow 1.14 to evaluate the architecture style in practice. The evaluation was performed with data from an office building in Finland and the evaluation consisted of following short-term load forecasting tasks: 96 h, 48 h, and 24 h forecasts with 6 years of training data, and 24 h forecast with 1 year and 6 months of training data to evaluate the data-efficiency of the model. As part of the evaluation, the SBN was compared to a state-of-the-art deep learning network [16], which it outperformed in all of the above mentioned forecasting tasks. In particular, the SBN provided significantly better results in terms of data-efficiency.

The SBN is general in the sense that it can be used for other similar time series forecasting domains. As a future work we will apply it to different type of energy forecasting problems and data sets. Moreover, approaches to make the SBN even more data-efficient will be investigated. An interesting idea to this end is related to the model pre-training with transfer learning. As is described in the paper, the SBN is composed of *Instant forecaster* and different boosters. Boosters solve very general univariate time series forecasting problem while Instant forecaster makes the data set specific prediction. Therefore, it could be possible to pre-train the boosters with totally different energy data sets and only train the *Instant forecaster* with the target data set. Since the booster would not need to be trained, this would make the SBN even more data-efficient with respect to training data from the specific building or energy system.

Author Contributions: conceptualization, T.S., J.K. and D.P.; methodology, T.S.; software, T.S.; writing–original draft preparation, T.S. and J.K.; writing–review and editing, T.S., J.K. and D.P.; visualization, T.S.; project administration, D.P.; funding acquisition, D.P. All authors have read and agreed to the published version of the manuscript.

Funding: This work has been funded by VTT Technical Research Centre of Finland and Business Finland.

Conflicts of Interest: The authors declare no conflict of interest.

References

1. Berardi, U. A cross-country comparison of the building energy consumptions and their trends. *Resour. Conserv. Recycl.* **2017**, *123*, 230–241.
2. Amasyali, K.; El-Gohary, N.M. A review of data-driven building energy consumption prediction studies. *Renew. Sustain. Energy Rev.* **2018**, *81*, 1192–1205, doi:10.1016/j.rser.2017.04.09.
3. Bagnasco, A.; Fresi, F.; Saviozzi, M.; Silvestro, F.; Vinci, A. Electrical consumption forecasting in hospital facilities: An application case. *Energy Build.* **2015**, *103*, 261–270, doi:10.1016/j.enbuild.2015.05.056.
4. Chae, Y.T.; Horesh, R.; Hwang, Y.; Lee, Y.M. Artificial neural network model for forecasting sub-hourly electricity usage in commercial buildings. *Energy Build.* **2016**, *111*, 184–194, doi:10.1016/j.enbuild.2015.11.045.
5. Wong, S.; Wan, K.; Lam, T. Artificial neural networks for energy analysis of office buildings with daylighting. *Appl. Energy* **2010**, *87*, 551–557, doi:10.1016/j.apenergy.2009.06.028.
6. Escrivá-Escrivá, G.; Álvarez Bel, C.; Roldán-Blay, C.; Alcázar-Ortega, M. New artificial neural network prediction method for electrical consumption forecasting based on building end-uses. *Energy Build.* **2011**, *43*, 3112–3119, doi:10.1016/j.enbuild.2011.08.008.
7. Zhang, Y.; Li, Q. A Regressive Convolution Neural network and Support Vector Regression Model for Electricity Consumption Forecasting. *arXiv* **2018**, arXiv:1810.08878.
8. Jain, R.K.; Smith, K.M.; Culligan, P.J.; Taylor, J.E. Forecasting energy consumption of multi-family residential buildings using support vector regression: Investigating the impact of temporal and spatial monitoring granularity on performance accuracy. *Appl. Energy* **2014**, *123*, 168–178, doi:10.1016/j.apenergy.2014.02.057.

9. Jain, R.K.; Damoulas, T.; Kontokosta, C.E. Towards Data-Driven Energy Consumption Forecasting of Multi-Family Residential Buildings: Feature Selection via The Lasso. In Proceedings of the Computing in Civil and Building Engineering, Orlando, FL, USA, 23–25 June 2014; pp. 1675–1682, doi:10.1061/9780784413616.208.
10. Marino, D.L.; Amarasinghe, K.; Manic, M. Building Energy Load Forecasting using Deep Neural Networks. *arXiv* **2016**, arXiv:1610.09460.
11. Kong, W.; Dong, Z.Y.; Hill, D.J.; Luo, F.; Xu, Y. Short-Term Residential Load Forecasting Based on Resident Behaviour Learning. *IEEE Trans. Power Syst.* **2018**, *33*, 1087–1088.
12. Kong, W.; Dong, Z.Y.; Jia, Y.; Hill, D.J.; Xu, Y.; Zhang, Y. Short-Term Residential Load Forecasting Based on LSTM Recurrent Neural Network. *IEEE Trans. Smart Grid* **2019**, *10*, 841–851, doi:10.1109/TSG.2017.2753802.
13. Amarasinghe, K.; Marino, D.; Manic, M. Deep neural networks for energy load forecasting. In Proceedings of the 2017 IEEE 26th International Symposium on Industrial Electronics (ISIE), Edinburgh, UK, 19–21 June 2017; pp. 1483–1488, doi:10.1109/ISIE.2017.8001465.
14. Yan, K.; Wang, X.; Du, Y.; Jin, N.; Huang, H.; Zhou, H. Multi-Step Short-Term Power Consumption Forecasting with a Hybrid Deep Learning Strategy. *Energies* **2018**, *11*, 3089, doi:10.3390/en11113089.
15. Mocanu, E.; Nguyen, H.; Gibescu, M.; Kling, W. Deep learning for estimating building energy consumption. *Sustain. Energy Grids Netw.* **2016**, *6*, 91–99, doi:10.1016/j.segan.2016.02.005.
16. Chen, K.; Chen, K.; Wang, Q.; He, Z.; Hu, J.; He, J. Short-Term Load Forecasting With Deep Residual Networks. *IEEE Trans. Smart Grid* **2019**, *10*, 3943–3952.
17. He, K.; Zhang, X.; Ren, S.; Sun, J. Deep Residual Learning for Image Recognition. *arXiv* **2015**, arXiv:1512.03385.
18. Kingma, D.P.; Ba, J. Adam: A Method for Stochastic Optimization. *arXiv* **2014**, arXiv:cs.LG/1412.6980.
19. Chollet, F. Keras. 2015. Available online: https://keras.io (accessed on 7 May 2020).
20. Abadi, M.; Agarwal, A.; Barham, P.; Brevdo, E.; Chen, Z.; Citro, C.; Corrado, G.S.; Davis, A.; Dean, J.; Devin, M.; et al. TensorFlow: Large-Scale Machine Learning on Heterogeneous Systems. 2015. Available online: tensorflow.org (accessed on 7 May 2020).

Article

Smart Grid Monitoring by Wireless Sensors Using Binary Logistic Regression

Hariprasath Manoharan [1], Yuvaraja Teekaraman [2], Irina Kirpichnikova [2], Ramya Kuppusamy [3], Srete Nikolovski [4,*] and Hamid Reza Baghaee [5]

1. Department of Electronics and Communication Engineering, Audisankara College of Engineering and Technology, Gudur 524 101, India; hari13prasath@gmail.com
2. Faculty of Energy and Power Engineering, South Ural State University, Chelyabinsk 454 080, Russia; teekaramani@susu.ru (Y.T.); kirpichnikovaim@susu.ac.ru (I.K.)
3. Department of Electrical & Electronics Engineering, Sri Sairam College of Engineering, Bangalore 562106, India; ramyapks26@ieee.org
4. Power Engineering Department, Faculty of Electrical Engineering, Computer Science and Information Technology, University of Osijek, 31000 Osijek, Croatia
5. Department of Electrical Engineering, Amirkabir University of Technology, Tehran 15875–4413, Iran; hrbaghaee@aut.ac.ir

* Correspondence: srete.nikolovski@ferit.hr

Received: 19 June 2020; Accepted: 18 July 2020; Published: 2 August 2020

Abstract: This article focuses on addressing the data aggregation faults caused by the Phasor Measuring Unit (PMU) by installing Wireless Sensor Networks (WSN) in the grid. All data that is monitored by PMU should be sent to the base station for further action. But the data that is sent from PMU does not reach the main server properly in many situations. To avoid this situation, a sensor-based technology has been introduced in the proposed method for sensing the values that are monitored by PMU. Also, the basic parameters that are necessary for determining optimal solutions like energy consumption, distance and cost have been calculated for wireless sensors, whereas, for PMU optimal placements with cost analysis have been restrained. For analyzing and improving the accuracy of the proposed method, an effective Binary Logistic Regression (BLR) algorithm has been integrated with an objective function. The sensor will report all measured PMU values to an Online Monitoring System (OMS). To examine the effectiveness of the proposed method, the examined values are visualized in MATLAB and results prove that the proposed method using BLR is more effective than existing methods in terms of all parametric values and the much improved results have been obtained at a rate of 81.2%.

Keywords: smart grids (intelligent networks); phasor machine learning; binary logistic regression; wireless network; Sensors

1. Introduction

In recent days, the application of Phasor Measuring Units (PMUs) and Wireless Sensor Networks (WSN) for monitoring electrical grid parameters has been observed as a developing area. For making the operation of the grid smarter and smoother, a sensing device that observes the functioning of PMU and intimates it to the Online Monitoring System (OMS) is essential. Therefore, this way of smart integration paves the way for preventing critical grid situations such as excessive loss of power consumption. Even though the global positioning system (GPS) receiver is present, it can only track the position of PMUs. In addition, all the measurements which are provided by the PMU will be transferred to the satellite and then to the control center. While transferring the parametric data, the occurrence of a fault may be present, and there is no way to detect faults such as wrong measurement data with noise. Moreover, the errors in measurement can only be controlled after detecting the type of problem that persists in PMU. This requires the help of a large amount of manpower, and immediate remedial actions cannot be taken. As a result, there might be a huge loss of energy where more blackouts can occur within a short period of time.

Therefore, the communicating PMU device requires a wireless device for supervision which will be able to detect the faults quickly and report them to the control center. In a modern trend, the wireless sensors which can sense the fault data in PMUs with less energy consumption can be integrated with PMUs for better control operations. The block diagram of the proposed method is indicated in Figure 1.

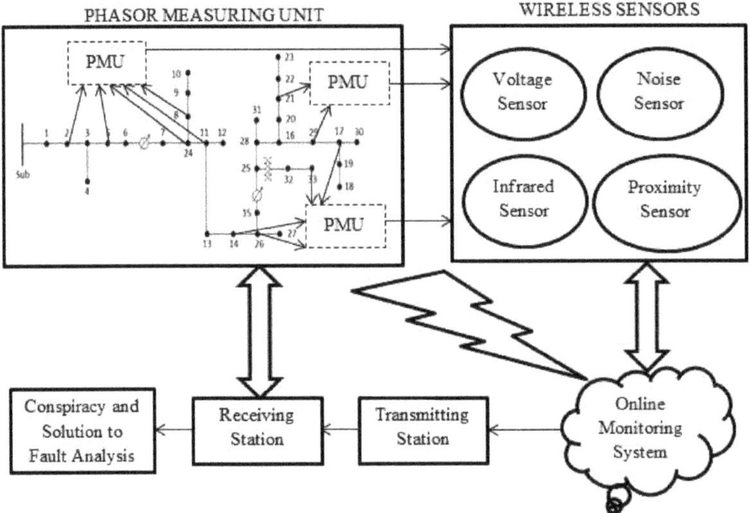

Figure 1. Incorporation of sensors with Phasor Measuring Units (PMU) for fault detection.

1.1. Motivation of Proposed Work

The primary motivation for deliberating this research work is to monitor the level of voltage, the status of PMU in the grid and to overcome the blackouts that occur in case of PMU failure within a short period. Therefore, the communicating PMU device requires an intelligent wireless device for controlling it. The installed group of sensors will be able to detect the faults quickly and reports it to the control center. The wireless sensors which can sense the fault data in PMUs with less energy consumption can be integrated with PMUs for better control operations.

1.2. Short Review on Existing Literature

There are numerous gaps that exist in the prevailing literature [1–12] when PMUs are connected in grid. In the following literature, all of the authors have installed PMU for monitoring the grid but fail to detect any of the important parameters. In many cases, the existing literature lacks discussion on the installation of sensors in the grid for making it smart, although the importance of placing sensors in the grid has been analyzed by several researchers where the framework consists of placement of PMUs at suitable locations for achieving full observability [1].

1.3. Research Gap and Motivation

All the existing literature has not provided any discussion related to the implementation of an additional communication device for monitoring the status of PMU, which includes parameters like energy consumption, voltage, and angle measurements. Also, no researchers have examined the detection of data faults in PMU, which provides necessary voltage values for operating the electric grids. Therefore, to overcome the drawbacks mentioned above, a sensor device has been integrated with all optimal PMUs that are installed in the grid. The sensor builds the gap (Table 1) by monitoring and reporting the status and fault information of PMU to the receiver (OMS). This will make the work station take remedial actions for preventing driving under emergencies, thus making the grid smart.

Table 1. Significant contributions associated with the proposed method.

Reference	**Optimization Method**	**Monitored Parameters**	
		For PMU	**For Sensors**
[1]	Frequency Distribution Recorder (FDR) using optical sensors	PMU placements and cost of installation	-
[3]	Binary Coded Genetic Algorithm	Observability and cost of installation	-
[6]	Virtualized PMU approach	-	Energy consumption and cost of installation
Proposed	Improved Binary Logistic Regression (BLR)	PMU placements, observability and cost of installation	Energy consumption, distance covered and cost of installation

1.4. Outline of Article

The remaining sections of the article is organized as follows: Section 2 describes a brief literature review where the fundamentals of the proposed work is designed. Section 3 formulates the proposed model on PMU with sensors. Section 4 integrates the optimization algorithm with proposed formulations. Section 5 validates the efficiency and parametric values of the projected method and finally, Section 6 concludes the research discussion with discussion on future revisions to be made.

2. Literature Survey

The importance of placing sensors in the grid has been analyzed by several researchers where the framework consists of placement of PMUs at suitable locations for achieving full observability [1]. Since there is a usage of traditional transformers across different areas around the world, a sensor must be needed for acquiring signals at the destination. Also, the control mechanisms in PMU are possible only when an accurate sensor is installed on the grid [1]. However, the major drawback in [1] is that only the placement locations of PMU are calculated where there is a possibility to replace the traditional transformer. Still, the same operation of the transformer has been performed in a digitalized way. Now, from the digital outcomes, which are indicated in [1], a challenging task on testing the bad data has been performed [2]. In this testing, the variation of data error is also detected and controlled by using a Kalman filter. The procedure starts by analyzing the previous state of PMU, which is under observation and compares it with the currently updated value where the PMU remains in a fully observable state. Even though this method produces less data error, it can only be applicable if there is any change in the voltage magnitude [2], and the sensor development procedure is not integrated into the place where voltage change occurs.

An optimal approach for wide-area systems has been developed [3] at a later phase for analyzing the impact of communication infrastructure that is integrated with PMU. In this case, the importance of communication infrastructure in the grid has been well-defined by using a genetic algorithm. This method uses a high number of sensor nodes where the cost of sensor placement will be much higher than traditional methods [3]. The method described in [3] for achieving complete observability is not possible in all situations because, in real-time, a mesh network cannot be formed under different situations. This limitation has been solved by using a phase-detection model [4] where the data attack procedure is evaluated using the Global Positioning System (GPS). Here, the clock precision error is also calculated, and the potential risk of placing the PMU data using time synchronization standard is observed [4]. But, even though the method provides a solution to data attack methods, a detailed procedure of sensor integration with important parametric analysis like less energy consumption of nodes, proper data aggregation has not been described [4]. If the data is aggregated correctly, then there is no need to detect the phase of PMUs.

An image acquisition-based sensor visionary system has been developed [5] for real-time monitoring of grids. In this method, a line detection method has been developed where the three-dimensional image of the damaged line in a particular area will be identified. Therefore, once the image arrives at the receiver, the appropriate necessary actions will be taken. This method involves a high cost of computation, and there is a discussion-based in smart meter investigation [5]. Still, the sensors are not integrated with the PMU-based rectification system. Also, the power failures can be easily identified by the developed method [5], where a cloud storage system has been built. However, these methods will require manpower, and there is no proficiency in automatic rescue operation [5] if large scale systems are integrated. For solving the drawback mentioned above, a virtual method of PMU analysis has been carried out [6]. In this method, traffic will be generated where each region will be divided into different zones. A latency test with a communication overlay has been designed for locally generated data. Also, an online monitoring system using Google application engines has been integrated for data aggregation. However, in the outcome analysis, the only percent of power-saving with bandwidth has been discussed, and there is no clear information whether the distance covered by the communication device in the zonal areas are higher or lesser [6].

The method mentioned above [6] has been simplified using software-defined networks [7] for making the communication much easier through the channel. Since the demand for electricity increases day-by-day, the requirement for monitoring the individual grid using network software has also been increased. If the software-defined networks are used then, the automatic control process can be easily completed [7]. This paves the way for creating sensor-based technological solutions in analyzing the network grid. However, there is no practical information regarding control information and parametric monitoring of PMU when it is placed in different areas of grid networks [7]. The usage of multiple sensors in the grid paves the way for monitoring different parameters, which even includes the temperature and humidity of PMU [8]. This has been indicated as a power management unit, which in turn gives rise to energy harvesting systems. In this method, a battery and automated power platform have been designed for cost-efficient implementation [8]. It is observed that in this method, the only microcontroller has been used for effective operation with transmitting and receiving antennas. The usage of both microcontrollers and multiple sensors will increase cost, and thus the energy consumption of each node will be much higher [8].

A survey has been showing [9] on the generalization of PMU with reliable models where the procedure for data uncertainty, fault analysis in the grid has been discussed. The authors describe that for reliability evaluation methods such as Markov based models can be integrated, but the application of PMU has been advanced in different areas, and the only solution to overcome the defects in PMU is by integrating real-time communication equipment, which identifies fault even in remote locations [9]. While designing the sensor technologies for integration with PMU, the standards suggested by IEEE C37.118 should always be followed [10]. The merging unit for interoperability test has been suggested where the signal identification and transmission test has been showing, which are kept as per the standards of IEEE. This is followed while placing the PMUs at desired locations. However, the sensors can be integrated as per the requirement by considering the optimal placement of PMUs [10].

Further, the sensors should be feasible for integrating with PMUs [11], and the feasibility solutions have been described by using cloud-based networks. The process has been executed by considering synchrophasor systems, and a high standard of encryption is also provided for ensuring the safe process of data transfer [11]. Even though the best security solutions are provided in [11], only the data security cost has been analyzed without any discussion regarding the implementation cost. Finally, a discussion on self-healing smart grids has been offered [12] with different emerging techniques where network communication with middleware solutions has been provided. To date, the aforementioned existing methods are the only available approaches for PMU integration with sensor-based networks, and all the methodologies have been analyzed for building a novel approach.

Objectives

The main objective of this study is to integrate a communicating device (Sensors) for monitoring the status and to report the fault information to the control center (OMS) that is observed by PMUs. This new flanged objective will be more efficient for building a smart grid environment by satisfying the constraints of energy consumption, range, and distance. The implementation design of the proposed objective which is formulated in Equation (7) is applied to a large-scale transmission system and small-scale distribution system.

3. Problem Formulation

Since the prime objective of the proposed method is to achieve complete communication observability in the presence of sensors, it is necessary to calculate the range covered by each individual sensor. For achieving long-distance communication, the entire area where the PMU is placed in the grid will be divided separately and will be marked as individual groups (Circles). Therefore, the maximum range will be achieved, and it can be calculated mathematically using (1).

$$d(i) = max \sum_{i=1}^{n} a_i \tag{1}$$

where, $d(i)$ represents the distance of the sensor to be maximized; and a_i denotes the possible arrangements of PMU that are placed in the grid.

The above equation indicates that if the distance covered by the sensor is maximized then, the cost of installation will be automatically reduced, and the energy consumption of the installed nodes will be minimized as much as possible. All Equations (1)–(6) have been derived from [2,3] in a unique way where the objective function indicated in (7) has not been stated in any existing literature, which forms the novelty of the projected methodology. The possible arrangements represent the accurate connection that is provided by the sensors to the PMUs, which are denoted using (2).

$$a_i = (a_1 + a_2 + \dots + a_n) = e(i) \tag{2}$$

where, $e(i)$ represents the exact value of the sensors that are placed near the PMUs.

Since the range covered by sensors in the grid which are located at different regions are in circles, the exact range can be calculated by combining (1) and (2) as,

$$e_r(i) = 0.47 \times d(i) \tag{3}$$

where, for precise monitoring each sensor has been divided into separate groups where PMUs are installed in the grid. There are several areas where the installed sensors have to monitor, and after monitoring the precise value it will be uploaded to OMS. This is not possible by implementing a single sensor as the range covered by PMU is much higher. Therefore, sensors are deployed as groups in the same areas where PMUs are installed. Thus, 0.47 is the exact value which is covered by each sensor in each sub-region $a_1, a_2 \dots a_n$ where the PMUs are placed. Once the range covered by the sensor is maximized then, the energy consumption should be minimized. This is an important concern in the case of PMUs because the energy of PMUs is already very high. If the energy of the sensor is also higher, then it is not logical to place the sensor in different regions.

In addition, these grouped sensors will determine important parameters which include energy consumption and cost. For example, if the number of PMUs in a particular area is much lesser, the installed sensor group will also be much lesser, and, even if any additional group is installed then, it will be switched off for a particular period of time thus resulting in energy saving processes.

Therefore, energy consumption can be mathematically expressed using (4). In line with the above concern, if a group of sensors are placed in one region and if any sensor fails in one particular group then there is no need for replacing the sensor. Instead, the energy supplied to other sensors in the same group will be increased for equalizing the total supplied energy and the cost of installation will be saved.

$$T_i(PMU_i, S_i) = min \sum_{i=1}^{n} [V_i(PMU_i) - \Delta d_i(PMU_i)] + [V_i(S_i) - \Delta d_i(S_i)] \quad (4)$$

where, V_i denotes the input voltage of both PMU and sensors as per IEEEC37.118 standard; and Δd_i represents the cut off voltage of both PMU and sensors, respectively.

In the proposed work, only the voltage data of PMUs are integrated, so the energy consumption of both sensors and PMU will be determined based on the values of voltage only. Here, the maximum voltage (230V) will be converted to 5V inside the power supply module for suitable installation, and if the value is less, then cut-off voltage will be indicated. If the major parametric inputs such as distance and energy consumption are given properly, then the data can be aggregated appropriately. This is a major issue in several situations because the areas are divided into circles where the data aggregation should happen in sequence order, and it will be examined using an online monitoring system. Therefore, the data combination from different areas can be given as,

$$DC_i = \sum_{i=1}^{n} O_i^{PMU} \times d(i) \quad (5)$$

where, O_i^{PMU} represents the total number of observable PMUs.

It can be seen from (5) that once the distance of the sensor is multiplied with the maximum number of observable PMUs, then the data will be sent from transmitter to receiver in sequence order, and the same will follow for large-scale systems to reduce the data similarity index. Implementation of (5) will provide the data clearly (i.e.,) the data of the first area will be obtained, and once the data is stored, then the data of the next area will be attained. This will reduce the confusion among workers, and they can easily detect the areas where the problem has occurred. Analyzing the implementation cost of the proposed method is an important constraint because here, two different communication devices are combined (PMU and sensors). Therefore, the cost should be minimized as low as possible, and it is calculated mathematically by using (6).

$$TC_i = min \sum_{i=1}^{n} (C_i O_i^{PMU}) + (C_i N_i^S) \quad (6)$$

where, N_i^S denotes the number of sensors installed for observing the features of PMU.

The objective function can be represented using (7) as,

$$OB_i = min \sum_{i=1}^{n} T_i(PMU_i, S_i), TC_i \ \& \ max \sum_{i=1}^{n} d(i) \quad (7)$$

The above equation indicates that if the distance covered by the sensor is maximized, then the cost of installation will be automatically reduced, and the energy consumption of the installed nodes will be minimized as much as possible. This objective function indicated in (7) has not been stated in any existing literature, which forms the novelty of the projected methodology.

4. Optimization Algorithm

An optimization algorithm plays a vital role in deciding the performance of the proposed methodology. Genetic algorithms are incorporated in existing literature [3] for monitoring PMU because of its natural selection process and ability to determine optimal solutions for subsequent iterations. Also, the basic setup using genetic algorithms has been followed [3], where the observability which is detected by binary values (0 or 1) will be different when genetic algorithms are introduced. In this process, unobservable PMUs will be automatically fixed thus allowing the algorithm to converge earlier. Even though the data transmission process is much higher when genetic algorithms are implemented, the same strategy cannot be followed when sensors are implemented, because sensors have unique characteristics for identifying erroneous values of PMU. Thus, the location of PMUs and the occurrence of faults are detected with binary numbers (0 or 1) by integrating a new improved Binary Logistic Regression (BLR) [13] to improve the performance of communication equipment that is implemented in the projected method. The major reason for choosing BLR is that it is one type of machine learning algorithm that exists in recent trends, and the PMU placement is a non-linear problem where the best solutions to non-linear equations will be provided only by BLR. Also, the integration of formulations with BLR has been derived from existing literatures [14,15].

For the projected method, accuracy in determining the fault that exists in PMUs through sensors is an important task; therefore, BLR is a perfect choice where the accuracy rate will be much higher when compared with other algorithms. The prediction of real value PMU using sensors will be mapped to binary values using the exponential function, which is given in (8).

$$b(i) = \frac{1}{1 + e^{TC_i}} \tag{8}$$

where, TC_i denotes the total cost of communication device which consists of both PMU and sensors.

The equivalent proposition for (8) can be defined using a sigmoid function and is given as,

$$y(i) = \sum_{i=1}^{n} \sigma_i(n) \tag{9}$$

where, $\sigma_i(n)$ represents the matrix equation for all the PMUs in logistic scale.

Therefore, (8) can be modified as,

$$b(i) = \frac{1}{1 + \sum_{i=1}^{n} e^{(C_i O_i^{PMU}) + (C_i N_i^S)}} \tag{10}$$

The cost function calculation of linear regression is easy to implement, and it is much similar to logistic regression. But if the cost function of linear terms is integrated, then a non-convex solution will be obtained where the major objective of minimizing the cost will result in failure. For a good communication system implementation, the cost function should be minimum, and it can be calculated using BLR as shown in Figure 2,

$$Cost(i) = \begin{cases} -\log(b(i)) & \text{if } b = 1 \\ -\log(1 - b(i)) & \text{if } b = 0 \end{cases} \tag{11}$$

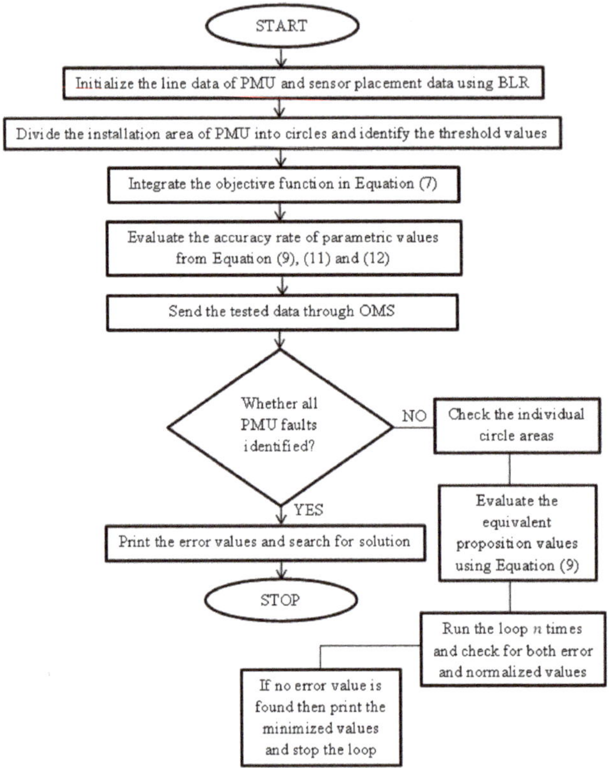

Figure 2. Flow chart of Binary Logistic Regression (BLR).

Equation (11) indicates that, for calculating the cost of all installed PMUs in the grid, both the binary values will be taken into account; whereas, for the optimum cost of installation (TC_i) only the binary value 1 will be considered. For other communication devices (Sensors), the total cost of implementation will be calculated separately, and it depends on base cost and the number of sensors installed. For each parametric calculation, the gradient descent function will be calculated, which is given in Equation (12).

$$g_i = min \sum_{i=1}^{n} \varphi_i - \tau \frac{\partial}{\partial \varphi_i} \tag{12}$$

where, φ_i and $\partial \varphi_i$ represents the first and second-order derivative of gradient descent function.

In (12), all the parameters to be monitored will be implemented based on their derivative function values. From (12), the entire parametric convergence rate will be calculated where BLR has converged at a faster rate when compared to other algorithms. The flow chart of proposed algorithm is shown in Figure 2.

5. Results and Discussion

In this section, the effect of integrating the proposed formulations with BLR has been elaborated. It is necessary to analyze the impact of PMU on both transmission and distribution grids. An OMS will monitor all the updated status of PMU, which is provided by sensors. For better understanding, the OMS provides the simulation results in MATLAB for a specific time period. Therefore, to check the efficiency of the projected method, the following case studies are performed.

Case 1: Effect of energy consumption
Case 2: Maximization of range
Case 3: Minimization of cost

For each of the test cases, the following two test systems are taken under consideration (i.e.,); for transmission systems, a large-scale IEEE 118 bus [14] has been examined. For the distribution system, an IEEE 33 node [15] has been tested. All the test systems consist of only voltage measurements which are obtained from line data for detecting the occurrence of a fault in PMU, and the sensor will report the faults to the control center.

5.1. Case 1

In this case, the energy consumption of sensors that are integrated with PMU has been analyzed. It is essential that the communication devices should always provide low energy with enhanced performance. Particularly, for the case of PMU, it is very important because the energy consumed by PMU in the grid will be much higher. So, if there is a need for another communication device, then the energy acquired by the device should be as low as possible. Therefore, in this important case, it is necessary to minimize the energy consumption of nodes and is calculated using (4). Since voltage measurements are considered, the sensor will make use of the degenerated voltage measurements for accurate calculations. The power supply will be converted suitably to 5V for integrating the sensors with PMU.

The simulation result of the test systems for this case has been projected in Figure 3. It can be observed that the data transfer rate of PMU and sensor is much higher, which always varies between 100 to 500 Mbps. Even for much higher data rates, the energy consumption of the sensor is minimized. For example, in the IEEE 118 bus system (Figure 3a), if the transfer rate is 300 Mbps, then the corresponding energy consumed by the nodes in the proposed BLR will be 76 mW. In contrast, for the existing method [3] (GA), it is 89 mW. Similarly, for the IEEE 33 node system (Figure 3b), if the transfer rate is 400 Mbps, then the energy consumed by the nodes for the proposed system (BLR) will be 27 mW. But, for the same data transfer rate, the energy consumed by GA [3] is 42 mW. This proves that BLR is much more efficient than existing systems, and it consumes less energy for both transmission and distribution systems.

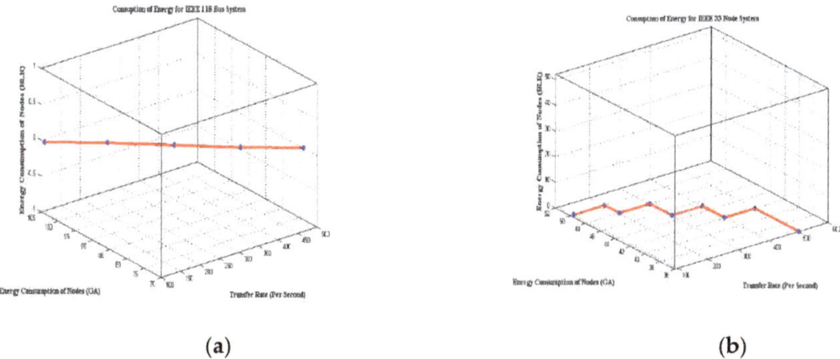

(a) (b)

Figure 3. Energy consumption of test systems: (**a**) IEEE 118 bus and (**b**) IEEE 33 node.

5.2. Case 2

In this case, the fundamental parameter, which is the most important for all sensor integrations in the grid called a range of data measurement, is examined. It is always necessary that the distance covered by a distinct sensor should always be wider. Already, the communicating PMU device will provide a wide range of data measurements where the installed sensors should be able to cover the wide-ranging distance of PMUs. If this constraint is satisfied, then it is possible to measure all the data faults that are measured by PMU within a short interval of time.

The sensor should be calibrated in such a way that it should maximize the distance; therefore, in the proposed method for security protections, the regions covered by PMUs are divided into different zones (Circles). Therefore, accurate distant sensors can be placed for measuring the data of PMUs, and it will be updated in the OMS. Figure 4 shows the distance covered by sensors in both transmission and distribution test systems. It can be seen from Figure 4 that the distance covered by the proposed BLR is much higher than the existing GA method [3]. For example, in the IEEE 118 bus system, the number of PMUs for achieving full observability is 68, and the maximum distance covered by a distinct sensor in a particular area is around 11.36 km, whereas, for the same test system, the distance covered by GA is only 6.8 km. Similarly, for the IEEE 33 node, the number of optimal PMUs will be 11, and the appropriate distance covered by the sensor is 2.1 km. But, with the same number of PMUs, the GA method covers only 1.5 km per area, which are divided into circles.

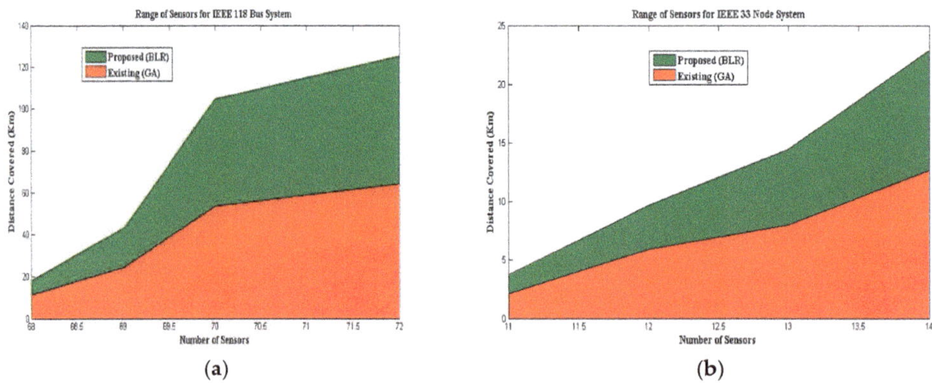

Figure 4. Distance covered by sensors for test systems: (**a**) IEEE 118 bus; and (**b**) IEEE 33 node.

5.3. Case 3

After achieving maximum distance, the next case study relies on the calculation of the cost of installed sensors and PMUs from Equation (6). In the proposed method, two communicating devices are installed; therefore, the cost for both sensors and PMU should be minimized. For each area, the required maximum number of sensors should be integrated for proper data aggregation. Also, if it is possible, two additional low-cost sensors can be added in the area because sometimes there might be a failure in the data detection operation of sensors. If this case persists, the additional sensors can be switched on immediately. The base cost of PMU and sensor is about 3046200 and 11421.56 INR, respectively [16,17]. Figure 5 shows the maximum number of sensors necessary for monitoring the optimal PMUs with the equivalent installation cost.

From Figure 5, it can be realized that, for the proposed BLR implementation in IEEE 118 bus and IEEE 33 node systems, the maximum number of sensors in a specific area is 15 (Figure 5a) and 9 (Figure 5b), respectively. But for the existing GA [3], the number of sensors in a distinct area for IEEE 118 bus and IEEE 33 node systems is 16 and 11, respectively. Similarly, the equivalent total cost of the projected method is minimized to 1019870 INR for IEEE 118 bus system (Figure 5c), and for IEEE 33 nodes, the cost of implementation is around 154069 INR (Figure 5d). The total cost of communication device placement is compared with GA [3] for both IEEE 118 bus and IEEE 33 node systems, which affords nearly 1147912 INR and 166890 INR, respectively. This comparison statement proves that BLR is a cost-effective method when compared with other methodologies.

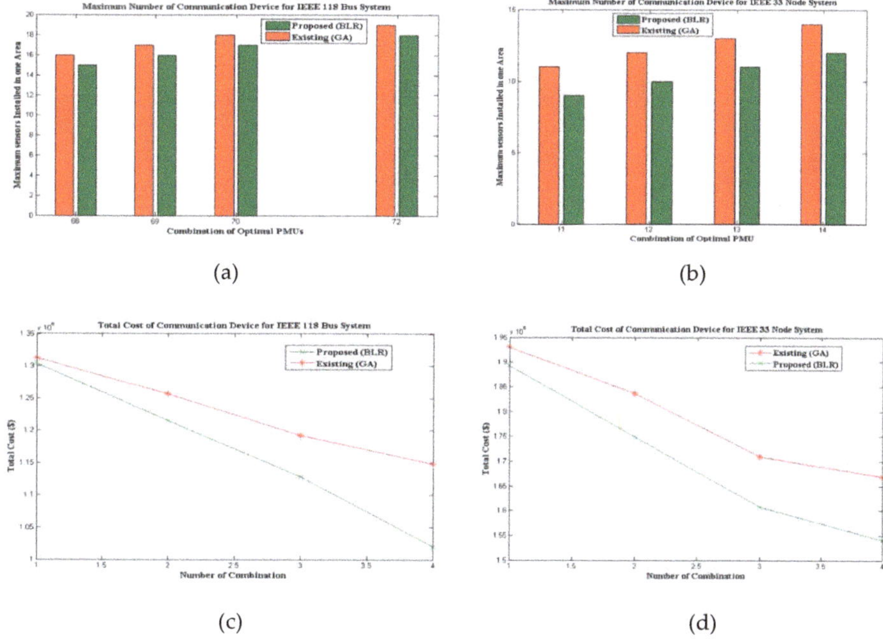

Figure 5. Concentrated installation and cost of communication devices for IEEE 118 bus and IEEE 33 node systems.

6. Conclusions

In this article, a new flanged method of detecting the data fault error in PMU by sensors has been addressed with the minimization of installation cost and energy consumption. Typically, PMU is a communicating device which measures the voltage magnitude and phase angle of all the corresponding buses. This is an important measurement relative to the grids because the voltage plays a major role in the supply of electricity. If there is any failure that is associated with PMUs, then it requires a large amount of manpower for resolving it. Therefore, an automated measuring device is necessary for monitoring the PMUs, which is provided by integrating the wireless sensors. Also, the basic parametric values such as maximum distance with minimization of cost and energy are also fulfilled by the proposed methodology by integrating BLR with OMS. Further, it is important to test the efficiency of the proposed method; therefore, the proposed arrangement has been applied for both transmission (IEEE 118 bus) and distribution systems (IEEE 33 node), and the results prove to be more effective when compared with the GA method.

Directions for Future Work

The proposed work on PMU with sensors can be extended to large-scale and practical real-time systems for observing the complete outcome of grids. Even the extension work on large-scale systems should be introduced for monitoring wide areas. The main challenge for incorporating this future work resides on placing the group of sensors and dividing them appropriately at areas where PMUs are going to be installed.

Author Contributions: Conceptualization, and methodology, H.M.; software, Y.T.; validation, R.K.; formal analysis, S.N.; investigation, H.R.B.; data curation, I.K.; writing—original draft preparation, H.M.; writing—review and editing, Y.T.; visualization, R.K. All authors have read and agreed to the published version of the manuscript.

Funding: This research received no external funding.

Conflicts of Interest: The authors declare no conflict of interest.

References

1. Yao, W.; Wells, D.; King, D.; Herron, A.; King, T.; Liu, Y. Utilization of optical sensors for phasor measurement units. *Electr. Power Syst. Res.* **2018**, *156*, 12–14. [CrossRef]
2. Paudel, S.; Smith, P.; Zseby, T. Stealthy attacks on smart grid PMU state estimation. *ACM Int. Conf. Proceeding Ser.* **2018**, *2018*, 1–10.
3. Bashian, A.; Assili, M.; Anvari-Moghaddam, A.; Catalão, J.P.S. Optimal design of a wide area measurement system using hybrid wireless sensors and phasor measurement units. *Electronics* **2019**, *8*, 1085. [CrossRef]
4. Akkaya, I.; Lee, E.A.; Derler, P. Model-based evaluation of GPS spoofing attacks on power grid sensors. In Proceedings of the 2013 Workshop on Modeling and Simulation of Cyber-Physical Energy Systems (MSCPES), Berkeley, CA, USA, 20–20 May 2013.
5. Subair, S.; Sanjeev, J.; Durga, M.S. Application of Phasor Measurement Units and Internet of Things for Real Time Monitoring of Smart Grid Using 3d Imagery. *IOSR J. Eng.* **2018**, *2018*, 84–87.
6. Meloni, A.; Pegoraro, P.A.; Atzori, L.; Sulis, S. An IoT architecture for wide area measurement systems: A virtualized PMU based approach. In Proceedings of the 2016 IEEE International Energy Conference (ENERGYCON), Leuven, Belgium, 4–8 April 2016.
7. Rehmani, M.H.; Davy, A.; Jennings, B.; Assi, C. Software Defined Networks-Based Smart Grid Communication: A Comprehensive Survey. *IEEE Commun. Surv. Tutor.* **2019**, *21*, 2637–2670. [CrossRef]
8. Bellier, P.; Laurent, P.; Stoukatch, S.; Dupont, F.; Joris, L.; Kraft, M. Autonomous micro-platform for multisensors with an advanced power management unit (PMU). *J. Sens. Sens. Syst.* **2018**, *7*, 299–308. [CrossRef]
9. Mohanta, D.K.; Murthy, C.; Sinha Roy, D. A Brief Review of Phasor Measurement Units as Sensors for Smart Grid. *Electr. Power Components Syst.* **2016**, *44*, 411–425. [CrossRef]
10. Song, E.Y.; Fitzpatrick, G.J.; Lee, K.B. Smart Sensors and Standard-Based Interoperability in Smart Grids. *IEEE Sens. J.* **2017**, *17*, 7723–7730. [CrossRef] [PubMed]
11. Coppolino, L.; D'Antonio, S.; Mazzeo, G.; Romano, L.; Sgaglione, L. An Approach for Securing Cloud-Based Wide Area Monitoring of Smart Grid Systems. *Lect. Notes Data Eng. Commun. Technol.* **2019**, *22*, 952–959.
12. Xu, G.X.; Wu, Q.; Daneshmand, M.; Liu, Y.; Wang, M. A data privacy protective mechanism for WBAN. *Wirel. Commun. Mob. Comput.* **2015**, *14*, 421–430.
13. Liu, L. Research on logistic regression algorithm of breast cancer diagnose data by machine learning. In Proceedings of the 2018 International Conference on Robots & Intelligent System (ICRIS), Changsha, China, 26–27 May 2018; pp. 157–160.
14. Pasqualetti, F.; Dorfler, F.; Bullo, F. Cyber-physical security via geometric control: Distributed monitoring and malicious attacks. *Proc. IEEE Conf. Decis. Control* **2012**, 3418–3425.
15. Bahramipanah, M.; Cherkaoui, R.; Paolone, M. Decentralized voltage control of clustered active distribution network by means of energy storage systems. *Electr. Power Syst. Res.* **2016**, *136*, 370–382. [CrossRef]
16. Young, M.; Silverstein, A. Factors Affecting PMU Installation Costs. In *U.S. Department of Energy Office of Electricity Delivery and Energy Reliability (DOE-OE)*; U.S Department of Energy: Washington, DC, USA, 2014; p. 12.
17. Lai, C.S.; McCulloch, M.D. Levelized cost of electricity for solar photovoltaic and electrical energy storage. *Appl. Energy* **2017**, *190*, 191–203. [CrossRef]

Article

A Novel Intrusion Mitigation Unit for Interconnected Power Systems in Frequency Regulation to Enhance Cybersecurity

Faisal R. Badal [1], Zannatun Nayem [2], Subrata K. Sarker [3], Dristi Datta [3], Shahriar Rahman Fahim [4], S. M. Muyeen [5,*], Md. Rafiqul Islam Sheikh [4] and Sajal K. Das [1]

1 Department of Mechatronics Engineering, Rajshahi University of Engineering & Technology, Rajshahi 6204, Bangladesh; faisalrahman@mte.ruet.ac.bd (F.R.B.); sajal.das@mte.ruet.ac.bd (S.K.D.)
2 Department of Computer Science and Engineering, Rajshahi University of Engineering & Technology, Rajshahi 6204, Bangladesh; zannatsumaiya.bauet36@gmail.com
3 Department of Electrical and Electronic Engineering, Varendra University, Rajshahi 6204, Bangladesh; subrata@vu.edu.bd (S.K.S.); dristi@vu.edu.bd (D.D.)
4 Department of Electrical and Electronic Engineering, Rajshahi University of Engineering & Technology, Rajshahi 6204, Bangladesh; fahim.fact@gmail.com (S.R.F.); mri.sheikh@eee.ruet.ac.bd (M.R.I.S.)
5 School of Electrical Engineering, Computing and Mathematical Sciences, Curtin University, Perth 6845, Australia
* Correspondence: sm.muyeen@curtin.edu.au

Abstract: Cyberattacks (CAs) on modern interconnected power systems are currently a primary concern. The development of information and communication technology (ICT) has increased the possibility of unauthorized access to power system networks for data manipulation. Unauthorized data manipulation may lead to the partial or complete shutdown of a power network. In this paper, we propose a novel security unit that mitigates intrusion for an interconnected power system and compensates for data manipulation to augment cybersecurity. The studied two-area interconnected power system is first stabilized to alleviate frequency deviation and tie-line power between the areas by designing a fractional-order proportional integral derivative (*FPID*) controller. Since the parameters of the *FPID* controller can also be influenced by a CA, the proposed security unit, named the automatic intrusion mitigation unit (AIMU), guarantees control over such changes. The effectiveness of the AIMU is inspected against a CA, load variations, and unknown noises, and the results show that the proposed unit guarantees reliable performance in all circumstances.

Keywords: interconnected power system; cybersecurity; *FPID* controller; automatic intrusion mitigation unit

Citation: Badal, F.R.; Nayem, Z.; Sarker, S.K.; Datta, D.; Rahman Fahim, S.; Muyeen, S.M.; Islam Sheikh, M.R.; Das, S.K. A Novel Intrusion Mitigation Unit for Interconnected Power Systems in Frequency Regulation to Enhance Cybersecurity. *Energies* **2021**, *14*, 1401. https://doi.org/10.3390/en14051401

Academic Editor: Michael Short

Received: 12 January 2021
Accepted: 22 February 2021
Published: 4 March 2021

Publisher's Note: MDPI stays neutral with regard to jurisdictional claims in published maps and institutional affiliations.

1. Introduction

The increasing use of modern devices has increased the energy crisis to meet the energy demands of the world's growing population [1–5]. Modern power systems integrated with information and communication technology (ICT) add flexibility for consumers but more challenges for stockholders due to their operation and control [6–8]. The reliable operation of modern power systems depends on many factors, such as accurate system modeling, control, disturbance-handling capabilities, and so forth [9–12].

Interconnected power systems (INPSs) consist of several subgeneration units or areas that can operate individually. Each area of an INPS has separate generation units connected by a tie line. The function of this line is to manage the flow of power between the areas [13–15].

Maintaining the tie-line power at its nominal value is a challenge and requires a great deal of attention [16–19]. Frequency deviation due to improper matching between the generation and demand may result in the loss of harmony and a complete blackout. Frequency deviation and tie-line power mismatch may introduce load damage, a reduction in production, overheating, and so on [12,20,21].

Experts have reported several methods to compensate for the frequency deviation of INPSs [17,22–26]. These methods mitigate deviation against the uncertainty, disturbance, and nonlinear behavior of the system but not against a cyberattack (CA). The mitigation of a CA on power systems has brought a new research dimension for researchers. If the CA is realistic, the malefactors can easily provide false data through the communication link to the power system control station. The injection of false data may increase the possibility of unrealistic system performance [27–30].

The rate of CAs on cyberphysical systems (CPSs) has increased significantly over the last decade. Recent news has attracted the attention of the researchers to develop strong security for CPSs. The CA on the U.S. gas pipeline and smart grid, as well as on the Ukraine power system, indicates the frequent and regular occurrence of attacks on power systems [31–33]. Grid reliability and resiliency were largely affected by this CA. Unauthorized access, false data injection, data integrity, sensitive data collection, etc., are some of the objectives of attackers to hamper power systems [34–36].

Although several load frequency controllers (LFCs) are designed to minimize frequency deviation, they may also be largely affected by CAs. The function of LFCs is to collect data from the frequency sensor and regulate the speed of the generation units. Any changes in the speed regulator by the governor make the governor's speed uncontrollable and produce a large frequency as well as a tie-line power deviation of the INPS. Again, any changes in the associated control parameters are also responsible for reducing the stability margin and reliability of the system. Thus, the speed regulator and control variables are critical parameters that attract the attackers' attention [37–39].

Conventional control methods can only control the system performance against system dynamics and different disturbances. The control of a CA is rarely possible with the help of these control methods. The parameters of the system and the controller are sensitive data that can easily be modified by attackers. Thus, alone, the controller cannot assure the stability of the system during a CA, which ensures the requirement of a cybersecurity-based control algorithm [40].

A CA on the parameters and settings of the system can be mitigated by analyzing the behavior of the circuit breaker, protection devices, and logics [41]. The detection of false data injection based on the measurement variation method is proposed in [42]. The control algorithm detects the presence of a CA by measuring the distances between two probability distributions. Reference [43] proposes a transformation-based scheme used to find the modification or injection of false data into the smart grid. The investigation of a CA is carried out by proper estimation of the measurement variation based on the Kullback–Leibler distance algorithm. The power system has a quasistatic nature that reduces the detection capability of this method.

Reference [38] considers a technique to provide false data to the power system and investigates its impact on production costs and demands. Positively and negatively biased CAs and their impacts on the power system are discussed in [40]. This work proposes a switching technique to control a LFC with an integrator controller. Another switching algorithm is investigated in [44]. The authors of [45] analyzed the frequency deviation of a single-area power system with a LFC and an automatic generation control (AGC) against positively and negatively biased CAs. The positively biased CA produced a greater impact on both the LFC and AGC compared to the negatively biased CA.

The aforementioned research articles provide control structures against CAs for a single-area power system, although they are unable to protect the operation of the INPS during a CA. Although the integrator-based AGC confirms the reliable operation for a single-unit LFC, the use of INPSs can be questionable due to their multiple control parameters, and these controllers are also sensitive to change in operating conditions. Therefore, a separate protection unit for INPS must be considered during a CA. Inspired by the above issues, this work proposes a novel security unit, called the automatic intrusion mitigation unit (AIMU), to protect INPSs during a CA. The contributions of this article are:

- The design of a fractional-order *PID* (*FPID*) controller to compensate for frequency deviation and the tie-line power mismatch of an interconnected power system.
- The investigation of the closed-loop interconnected power system's behavior due to changes in sensitive parameters such as the speed regulator, biasing factor, and parameters of the controller.
- The design of a novel security unit for the INPS to mitigate the frequency deviation and tie-line power discrepancy against the parameter changes of both the system and controller due to a CA.

This paper is organized as follows. Section 2 models the INPS, and Section 3 discusses the load frequency controller design. The impacts of a CA, the proposed switching technique, and the performance of the AIMU are investigated in Sections 4–6. The conclusions are presented in Section 7.

2. Interconnected Power System Modeling

The control structure of the INPS is constructed by using two control loops; e.g., the primary and secondary loops for controlling frequency. The primary loop acts as a proportional loop that decreases the frequency deviation produced due to load and demand change, but it is unable to regain zero frequency deviation. The frequency deviation returns to a nominal value (i.e., zero deviation) by implementing another control loop, known as the secondary frequency control loop.

Figure 1 presents a model of an INPS that has three main components: a governor, a turbine, and a mass-load system. The mass-load system produces an impact on the INPS, the model of which can be represented by [12,14,16,46,47]

$$X_{ml}(s) = \frac{1}{Ms + D} \tag{1}$$

where M and D are the inertia and damping constant. The governor can be modeled as

$$X_g(s) = \frac{1}{1 + Y_g s} \tag{2}$$

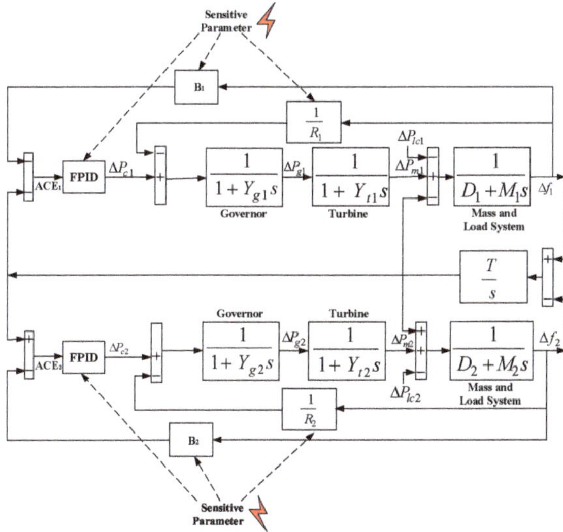

Figure 1. Model of an interconnected power system.

The performance of the governor largely depends on the parameter known as the speed regulator (R). Any variation of this parameter is responsible for deviating the speed of the generation unit and increasing the frequency and power deviation. Another element known as the turbine can be modeled as

$$X_t(s) = \frac{1}{1 + Y_t s} \tag{3}$$

For an INPS, two or more areas are connected to each other and deliver power through the tie line, the model of which can be represented by

$$\Delta P_{12} = \frac{|V_1||V_2|}{X_{12}} \sin(\phi_1 - \phi_2) \tag{4}$$

where X_{12} is the reluctance of the tie line and areas. The load change is usually kept small; e.g., 2%. Thus, the delivered tie-line power due to the small perturbation of load where $\phi = \phi^0$ can be represented by

$$\Delta P_{12} = T(\Delta f_1 - \Delta f_2) \tag{5}$$

The reference signal and feedback construct the area control error (*ACE*) that feeds the error measurement to the controller to minimize the frequency and power deviation. The model of *ACE* can be given as

$$ACE = \Delta P_{12} + B\Delta f \tag{6}$$

where B is a biasing factor that can be represented by

$$B = \frac{1}{R} + D \tag{7}$$

In an INPS, there are nine total state variables. There are three state variables for each area; i.e., Δf_1, ΔP_{g1}, ΔP_{m1}, and Δf_2, ΔP_{g2}, and ΔP_{m2}. There is one state variable for the tie line; i.e., ΔP_{12}. Finally, there are two state variables for the inputs ΔP_{c1} and ΔP_{c2}. Thus, the dynamic model of the INPS can be represented by

$$\begin{aligned}
\Delta \dot{f_1} &= \tfrac{D_1}{M_1}\left\{\tfrac{1}{D_1}\Delta P_{m1} - \tfrac{1}{D_1}\Delta P_{lc1} - \Delta f_1 - \tfrac{1}{D_1}\Delta P_{12}\right\} \\
\Delta \dot{P}g_1 &= \tfrac{1}{Y_{g1}}\left\{\Delta P_{c1} - \Delta P_{g1} - \tfrac{1}{R_1}\Delta f_1\right\} \\
\Delta \dot{P}_{m1} &= \tfrac{1}{Y_{t1}}\{\Delta P_{g1} - \Delta P_{m1}\} \\
\Delta \dot{P}_{12} &= T\{\Delta f_1 - \Delta f_2\} \\
\Delta \dot{f_2} &= \tfrac{D_2}{M_2}\left\{\tfrac{1}{D_2}\Delta P_{m2} - \tfrac{1}{D_2}\Delta P_{lc2} - \Delta f_2 + \tfrac{1}{D_2}\Delta P_{12}\right\} \\
\Delta \dot{P}g_2 &= \tfrac{1}{Y_{g2}}\left\{\Delta P_{c2} - \Delta P_{g2} - \tfrac{1}{R_2}\Delta f_2\right\} \\
\Delta \dot{P}_{m2} &= \tfrac{1}{Y_{t2}}\{\Delta P_{g2} - \Delta P_{m2}\} \\
\Delta \dot{P}_{c1} &= B_1 f_1 + \Delta P_{12} \\
\Delta \dot{P}_{c2} &= B_2 f_2 - \Delta P_{12}
\end{aligned}$$

Thus, the INPS in the form of state space can be written as

$$\begin{bmatrix} \Delta\dot{f}_1 \\ \Delta\dot{P}_{m1} \\ \Delta\dot{P}_{g1} \\ \Delta\dot{f}_2 \\ \Delta\dot{P}_{m2} \\ \Delta\dot{P}_{g2} \\ \Delta\dot{P}_{12} \\ \Delta\dot{P}_{c1} \\ \Delta\dot{P}_{c2} \end{bmatrix} = \begin{bmatrix} -\frac{D_1}{M_1} & \frac{1}{M_1} & 0 & 0 & 0 & 0 & -\frac{1}{M_1} & 0 & 0 \\ 0 & -\frac{1}{Y_{t1}} & \frac{1}{Y_{t1}} & 0 & 0 & 0 & 0 & 0 & 0 \\ -\frac{1}{R_1 Y_{g1}} & 0 & -\frac{1}{Y_{g1}} & 0 & 0 & 0 & 0 & 0 & 0 \\ 0 & 0 & 0 & -\frac{D_2}{M_2} & \frac{1}{M_2} & 0 & \frac{1}{M_2} & 0 & 0 \\ 0 & 0 & 0 & 0 & -\frac{1}{Y_{t2}} & \frac{1}{Y_{t2}} & 0 & 0 & 0 \\ 0 & 0 & 0 & -\frac{1}{R_2 Y_{g2}} & 0 & -\frac{1}{Y_{g2}} & 0 & 0 & 0 \\ T & 0 & 0 & -T & 0 & 0 & 0 & 0 & 0 \\ B_1 & 0 & 0 & 0 & 0 & 0 & 1 & 0 & 0 \\ 0 & 0 & 0 & B_2 & 0 & 0 & -1 & 0 & 0 \end{bmatrix} \begin{bmatrix} \Delta f_1 \\ \Delta P_{m1} \\ \Delta P_{g1} \\ \Delta f_2 \\ \Delta P_{m2} \\ \Delta P_{g2} \\ \Delta P_{12} \\ \Delta P_{c1} \\ \Delta P_{c2} \end{bmatrix}$$

$$+ \begin{bmatrix} 0 & 0 \\ 0 & 0 \\ \frac{1}{Y_{g1}} & 0 \\ 0 & 0 \\ 0 & 0 \\ 0 & \frac{1}{Y_{g2}} \\ 0 & 0 \\ 0 & 0 \\ 0 & 0 \end{bmatrix} \begin{bmatrix} \Delta P_{c1} \\ \Delta P_{c2} \end{bmatrix} + \begin{bmatrix} -\frac{1}{M_1} & 0 \\ 0 & 0 \\ 0 & 0 \\ 0 & -\frac{1}{M_2} \\ 0 & 0 \\ 0 & 0 \\ 0 & 0 \\ 0 & 0 \\ 0 & 0 \end{bmatrix} \begin{bmatrix} \Delta P_{lc1} \\ \Delta P_{lc2} \end{bmatrix}$$

Table 1 lists the values of the parameters of the INPS.

Table 1. Description of the interconnected power system (INPS) parameters.

Description	Notation	Unit	Area 1	Area 2
Speed regulator	R	pu	0.05	0.0625
Biasing factor for frequency	B	pu/Hz	20.6	16.9
Generator inertia constant	M	s	10	8
Load-damping factor	D	-	0.6	0.9
Governor time constant	Y_g	s	0.2	0.3
Turbine time constant	Y_t	s	0.5	0.6
Synchronizing power coefficient	T	pu	-	2
Load change	ΔP_{12}	pu	0.2	0

3. Load Frequency Controller Design

In an INPS, the frequency and *ACE* are two crucial pieces of data that need to be controlled to increase the level of protection of the power system. Any change in the supply-to-demand ratio or the speed regulator of the governor is responsible for deviations from the nominal frequency. A frequency controller is implemented to control these frequency deviations in the power system with an additional control approach. In this work, we designed a *FPID* controller for load frequency control.

The *FPID* controller is a modified conventional integer-order *PID* controller based on the fractional calculus used to design the integral and derivative mode of the *FPID* controller [48–50]. Thus, the construction of the *FPID* controller provides two more controlling parameters that enhance the tuning performance of the *FPID* controller.

The dynamic model of an integral part of the *FPID* controller can be written as

$$_{T}I_t^{\omega} f(t) = \frac{1}{\Xi(\omega)} \int_T^t \frac{f(\sigma)}{(t-\sigma)^{1-\omega}} d\sigma \tag{8}$$

Here, $\Xi(\bullet)$ is the function of the FI controller having fractional order σ.

Similarly, the derivative part of this controller can be defined as

$$ {}_{T}D_t^{\omega} f(t) = \frac{1}{\Xi(n-\omega)} \frac{d^n}{d^n t} \int_T^t (t-\sigma)^{n-\omega-1} f(\sigma) d\sigma \tag{9} $$

From the above two equations, the output of the *FPID* controller can be given as

$$ y(t) = K_p e(t) + K_i D_t^{-a} e(t) dt + K_d D_t^b e(t) dt \tag{10} $$

Thus, the *FPID* controller in the frequency domain can be represented by

$$ C_{FPID} = K_p + K_i S^{-a} + K_d S^b \tag{11} $$

where a and b are the power of the I and D of the *FPID* controller. The *FPID* controller has five modes of operation:

(i) The *FPID* acts as a *FPID* controller when a is a fractional number and b is a fractional number;
(ii) The *FPID* acts as a P controller when $a = 0$ and $b = 0$;
(iii) The *FPID* acts as a *PI* controller when $a = 1$ and $b = 0$;
(iv) The *FPID* acts as a *PD* controller when $a = 0$ and $b = 1$;
(v) The *FPID* acts as a *PID* controller when $a = 1$ and $b = 1$.

a and b of the *FPID* controller enhance the system quality and performance compared to the *PID* and *PI* controllers in the case of a smaller processing time and larger bandwidth, as well as being suitable for higher-order systems [51–53]. In this work, we selected $a_1 = 0.995$, $b_1 = 1.5$ and $a_2 = 2$, $b_2 = 1.5$ to design the *FPID* controller and compared the results with those of the integer-order *PI* and *PID* controllers, considering $a = 1$, $b = 0$ and $a = 1$, $b = 1$, respectively.

The high performance of the *FPID* controller compared to the *PI* and *PID* controllers can be investigated from Figure 2. As presented in Figure 2, the studied open-loop system indicates a high level of oscillation against a step input and deviates by a large frequency and tie-line power from the reference, whereas the *FPID* controller has approximately zero deviation of frequency and tie-line power from the reference, which is the lowest compared to the integer-order *PI* and *PID* controllers.

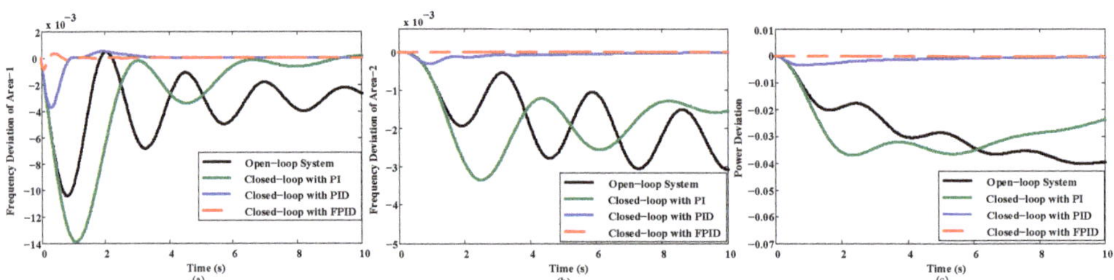

Figure 2. Comparison of frequency deviations and tie-line power for open- and closed-loop systems. (**a**) Frequency deviation for Area 1; (**b**) frequency deviation for Area 2; (**c**) tie-line power.

4. Effects of a Cyberattack

The design of the LFCs only for the INPS is not enough to guarantee protection against a CA. This is because the control parameters can also be targeted by attackers. Figure 2 shows that although the system is efficiently controlled by implementing different control approaches, any change in the controller or system parameters may lead to deviation from the nominal operation of the system. This work aimed to address the causes of parameter changes and present a solution to overcome these problems. Misleading information about the frequency measurement, speed regulator, or controller parameters due to a CA are responsible for generating false *ACE* signals, and the power system may produce or supply an undesirable amount of power. Thus, system frequency, as well as the power of the system, will be greatly affected [28].

The adverse effect on the INPS due to a parameter variation was investigated. The variation of sensitive parameters, such as the speed regulator, the biasing factor of the system, and K_p, K_i, K_d, a, and b of the *FPID*, *PID*, and *PI* controllers is considered. This case study is essential to address the need for INPS security and mitigate the effects of the unwanted changing of the parameters.

4.1. Case Study 1: Variation of the Speed Regulator

The speed regulator (R) is a sensitive parameter. The change of *R* produces an impact on the speed of the governor responsible for deviating the system frequency and power. Figure 3 shows that any small change of *R* is responsible for deviating the performance of the INPS and making the system operation unreliable.

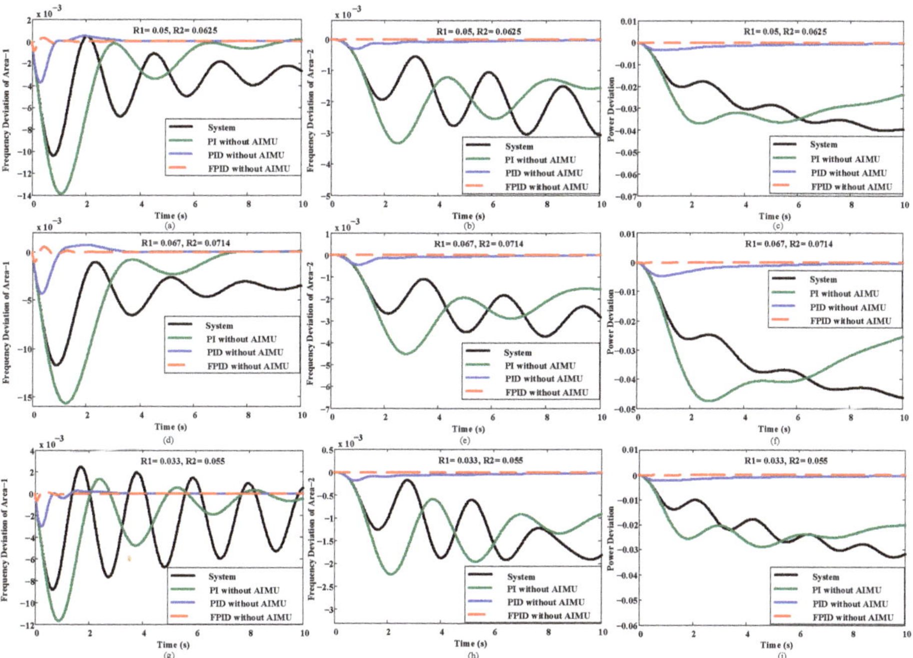

Figure 3. Performance of the interconnected power system against the speed regulator variation without an automatic intrusion mitigation unit (AIMU).

4.2. Case Study 2: Variation of the Biasing Factor

The biasing factor (B) is another critical parameter that produces an impact on the construction of the *ACE* signal. The change of *B*1 and *B*2 is responsible for producing the wrong error signal fed to the controller. Thus, a wrong control signal is produced by the associated controller and may increase fluctuations in the INPS performance. The impact due to the change of *B* is shown in Figure 4. Figure 4 shows a huge change in the performance of the INPS due to the variable value of *B* that may cause the system to become unstable and damage the load.

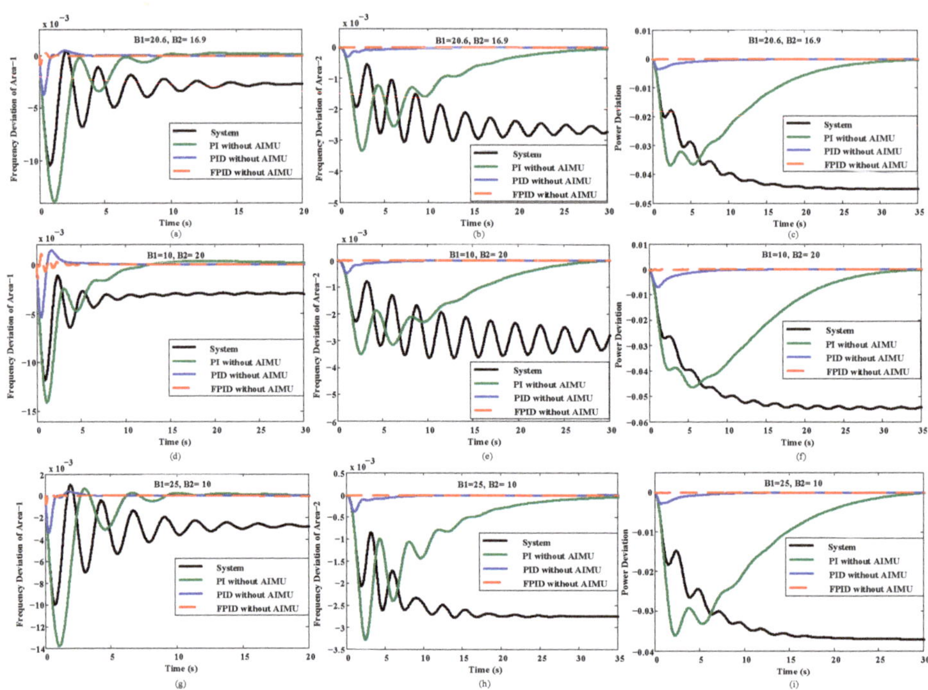

Figure 4. Performance of the interconnected power system against the biasing factor variation without an AIMU.

4.3. Case Study 3: Variation of the Controller Parameter

The performance of the INPS largely depends on the efficient design of the controller. The stable and reliable operation of the whole system and loads is directly related to the control efficiency and structure. Any change of this control structure will make the controller unable to produce the required control signals. Table 2 was used to test the controller performance without an AIMU. Figure 5 presents the variation of the controller parameters and their impacts on the deviation of area frequency and power. Table 3 presents the impacts of the parameter changes on the power system.

Table 2. Parameter values for Case Study 3.

Data Type	K_{p1}	K_{p2}	K_{i1}	K_{i2}	K_{d1}	K_{d2}	a_1	a_2	b_1	b_2
*FPID*1	100	15	150	0.01	5	5	0.995	2	1.5	1.5
*FPID*2	80	20	130	0.1	7	5	0.1	1	1.5	1.5
*FPID*3	70	150	80	10	8	4	1.6	0.001	1.5	1.7
*PID*1	3.5	4	6	2	2	2	-	-	-	-
*PID*2	5	1	1	4	1	4	-	-	-	-
*PID*3	1	10	3	2	9	2	-	-	-	-
*PI*1	0.001	0.0005	0.2	0.01	-	-	-	-	-	-
*PI*2	1	0.3	0.2	0.01	-	-	-	-	-	-
*PI*3	0.11	0.0005	0.4	0.1	-	-	-	-	-	-

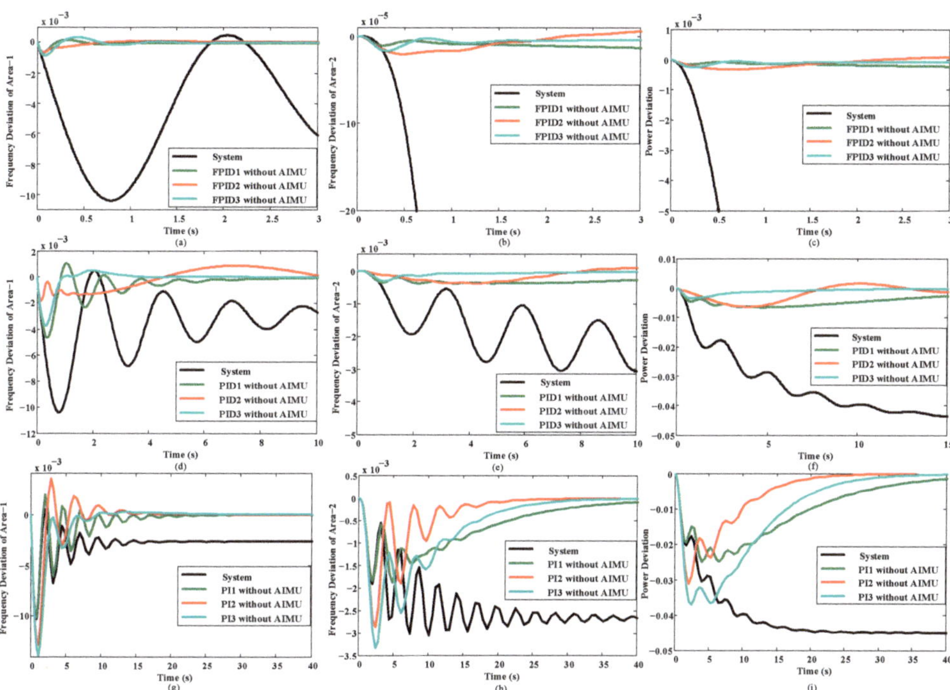

Figure 5. Performance of the interconnected power system against the controller parameter variation without an AIMU.

Table 3. Impacts of a cyberattack (CA) on the INPS.

Parameters	Increased Value	Decreased Value
R	(i) Increased overshoot (OS), (ii) decreased oscillation, (iii) increased steady-state error (SSE)	(i) Decreased OS, (ii) increased oscillation, (iii) decreased SSE
B	(i) Decreased OS, (ii) increased oscillation, (iii) decreased SSE	(i) Increased OS, (ii) decreased oscillation, (iii) increased SSE
K_p	(i) Decreased OS, (ii) decreased SSE, (iii) increased settling time (ST)	(i) Increased OS, (ii) increased SSE, (iii) decreased ST
K_i	(i) Increased OS, (ii) decreased SSE, (iii) increased ST	(i) Decreased OS, (ii) increased SSE, (iii) decreased ST
K_d	(i) Decreased OS, (ii) decreased ST	(i) Increased OS, (ii) increased ST
a	(i) Decreased SSE	(i) Increased SSE
b	(i) Decreased oscillation	(i) Increased oscillation

The findings of these case studies present the evidence of the lack of system security in conventional power systems. Any change of the sensitive parameters is responsible for deviating system performance, as shown in Figures 3 and 4. Researchers are always interested in developing different control algorithms to control system performance against any disturbance, uncertainties, or system dynamics. *PI*, *PID*, *FPID*, or other control algorithms are inventions of researchers to control system stability under system parameter variation or disturbance.

Today's research not only focuses on the design of any controller but also accounts for the security issue because it is rarely possible to control modern devices with a single controller due to the development of ICT. At present, ICT makes it possible to easily access any modern system. Controllers such as *PI*, *PID*, or *FPID* are able to control system performance but are unable to ensure the cybersecurity of the system. Thus, choosing only the controller without considering a CA hampers both the system and controller parameters because attackers can attack both the system and controller, as shown in Figures 3–5. To address these issues, it is essential to design a cybersecurity-based controller that not only stabilizes system performance but also diminishes the adverse effects of false data injection into both the system and controller.

5. Design Methodology of the Proposed Automatic Intrusion Mitigation Unit

It is obvious from the aforementioned discussion that the biasing factor and speed regulator of the governor, as well as the variables of the associated control unit of the INPS, are attracting parameters to attackers. By changing these parameters through the CA, they can destroy the power system and interrupt the delivery of power to the customer.

In this paper, an automatic intrusion mitigation unit (AIMU), as shown in Figure 6, is proposed that mitigates the effects of a CA and enhances the security of the INPS, the construction of which is shown in Figure 7 and is the main contribution of this paper. This control technique has both a network and hardware connection. The control structure of the proposed AIMU has four input terminals that account for the speed regulator, control signal from the *FPID* controller, current, and fixed information of the sensitive data collected from the network and hardware. The fixed data stored in the AIMU are termed hardware data. The data information of the system may change against the system dynamics, uncertainties, and disturbance, as well as the CA. Thus, the present data information of the system and controller at every moment is considered networked data.

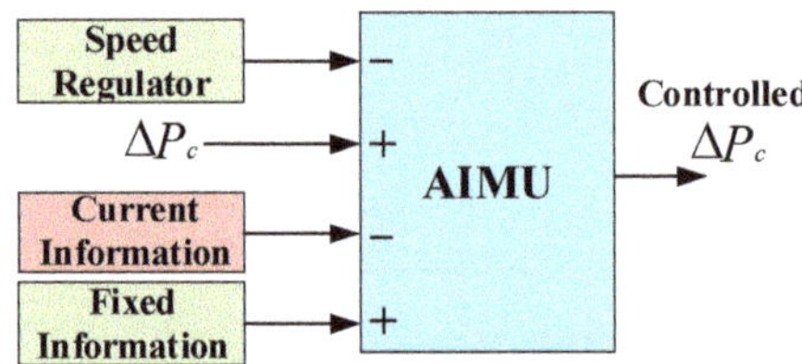

Figure 6. Construction of an automatic intrusion mitigation unit.

The detection of the current information about sensitive data is important to identify the presence of a CA. The collection of information from both the hardware and network is essential to maintain the safe and reliable operation of a power system. To regulate the balance between supply and demand, the service provider manually modifies hardware data such as the parameters and setup of the system and controller. Once the data are modified, the service provider will make the data fixed to overcome illegal data modification or injection.

The data of a nominal system and the parameter values of a controller at which it provides better performance are considered fixed data. The fixed data for both the system and controller are listed in Tables 1 and 4. The speed regulator and biasing factor of the INPS, as well as K_p, K_i, K_d, a, and b of the *FPID*, *PID*, and *PI* controllers, are sensitive parameters usually affected by a CA. Thus, it is necessary to protect these parameters from attackers. To secure these parameters from attackers, hardware with an offline connection can be used to store the data, which acts as an input to the AIMU. The AIMU can only read data from this hardware for its further comparison process. This control unit cannot write or update these fixed hardware data without the permission of the service providers. This way, the proposed AIMU prevents the unauthorized access of attackers to the fixed data.

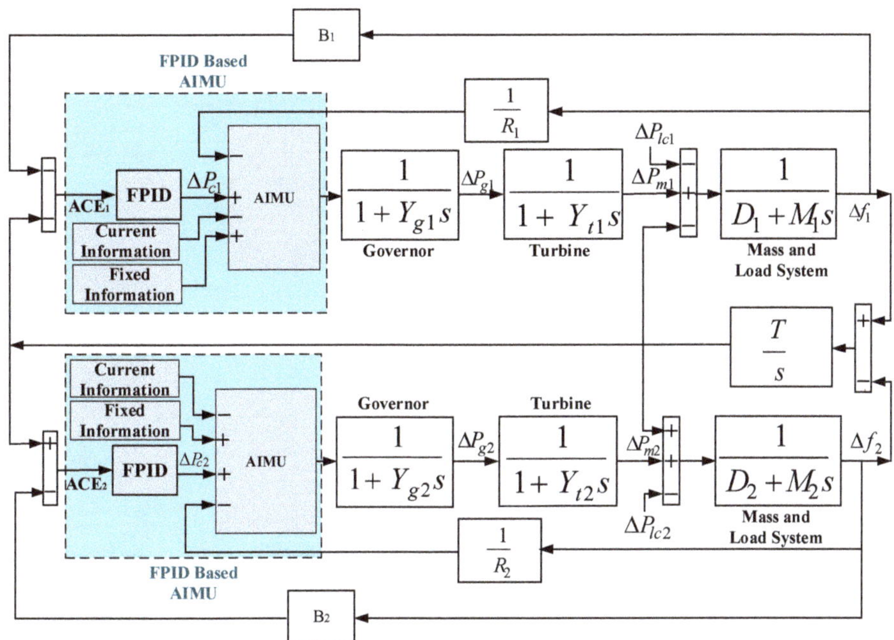

Figure 7. Model of an AIMU-based INPS.

Table 4. Fixed and chosen affected parameters of R and the controller for the interconnected power system.

Data Type	R_1	R_2	K_{p1}	K_{p2}	K_{i1}	K_{i2}	K_{d1}	K_{d2}	a_1	a_2	b_1	b_2
Affected data	0.055	0.0714	200	50	100	1	6	10	0.1	0.01	1.4	2
Fixed *FPID*	0.05	0.0625	100	15	150	0.01	5	5	0.995	2	1.5	1.5
Fixed *PID*	0.05	0.0625	3.5	4	6	2	2	2	-	-	-	-
Fixed *PI*	0.05	0.0625	0.001	0.0005	0.2	0.01	-	-	-	-	-	-

The AIMU then continuously monitors the present situation of these sensitive networked data of the INPS, and the controller and compares these networked data with the fixed hardware data at every moment to calculate the difference between the monitored data and fixed data.

The AIMU acts as a control unit that performs a mathematical calculation between the hardware and networked data. The polarity of the four input terminals such as R, ΔP_c, the fixed, and the current information of the AIMU is selected in such a way that all the monitored network data such as R, ΔP_c, and the current information are subtracted from the fixed hardware data to find any change between them, and their algebraic summation always generates the required constant value of the parameters for the system and controller. If it finds no changes between the measured data and the fixed data, it ensures the absence of a CA on the system. If any changes of these parameters are noticed by the AIMU, it generates a control signal to indicate the presence of a CA on the system or controller, makes the algebraic summation of R, ΔP_c, and the current information equal zero, and allows the fixed hardware data to control the system. Thus, any change of the data can be overcome automatically, and the system always exhibits a fixed performance with the aid of the AIMU, which protects the system against the diverse effects of an unauthorized CA.

Thus, the AIMU is designed considering its hardware is protected from a cyberattack due to its fixed offline operation. This means attackers can only attack the network data, which may change the system or controller parameters. The proposed AIMU detects changes in the network data, minimizes the effects on the system performance due to the parameter variations, and reduces the error to zero, based on the fixed hardware data.

6. Performance Evaluation

The performances of the INPS with and without the implementation of the AIMU are inspected here. The open-loop performance of the INPS, as shown in Figure 2, demonstrates the necessity for the controller design to make the system stable and reduce the oscillations in the INPS performance. For the stable operation of the INPS, a *FPID* controller is designed in this paper, which is further compared with conventional *PI* and *PID* controllers. To understand the impact of a CA on the INPS, a mimic environment of a CA is created, and the performance of the INPS is inspected in Figures 3–5. For further improvement, the performance of the controller-based AIMU is investigated under load and wind power change.

6.1. Performance Analysis with the AIMU

The parameter variations and their effects were inspected in Case Studies 1–3, which are presented in Figures 3–5, and their impacts are listed in Table 3. These ensure the necessity for the implementation of the AIMU to protect the system from CA and confirm the constant behavior of the INPS.

To evaluate the reliable and fixed behavior of the AIMU, Table 4 is used to list some affected and fixed data of the speed regulator and controller parameters. The affected data of Table 4 are used to produce a mimic environment of a CA to evaluate the effectiveness of the proposed AIMU. With these affected data, the performance of the INPS deviates from its fixed operation, which may damage the system and load. Figures 8 and 9 show the performance of the affected and secured systems. The implementation of the AIMU completely eliminates the affected data from the power system and feeds the fixed data from the hardware to exhibit constant frequency and power deviation control similar to the responses before a cyberattack. Thus, Figures 8 and 9 ensure the robust performance of the INPS with any changes of the system and control unit. Table 5 shows the performance of the different control algorithms with the AIMU.

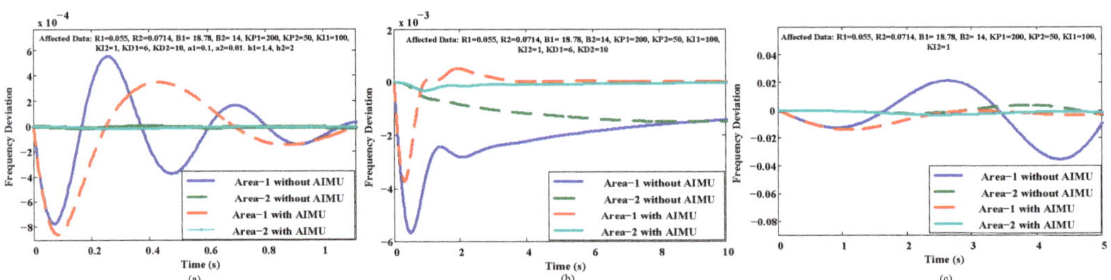

Figure 8. Control of the frequency of the AIMU-based INPS with (**a**) *FPID*, (**b**) *PID*, and (**c**) *PI* controllers.

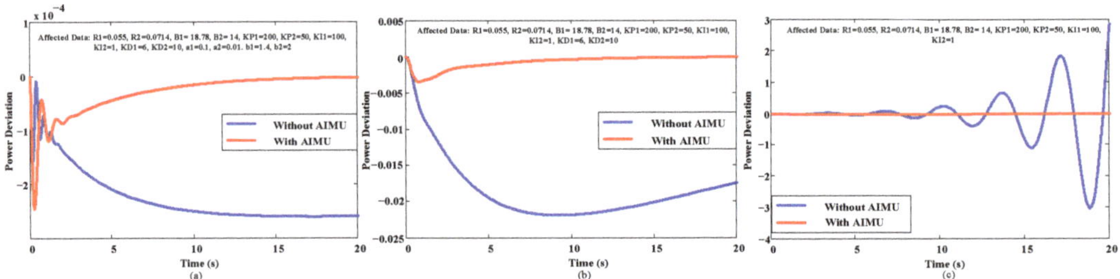

Figure 9. Control of the tie-line power of the AIMU-based INPS with (**a**) *FPID*, (**b**) *PID*, and (**c**) *PI* controllers.

Table 5. Performance of the INPS with *PI-*, *PID-*, and fractional order *PID* (*FPID*)-controller-based AIMUs. [rise time (RT), settling time (ST), and overshoot (OS)].

Controller	Δf_1			Δf_2			ΔP		
	RT (s)	ST (s)	OS	RT (s)	ST (s)	OS	RT (s)	ST (s)	OS
PI	7.1	17.3	−0.0133	27.3	29.01	−0.0033	30.5	34.02	−0.037
PID	1.1	3.9	−0.0038	3.9	10.2	−0.0003	7.02	12.01	−0.0031
FPID	0.2	0.7	−0.0009	0.01	0.2	−0.00001	0.011	0.1	−0.0001

6.2. Performance Evaluation under Load Change

The load change connected to the INPS is also responsible for diverging the performance of the INPS. The effects due to the change of ΔP_{lc} in the power system need to be controlled efficiently to maintain favorable INPS behavior.

The performance of the INPS in the case of load change is presented in Figure 10, exhibiting a huge deviation of INPS performance due to the load variation. The *FPID* controller efficiently minimizes the frequency and power deviation compared to the *PI* and *PID* controllers. Table 6 ensures the better performance of the *FPID* controller with AIMU with respect to rise time (RT), settling time (ST), and overshoot (OS).

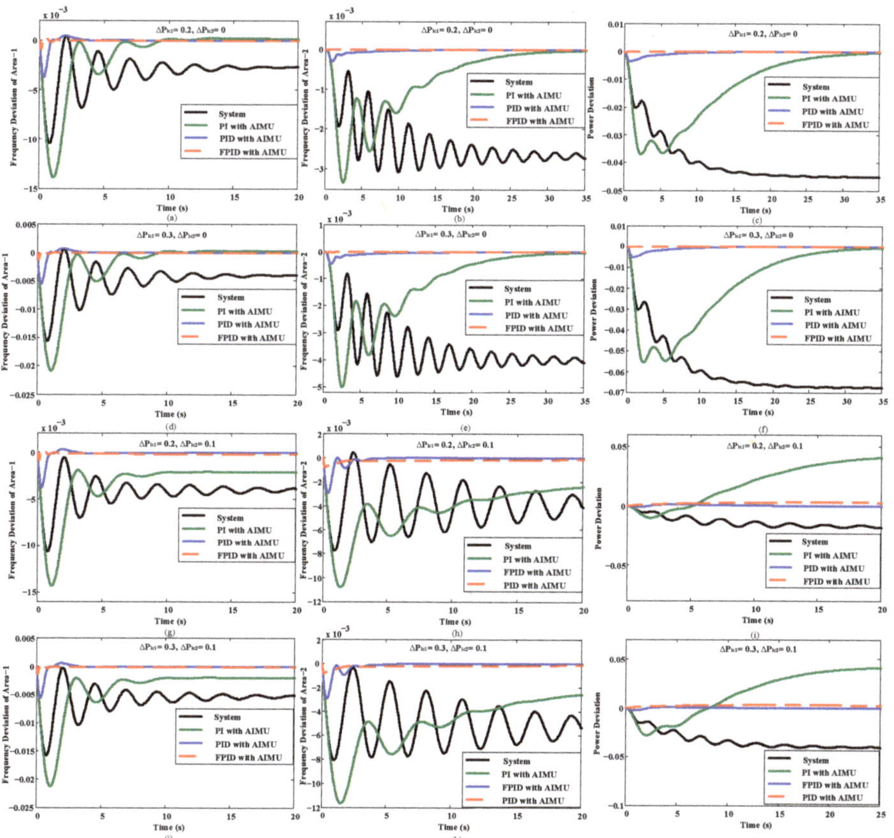

Figure 10. Performance of the interconnected power system with *PI*, *PID*, and *FPID* controllers against load change with the AIMU.

Table 6. Performance of the INPS under load change.

Load	Controller	Δf_1			Δf_2			ΔP		
		RT (s)	ST (s)	OS	RT (s)	ST (s)	OS	RT (s)	ST (s)	OS
$\Delta P_{lc1} = 0.2$, $\Delta P_{lc2} = 0$	*PI*	7.1	17.3	−0.0133	27.3	29.02	−0.0033	30.5	34.01	−0.037
	PID	1.1	3.9	−0.0038	3.9	10.2	−0.0003	7.01	12.02	−0.0031
	FPID	0.2	0.7	−0.0009	0.01	0.2	−0.0001	0.011	0.1	−0.0001
$\Delta P_{lc1} = 0.3$, $\Delta P_{lc2} = 0$	*PI*	6.5	12.7	−0.021	28.1	31.1	−0.005	28.5	31.9	−0.057
	PID	1.01	3.8	−0.0051	7.01	10.01	−0.0003	10.03	12.9	−0.005
	FPID	0.1	0.5	−0.001	0.01	0.011	−0.00002	0.01	0.11	−0.0001
$\Delta P_{lc1} = 0.2$, $\Delta P_{lc2} = 0.1$	*PI*	35.2	39.8	−0.0148	36.1	40.2	−0.015	5.2	45.7	−0.01
	PID	1.1	4.7	−0.0039	1.9	5.2	−0.0027	3.01	3.5	−0.0032
	FPID	0.2	0.9	−0.0008	1.7	2.3	−0.0007	2.7	3.1	−0.0001
$\Delta P_{lc1} = 0.3$, $\Delta P_{lc2} = 0.1$	*PI*	35.2	40.1	−0.022	38.3	45.1	−0.0119	9.02	36.9	−0.003
	PID	1.7	4.1	−0.0005	2.8	4.5	−0.0028	1.8	3.2	−0.0001
	FPID	0.23	0.95	−0.0001	1.8	3.5	−0.0009	1.2	3.01	−0.00001

6.3. Performance Evaluation under Wind Power Change

The performance of the INPS in this paper is also investigated under the presence of a renewable energy source. As a renewable energy source, wind energy is considered here. The variation of wind power, as presented in Figure 11, provides an inconsiderable impact on the frequency and power deviation and makes the system unstable [54]. Figure 12 presents the performance of the power system in the case of a wind power variation. The better performance of the *FPID* controller in the presence of the wind power variation makes it a high-performance control approach for INPS.

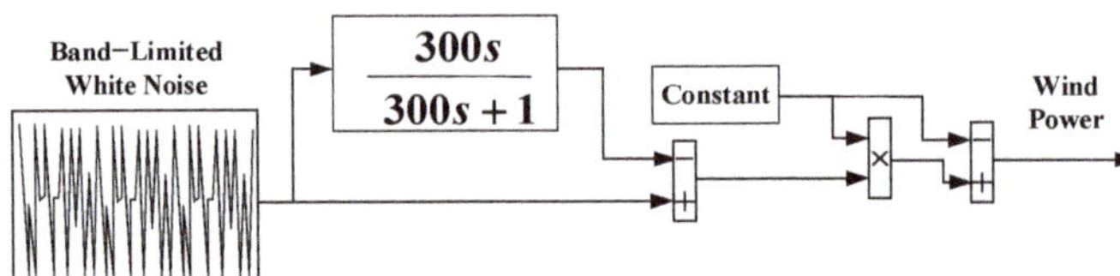

Figure 11. Wind speed model.

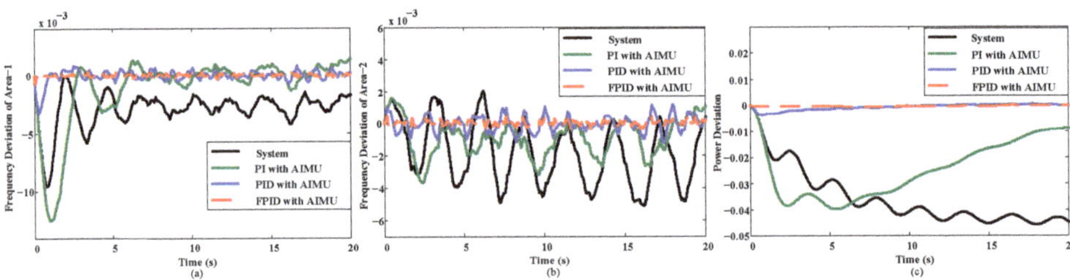

Figure 12. Performance of the AIMU-based interconnected power system with (**a**) *FPID*, (**b**) *PID*, and (**c**) *PI* controllers against a wind power variation.

7. Conclusions

The cybersecurity of modern power systems is crucial to empower their reliability, stability, and resilience. The dissimilarity of any element out of the reference increases the possibility of partially or completely damaging a power grid. In this paper, we proposed a novel intrusion mitigation unit to diminish the modification of all critical elements during the cyberattack of an interconnected power system. A two-area interconnected power system is modeled and stabilized by designing a *FPID* controller to compensate for frequency deviation and tie-line power mismatch. The closed-loop performance of the designed *FPID* controller over the conventional integer-order *PI* and *PID* controller remains superior in managing frequency deviation and tie-line power against no cyberattack. On the other hand, the *FPID* controller has no control over a cyberattack and executes undesirable performance. To limit and compensate for the effect of a cyberattack, the proposed security unit was designed using the status of the *FPID* controller and others. The effectiveness of the proposed AIMU is tested and compared with that of the *PI* and *PID* controller against a cyberattack, load variations, and the incorporation of noises. The proposed AIMU diminishes the diverse effects of a CA and efficiently minimizes the effect of load change and noise, the importance of which are listed in Table 7. The design of the proposed AIMU by AI techniques will be investigated further in the future.

Table 7. Comparison of power systems based on an AIMU.

Case	Advantages	Limitation
Power system without an AIMU	(i) Easy to install and operate, (ii) Easy to maintain	(i) Easily affected by a CA; (ii) Unauthorized access of the cyber attackers; (iii) Collect sensitive data; (iv) Injection of false data, (v) Change the sensitive parameters of the system and controller; (vi) Decreases the resilience, reliability, and stability of the system, (vii) Partially or completely damages the load and grid; (viii) Increases the utility costs; (ix) Largely deviates the frequency and power of the power system.
Power system with an AIMU	(i) Mitigates the diverse effects of unauthorized access to the power system; (ii) Creates a security wall to mitigate the collection and changing of sensitive data; (iii) Maintains a fixed level of system frequency and power; (iv) Increases system reliability, resilience, stability, and security; (v) Prevent sudden load shedding, power shortages, and a system breakdown.	(i) Requires advanced knowledge and experience to operate and control the system.

Author Contributions: Conceptualization, F.R.B. and S.K.S.; methodology, F.R.B., S.K.S. and S.R.F.; formal analysis, F.R.B., Z.N. and D.D.; investigation, F.R.B., S.K.S., S.K.D. and S.M.M.; writing—original draft preparation, F.R.B. and S.K.D.; writing—review and editing, F.R.B., S.K.S., S.K.D., S.M.M. and M.R.I.S.; supervision, S.K.D., S.M.M. and M.R.I.S. All authors have read and agreed to the published version of the manuscript.

Funding: This research received no external funding.

Institutional Review Board Statement: Not applicable.

Informed Consent Statement: Not applicable.

Data Availability Statement: Not applicable.

Conflicts of Interest: The authors declare no conflict of interest.

References

1. Gurrala, G.; Sen, I. Power System Stabilizers Design for Interconnected Power Systems. *IEEE Trans. Power Syst.* **2010**, *25*, 1042–1051. [CrossRef]
2. Sarkar, S.K.; Badal, F.R.; Das, S.K. A comparative study of high performance robust PID controller for grid voltage control of islanded microgrid. *Int. J. Dyn. Control.* **2017**, *6*, 1207–1217. [CrossRef]
3. Guha, D.; Roy, P.K.; Banerjee, S. Load frequency control of interconnected power system using grey wolf optimization. *Swarm Evol. Comput.* **2016**, *27*, 97–115. [CrossRef]
4. Sarker, S.K.; Badal, F.R.; Das, P.; Das, S.K. Multivariable integral linear quadratic Gaussian robust control of islanded microgrid to mitigate voltage oscillation for improving transient response. *Asian J. Control.* **2019**, *21*, 2114–2125. [CrossRef]
5. Badal, F.R.; Das, P.; Sarker, S.K.; Das, S.K. A survey on control issues in renewable energy integration and microgrid. *Prot. Control. Mod. Power Syst.* **2019**, *4*, 8. [CrossRef]
6. Armin, M.; Rahman, M.; Rahman, M.; Sarker, S.K.; Das, S.K.; Islam, R.; Kouzani, A.Z.; Mahmud, M.A.P. Robust Extended H∞ Control Strategy Using Linear Matrix Inequality Approach for Islanded Microgrid. *IEEE Access* **2020**, *8*, 135883–135896. [CrossRef]
7. Sarkar, S.K.; Badal, F.R.; Das, S.K.; Miao, Y. Discrete time model predictive controller design for voltage control of an islanded microgrid. In Proceedings of the 2017 3rd International Conference on Electrical Information and Communication Technology (EICT), Piscataway, NJ, USA, 7–9 December 2017; pp. 1–6.
8. Mohanty, P.K.; Sahu, B.K.; Pati, T.K.; Panda, S.; Kar, S.K. Design and analysis of fuzzy PID controller with derivative filter for AGC in multi-area interconnected power system. *IET Gener. Transm. Distrib.* **2016**, *10*, 3764–3776. [CrossRef]
9. Kiran, D.; Amani, G. Load Frequency Control of a Two-Area Power System Using FOPID with Harmony Search Algorithm. *Int. J. Adv. Eng. Res. Sci.* **2017**, *6495*, 12–17.
10. Badal, F.R.; Subrata, K.S.; Sajal, K.D. Transient Stabilization Improvement of Induction Generator Based Power System using Robust Integral Linear Quadratic Gaussian Approach. *Int. J. Smart Grid-ijSmartGrid* **2019**, *3*, 73–83.
11. Tasneem, Z.; Al Noman, A.; Das, S.K.; Saha, D.K.; Islam, R.; Ali, F.; Badal, F.R.; Ahamed, H.; Moyeen, S.I.; Alam, F. An analytical review on the evaluation of wind resource and wind turbine for urban application: Prospect and challenges. *Dev. Built Environ.* **2020**, *4*, 100033. [CrossRef]
12. Ahmed, S.H.A. Load Frequency Control for a Two Area Power System. Ph.D. Thesis, University of Khartoum, Khartoum, Sudan, 2017.
13. Pan, I.; Das, S. Fractional-order load-frequency control of interconnected power systems using chaotic multi-objective optimization. *Appl. Soft Comput.* **2015**, *29*, 328–344. [CrossRef]
14. Chuang, N. Robust H-infinite load-frequency control in interconnected power systems. *IET Control Theory Appl.* **2016**, *10*, 67–75. [CrossRef]
15. Thirukkovulur, A.K.; Nandagopal, H.; Parivallal, V. Decentralized control of multi-area power system restructuring for LFC optimization. In Proceedings of the 2012 IEEE International Conference on Computational Intelligence and Computing Research, Bengaluru, India, 16–19 December 2012; pp. 1–6.
16. Shah, N.N.; Kotwal, C.D. The state space modeling of single two and three ALFC of power system using integral control and optimal LQR control method. *IOSR J. Eng.* **2012**, *2*, 501–510. [CrossRef]
17. Farhangi, R.; Boroushaki, M.; Hosseini, S.H. Load–frequency control of interconnected power system using emotional learn-ing-based intelligent controller. *Int. J. Electr. Power Energy Syst.* **2012**, *36*, 76–83. [CrossRef]
18. Shabani, H.; Vahidi, B.; Ebrahimpour, M. A robust PID controller based on imperialist competitive algorithm for load-frequency control of power systems. *ISA Trans.* **2013**, *52*, 88–95. [CrossRef] [PubMed]
19. Fahim, S.R.; Yeahia, S.; Subrata, K.S.; Rafiqul, I.S.; Sajal, K.D. Self attention convolutional neural network with time series imaging based feature extraction for transmission line fault detection and classification. *Electr. Power Syst. Res.* **2020**, *187*, 106437. [CrossRef]
20. Mu, C.; Tang, Y.; He, H. Improved Sliding Mode Design for Load Frequency Control of Power System Integrated an Adaptive Learning Strategy. *IEEE Trans. Ind. Electron.* **2017**, *64*, 6742–6751. [CrossRef]
21. Saxena, S.; Hote, Y.V. Load Frequency Control in Power Systems via Internal Model Control Scheme and Model-Order Reduction. *IEEE Trans. Power Syst.* **2013**, *28*, 2749–2757. [CrossRef]
22. Shayeghi, H.; Jalili, A.; Shayanfar, H. Multi-stage fuzzy load frequency control using PSO. *Energy Convers. Manag.* **2008**, *49*, 2570–2580. [CrossRef]
23. Cam, E.; Gorel, G.; Mamur, H. Use of the Genetic Algorithm-Based Fuzzy Logic Controller for Load-Frequency Control in a Two Area Interconnected Power System. *Appl. Sci.* **2017**, *7*, 308. [CrossRef]
24. Prakash, S.; Sinha, S. Simulation based neuro-fuzzy hybrid intelligent PI control approach in four-area load frequency control of interconnected power system. *Appl. Soft Comput.* **2014**, *23*, 152–164. [CrossRef]
25. Giri, S.P.; Sinha, S.K. Four-Area Load Frequency Control of an Interconnected Power System Using Neuro-Fuzzy Hybrid Intelligent Proportional and Integral Control Approach. *J. Intell. Syst.* **2013**, *22*, 131–153. [CrossRef]

26. Prakash, S.; Sinha, S. Application of artificial intelligence in load frequency control of interconnected power system. *Int. J. Eng. Sci. Technol.* **2011**, *3*, 3. [CrossRef]
27. Teixeira, A.; Dan, G.; Sandberg, H.; Berthier, R.; Bobba, R.B.; Valdes, A. Security of smart distribution grids: Data integrity attacks on integrated volt/VAR control and countermeasures. In Proceedings of the 2014 American Control Conference, Portland, OR, USA, 4–6 June 2014; pp. 4372–4378.
28. Tan, R.; Nguyen, H.H.; Foo, E.Y.S.; Yau, D.K.Y.; Kalbarczyk, Z.; Iyer, R.K.; Gooi, H.B. Modeling and Mitigating Impact of False Data Injection Attacks on Automatic Generation Control. *IEEE Trans. Inf. Forensics Secur.* **2017**, *12*, 1609–1624. [CrossRef]
29. Esfahani, P.M.; Vrakopoulou, M.; Margellos, K.; Lygeros, J.; Andersson, G. Cyber-attack in a two-area power system: Impact identification using reachability. In Proceedings of the 2010 American Control Conference, Baltimore, MD, USA, 30 June–2 July 2010; pp. 962–967.
30. Sridhar, S.; Manimaran, G. Data integrity attack and its impacts on voltage control loop in power grid. In Proceedings of the IEEE PES General Meeting, Detroit, MI, USA, 24–28 July 2011; pp. 1–6.
31. Stuxnet Style Attack on US Smart Grid. 2018. Available online: https://www.scmagazineuk.com/stuxnet-style-attack-on-us-smart-gridcould-cost-government-1-trillion/article/535452/ (accessed on 12 January 2021).
32. *Ukraine's Power Outage Was a Cyber Attack*; Ukrenergo: Kyiv, Ucraina, 2017. Available online: http://www.reuters.com/article/.us-ukraine-cyber-attack-energyidUSKBN1521BA (accessed on 12 January 2021).
33. US Gas Pipeline Hit by Cyber Attack. 2018. Available online: https://www.infosecurity-magazine.com/news/us-gas-pipelines-hit-by-cyberattack/ (accessed on 12 January 2021).
34. Wadhawan, Y.; Almajali, A.; Neuman, C. A Comprehensive Analysis of Smart Grid Systems against Cyber-Physical Attacks. *Electronics* **2018**, *7*, 249. [CrossRef]
35. Wood, P.; Bagchi, S.; Hussain, A. Defending against strategic adversaries in dynamic pricing markets for smart grids. In Proceedings of the 2016 8th International Conference on Communication Systems and Networks (COMSNETS), Bangalore, India, 5–10 January 2016; pp. 1–8.
36. Wadhawan, Y.; Neuman, C.; Almajali, A. Analyzing cyber-physical attacks on smart grid systems. In Proceedings of the 2017 Workshop on Modeling and Simulation of Cyber-Physical Energy Systems (MSCPES), Pittsburgh, PA, USA, 21 April 2017; pp. 1–6. [CrossRef]
37. Ten, C.W.; Manimaran, G.; Liu, C.C. Cybersecurity for critical infrastructures: Attack and defense modeling. In *IEEE Transactions on Systems, Man, and Cybernetics-Part A: Systems and Humans*; IEEE: Piscataway, NJ, USA, 2010; Volume 40, pp. 853–865.
38. Deng, R.; Xiao, G.; Lu, R.; Liang, H.; Vasilakos, A.V. False data injection on state estimation in power systems—Attacks, im-pacts, and defense: A survey. *IEEE Trans. Ind. Inf.* **2016**, *13*, 411–423. [CrossRef]
39. Ten, C.-W.; Liu, C.-C.; Manimaran, G. Vulnerability Assessment of Cybersecurity for SCADA Systems. *IEEE Trans. Power Syst.* **2008**, *23*, 1836–1846. [CrossRef]
40. Hassan, M.; Roy, N.K.; Sahabuddin. Mitigation of frequency disturbance in power systems during cyber-attack. In Proceedings of the 2016 2nd International Conference on Electrical, Computer & Telecommunication Engineering (ICECTE), Rajshahi, Bangladesh, 8–10 December 2017; pp. 1–4.
41. Liu, X.; Shahidehpour, M.; Li, Z.; Liu, X.; Cao, Y.; Li, Z. Power System Risk Assessment in Cyber Attacks Considering the Role of Protection Systems. *IEEE Trans. Smart Grid.* **2017**, *8*, 572–580. [CrossRef]
42. Chaojun, G.; Jirutitijaroen, P.; Motani, M. Detecting False Data Injection Attacks in AC State Estimation. *IEEE Trans. Smart Grid* **2015**, *6*, 2476–2483. [CrossRef]
43. Singh, S.K.; Khanna, K.; Bose, R.; Panigrahi, B.K.; Joshi, A. Joint-Transformation-Based Detection of False Data Injection Attacks in Smart Grid. *IEEE Trans. Ind. Inform.* **2017**, *14*, 89–97. [CrossRef]
44. Yan, Y.; Qian, Y.; Sharif, H.; Tipper, D. A Survey on Cyber Security for Smart Grid Communications. *IEEE Commun. Surv. Tutorials* **2012**, *14*, 998–1010. [CrossRef]
45. Sahabuddin; Dutta, B.; Hassan, M. Impact of cyber-attack on isolated power system. In Proceedings of the 2016 3rd International Conference on Electrical Engineering and Information Communication Technology (ICEEICT), Dhaka, Bangladesh, 22–24 September 2016; pp. 1–4.
46. Mohamed, T.; Bevrani, H.; Hassan, A.; Hiyama, T. Decentralized model predictive based load frequency control in an inter-connected power system. *Energy Convers. Manag.* **2011**, *52*, 1208–1214. [CrossRef]
47. Alhelou, H.H.; Golshan, M.E.H.; Fini, M.H. Wind Driven Optimization Algorithm Application to Load Frequency Control in Interconnected Power Systems Considering GRC and GDB Nonlinearities. *Electr. Power Compon. Syst.* **2018**, *46*, 1223–1238. [CrossRef]
48. Shah, P.; Agashe, S. Review of fractional PID controller. *Mechatronics* **2016**, *38*, 29–41. [CrossRef]
49. Li, C.; Zhang, N.; Lai, X.; Zhou, J.; Xu, Y. Design of a fractional-order PID controller for a pumped storage unit using a grav-itational search algorithm based on the Cauchy and Gaussian mutation. *Inform. Sci.* **2017**, *396*, 162–181. [CrossRef]
50. Senberber, H.; Bagis, A. Fractional PID controller design for fractional order systems using ABC algorithm. In Proceedings of the 2017 Electronics, Palanga, Lithuania, 19–21 June 2017; pp. 1–7.
51. Khezri, R.; Oshnoei, A.; Hagh, M.T.; Muyeen, S. Coordination of Heat Pumps, Electric Vehicles and AGC for Efficient LFC in a Smart Hybrid Power System via SCA-Based Optimized FOPID Controllers. *Energies* **2018**, *11*, 420. [CrossRef]

52. Babaei, M.; Abazari, A.; Muyeen, S.M. Coordination between Demand Response Programming and Learning-based FOPID Controller for Alleviation of Frequency Excursion of Hybrid Microgrid. *Energies* **2020**, *13*, 442. [CrossRef]
53. Ahmed, S.M.; Khalid, H.M.; Muyeen, S.M.; Al-Durra, A. A Prediction Algorithm to Enhance Grid Resilience Towards Cyber Attacks in WACS Applications. *IEEE Syst. J.* **2019**, *13*, 710–719.
54. Li, X.; Hui, D.; Lai, X.; Yan, T. Power Quality Control in Wind/Fuel Cell/Battery/Hydrogen Electrolyzer Hybrid Micro-grid Power System. In *Applications and Experiences of Quality Control*; IntechOpen: London, UK, 2011. [CrossRef]

Article

Optimal Dispatch of Aggregated HVAC Units for Demand Response: An Industry 4.0 Approach

Michael Short [1,*], Sergio Rodriguez [1], Richard Charlesworth [2], Tracey Crosbie [1] and Nashwan Dawood [1]

[1] School of Computing, Engineering and Digital Technologies, Teesside University, Middlesbrough, Cleveland TS1 3BA, UK; S.Rodriguez@tees.ac.uk (S.R.); T.Crosbie@tees.ac.uk (T.C.); n.n.dawood@tees.ac.uk (N.D.)

[2] Energy Management Division, Siemens plc, Princess Road, Manchester M20 2UR, UK; Richard.Charlesworth@siemens.com

* Correspondence: m.short@tees.ac.uk

Received: 24 October 2019; Accepted: 11 November 2019; Published: 13 November 2019

Abstract: Demand response (DR) involves economic incentives aimed at balancing energy demand during critical demand periods. In doing so DR offers the potential to assist with grid balancing, integrate renewable energy generation and improve energy network security. Buildings account for roughly 40% of global energy consumption. Therefore, the potential for DR using building stock offers a largely untapped resource. Heating, ventilation and air conditioning (HVAC) systems provide one of the largest possible sources for DR in buildings. However, coordinating the real-time aggregated response of multiple HVAC units across large numbers of buildings and stakeholders poses a challenging problem. Leveraging upon the concepts of Industry 4.0, this paper presents a large-scale decentralized discrete optimization framework to address this problem. Specifically, the paper first focuses upon the real-time dispatch problem for individual HVAC units in the presence of a tertiary DR program. The dispatch problem is formulated as a non-linear constrained predictive control problem, and an efficient dynamic programming (DP) algorithm with fixed memory and computation time overheads is developed for its efficient solution in real-time on individual HVAC units. Subsequently, in order to coordinate dispatch among multiple HVAC units in parallel by a DR aggregator, a flexible and efficient allocation/reallocation DP algorithm is developed to extract the cost-optimal solution and generate dispatch instructions for individual units. Accurate baselining at individual unit and aggregated levels for post-settlement is considered as an integrated component of the presented algorithms. A number of calibrated simulation studies and practical experimental tests are described to verify and illustrate the performance of the proposed schemes. The results illustrate that the distributed optimization algorithm enables a scalable, flexible solution helping to deliver the provision of aggregated tertiary DR for HVAC systems for both aggregators and individual customers. The paper concludes with a discussion of future work.

Keywords: industry 4.0; digitalization; demand response; HVAC control; dynamic programming; nonlinear optimization

---

1. Introduction

1.1. Context and Motivation

With improved connectivity, dramatically increased access to low-cost computational power and recent advances in data science, many industries are currently on the verge of a second digital revolution known as 'Industry 4.0' [1]. Of the many challenges which can be addressed by Industry

4.0, concepts of sustainable, highly available plants and fully integrated renewable energy are prominent [1,2]. This involves aspects of industrial digitalization related to real-time, distributed and robust system-wide optimization methods coupled with the tighter integration of control, scheduling, planning and demand-side management for industrial production systems and energy networks [1,2]. The focus of this paper lies in the use of HVAC and/or refrigeration systems within buildings to provide aggregated energy-shifting services for demand response (DR) purposes within an Industry 4.0 framework. Building stock is estimated to account for around 40% of global energy consumption, exceeding that of other major sectors like industry and transportation [3,4]. As such, buildings represent the sector with both the largest energy saving and energy shifting potential for DR applications [3,4].

DR is the generic title given to a range of options customers have to reduce costs and/or generate profits by changing their pattern of consumption; economic benefits can be leveraged by deliberately moving consumption away from peak/critical times, or deliberately moving consumption onto off-peak/critical times [4–6]. Figure 1 illustrates the concept of DR. Heating, ventilation and air conditioning (HVAC) of buildings provides the largest possible source for this energy saving and energy shifting potential [4,5]. The opportunities for realizing demand response vary across Europe, as they are dependent on the particular regulatory, market and technical contexts in different European countries [4]. Nevertheless, successful DR programs are becoming increasingly common for large industrial customers. However, DR programs aimed at small and medium scale customers have mostly failed to meet their expected potential. Previous research has identified that blocks (or groups) of buildings offer more flexibility for the provision of DR than single buildings, the latter of which are generally too limited in capacity to operate individually in power markets [4–9]. As such, researchers and the energy industry are beginning to consider how blocks of buildings can operate collectively within energy networks to enhance the effectiveness of DR programs. The potential value of DR in blocks of buildings depends on the telemetry and control technologies embedded in the building management systems currently deployed at any given site and the potential revenue sources: both of which can vary significantly according to specific local and national conditions [4]. In this context, to encourage the growth of DR services and reap the potential benefits of DR, it is necessary for current research to demonstrate the economic and environmental benefits of DR for the different key actors required to bring DR services in blocks of buildings to market.

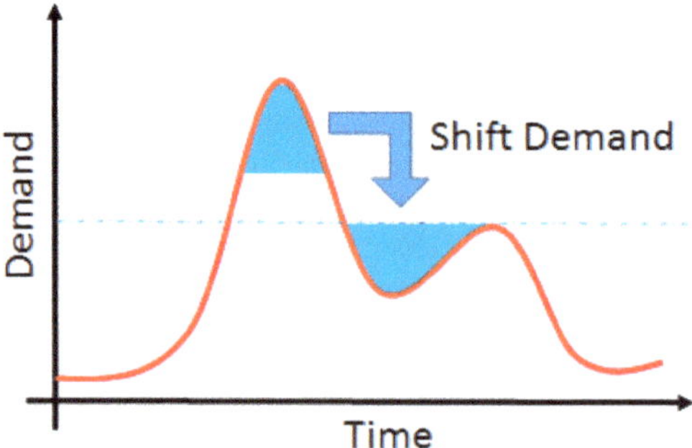

Figure 1. Demand response concept.

There are several established and emerging DR archetypes suitable for use with individual buildings and blocks or groups of buildings, e.g., see [4,6,7]. This paper is principally concerned with explicit (surgical) DR, in which prior notification is given to the DR participant of a pending event and its requirements for down-regulation or up-regulation of demand during a specified future time window. In the UK market, for example, the DR archetypes of short-term operating reserve (STOR) and demand turn up (DTU) are employed [6]. For both STOR and DTU events, enrolled participants are notified of an upcoming event and its requirements typically between 30–240 min before the required time of delivery (although day-ahead notice is sometimes given), and event durations are of at least 120 min [6]. STOR and DTU event windows occur at differing periods during the day (peak and off-peak, respectively) and require an aggregated minimum of three MW to be available. This latter restriction is also a motivating factor for considering blocks of buildings. In general, STOR and DTU can be considered as tertiary forms of reserve DR (ancillary services), which can be integrated alongside emerging fast-acting and very fast-acting decentralized approaches for providing secondary and primary frequency regulation services using HVAC systems, e.g., see [7,9].

1.2. Related Work

There is a growing body of work studying direct load control (DLC) techniques for HVAC units and other thermostatically controlled loads (TCLs) such as refrigerators. The fundamental requirements of DLC are outlined in an earlier work, which presents a general optimization framework for feeder-scale load reduction [10]. A priority-based control scheme for TCLs to participate in grid frequency regulation has also been developed [11]. A two-stage dispatch method has been demonstrated for TCLs in which a day-ahead scheduling model is solved to determine the optimal TCL dispatch, and then a real time control model allocates the desired setpoints to individual TCLs [12]. The use of a Markov transition matrix to model the populated TCLs to carry out non-disruptive load reduction underpins other work [13]. Previous work also makes use of thermal inertia models to control TCL in smart homes [14], and research has been conducted to propose a model for coordinated dispatch of TCLs and thermal generation units [15].

Motivated by the observation that users are not immediately convinced that replacing existing thermostats by an intelligent controller is worth the investment, a 'reference governor' approach for optimizing performance and energy consumption of building thermostatically controlled loads was developed in an earlier work [16]. In this research, a model predictive control (MPC) supervisory unit adjusts the temperature setpoints for a relay controlling the on/off load. The MPC problem is converted into a mixed-integer linear programing (MILP) problem, which is dispatched in real-time. Although the method is shown to be effective, it is restricted to a single heating/cooling zone, and little consideration is given in the earlier research to implementation details such as communication or computation overheads.

In all the above research and technical developments, the on/off control actions of individual TCLs are driven by the thermostat settings and DR actions are principally achieved by varying setpoints within pre-set temperature ranges. However, other work has sought to aggregate TCLs in numerous buildings. For example, Luo et al. [8] presented a framework for aggregating TCLs in a rolling-horizon framework, aimed principally at the Nordic electricity market. Standard thermal comfort models described by International Organization for Standardization (ISO) standard ISO 7730 are adopted to estimate occupant thermal comfort in a representative model for TCL dispatch. To avoid the technical intractability of dispatching TCLs individually in a large-scale centralized model, similar TCLs are first grouped based on feature similarities using a clustering technique and an averaged TCL model for each group is then created. A randomized meta-heuristic technique named the natural aggregation algorithm (NAA) is then employed to dispatch the averaged TCLs in each combined group. Since both the number of groups and the number of NAA iterations can be pre-selected, the computation time can be tailored to suit available resources; however, this is at the expense of solution sub-optimality. Although the technique is shown to give good results on practical examples, the optimal solution cannot be guaranteed [8].

Callaway [17] also describes a statistical model for aggregating TCLs such that frequency regulation services can be provided in the presence of generation variability. This work presents a new DR method to provide load following services by controlling the aggregated TCLs under Fokker–Planck diffusion models [17]. A minimum-variance (MV) statistical control law is derived to follow a one-step-ahead load signal and is demonstrated to perform well under simulation with 10,000 TCLs. Zhou et al. [18] developed a change-time-priority-list method to control power output taking into account customers' satisfaction for the aggregation of multiple HVAC systems to provide demand-side frequency regulation services, to support intermittency of renewable generation, e.g., in the presence of fluctuating wind power. The aggregation unit employs simple approximations for on/off change times in the HVAC units to prioritize their dispatch in order to follow load control signals. Simulations verify the tracking ability of the system when connected to 6000 TCLs.

In the earlier works discussed above, the proposed methods are shown to perform favorably, but large numbers of aggregated units (>1000) are needed for the statistical approximations to be accurate; this limits their down-scalability and they cannot be applied to situation of, say, ten HVAC units. In addition, the lack of a rigorous baselining scheme for participant settlement—and potentially high communication load between an aggregator (providing the means by which individual HVAC unit owners can participate in DR) and individual units—are major drawbacks from a practical perspective.

1.3. Contributions

In this paper, an efficient and partially decentralized approach to aggregated tertiary DR schemes for individual or blocks of buildings is proposed. As with other Internet-of-Things and Industry 4.0 applications, large-scale DR implementation across multiple HVAC units across different geographic locations results in large volumes of high-velocity real-time data, and very large optimization and scheduling problems that must be solved centrally. It is beneficial to run local processing of these data at the edge of the network (on 'edge devices') to reduce the frequency and real-time requirements of communication to core devices in such situations [2]. In addition, in an Industry 4.0 context, it makes sense to distribute the optimization and resource scheduling problems between the network edges and core as much as possible. This is the approach taken in the research presented here.

Initially, the paper focuses upon the dispatch problem for individual TCLs in the presence of a tertiary DR program. The dispatch problem for an individual HVAC zone is formulated as a non-linear constrained predictive control problem, which manipulates on/off controls to optimize thermal comfort and electricity consumption across a future time horizon. A dynamic programming (DP) algorithm is developed to solve the dispatch problem with fixed memory and computation time overheads, rendering it suitable to be implemented as an intelligent thermostat unit (ITU), which can be implemented on an edge device with IP communications. A key factor of the implementation is the exposure of adjustable constraints, which can be used to enforce a minimum and/or maximum level of control activity during a specified window of time. This allows electricity consumption to be increased (or decreased) during a specified DR window. In the absence of DR signals, the DP solution defaults to a simple thermostat with adjustable deadzone; this allows the ITU to calculate an accurate baseline in the presence of active DR signals to ensure consumption is surgically reduced (increased) to a specified level during a DR event, and post-settlement can be effectively handled.

The paper also presents an efficient unit allocation/reallocation scheme (UAS) for aggregated explicit DR employing multiple ITUs. The UAS is employed by an aggregator to allocate individual DR targets to individual zones based upon advertised price/energy relationships. It can also be employed to reallocate targets should an individual unit drop out from, or drop in to, an upcoming DR event, or should environmental conditions indicate that a previously allocated target for a unit cannot be achieved. The UAS is formulated as a modified knapsack algorithm, which can be efficiently solved using DP techniques, and is designed to be cloud-based and hosted in the network core. As such, the proposed ITU/UAS overall exhibits low bandwidth requirements—as only elementary data need to be periodically exchanged between the aggregator and ITUs. Additionally, the computational load is distributed evenly across the core and network edges and optimization workloads are executed in parallel for each of the ITUs.

A number of computational simulations and a large-scale experimental test are then described to verify and illustrate the performance of the proposed schemes. It is argued that the ITU optimization algorithm for individual HVAC zones, combined with the centralized UAS procedure, enables a scalable, flexible solution helping to deliver the provision of aggregated tertiary DR for HVAC systems in distributed environments for both aggregators and individual customers.

1.4. Structure

The remainder of this article is structured as follows. In Sections 2 and 3, the main technical developments are described in detail, including the assumed HVAC thermal dynamic model and the ITU/UAS optimization schemes. In Section 4, calibrated simulation studies and experimental results from practical tests are presented to illustrate the performance of the proposed schemes. Section 5 provides conclusions and suggestions for future areas of work.

2. Models and Assumptions

In this paper, the focus is principally upon explicit (surgical) DR events involving HVAC units. For ease of exposition, for the remainder of this paper it is assumed that the DR events to be handled are characterized by STOR-type events, in which prior notification of an upcoming event is given and the goal during the event is to reduce demand by a pre-agreed or negotiated amount. Since STOR-type events and DTU-type events are mutually exclusive, it is a relatively straightforward matter to modify the basic approach to handle DTU-type DR events, in which prior notification is again given but the goal during the event is to increase demand by a pre-agreed or negotiated amount. Figure 2 shows the assumed temporal relations between the key events in a STOR DR situation.

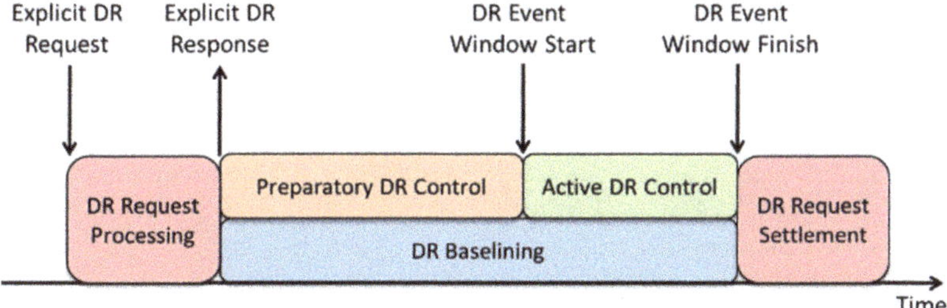

Figure 2. Timeline of events in a short term operating reserve (STOR) demand response (DR) event.

A variety of different modelling approaches (of varying complexity) for HVAC and building thermal dynamics can be found in the literature (see, e.g., [19]). Most models are based on physical principles related to mass, energy and momentum transfer and consist of complex partial differential equations that capture the building thermal and physical characteristics. Although the optimization to be developed in Section 3 allows for the use of complicated non-linear models, in practice simplified first-order models can typically perform just as well as more complicated models [17,20]. The descriptions and technical development henceforth consider the linear case for ease of exposition. In addition, for simplicity the assumption is also made that the unit in question delivers heated or cooled water to the building radiators, but may be applied to other configurations (e.g., hot air delivery) with an appropriate change of variables and/or model. Assuming continuous time indexed by the variable t, the linear dynamic relationship between the HVAC control signal $u(t) \in \{0, 1\}$, which switches electrical power delivery to the electro-thermal converter, the circulating water temperature delivered to the building $T(t)$ and the (ambient) inlet water temperature $T_A(t)$ is assumed to be well approximated by a first-order differential equation, as follows:

$$\dot{T}(t) = \frac{1}{\tau}(T(t) - T_A(t) + u(t-D) \cdot K_T), \quad (1)$$

where the dot above a variable represents its first derivative (d/dt), $D \geq 0$ represents any dead-time in the converter dynamics, τ is the thermal time constant of the converter and K_T is the equivalent thermal gain of the converter. A model such as this has previously been studied in the context of HVAC control for demand response applications, and time constants of 10–30 min and dead-times between 0–5 min are typical of most buildings [17–20]. In Equation (1), the sign of K_T distinguishes between a heating (positive) or cooling (negative) application. Neglecting the standby power consumption, the variable electrical power consumption of the HVAC unit $E(t)$ is determined by the state of the control signal and the electrical gain of the converter K_E, such that $E(t) = u(t)\ K_E$. Hence, as is well known, the HVAC unit may participate in STOR and DTU DR programs by modulating the electrical power consumption through deliberate under- or over-supply of thermal power, at the possible expense of some thermal comfort of the building occupants. Digitizing the Equation (1) for a time-step of T_s seconds, using k as the (integer) discrete-time index, yields:

$$T(k) = a \cdot T(k-1) + (1-a) \cdot [u(k-d) \cdot K_T - T_A(k-1)], \tag{2}$$

where the real parameter $a = exp(-T_s/\tau) > 0$ and the integer $d = 1 + (D/T_s)$, which if fractional is assumed rounded to the nearest whole integer. For the thermal model, Equation (1), it is suggested that the time step be chosen to be not more than 5% of the time constant τ. The reasoning behind this is that basic systems theory gives the open loop rise time of the model, Equation (1), as 2.2τ, and hence choosing T_s as suggested gives at least 44 time steps in the rise time, which should be adequate for good on/off control without leading to excessively short time steps (which could be problematic from an implementation viewpoint). For example, a 30-min time constant gives step times of not more than 90 s. The digitized model, Equation (2), provides a means in which the temperature evolution may easily be predicted over a future time horizon, enabling predictive optimization to be applied at each time step. The proposed optimization approaches are described in the following section.

3. Technical Development

The main technical development consists of three main elements: firstly, the ITU dispatch algorithm for an individual HVAC unit (Section 3.1); secondly, the approach to generate predictions of baseline and reduced consumption for an individual HVAC unit (Section 3.2); and thirdly, the UAS optimization approach for aggregated units (Section 3.3).

3.1. Individual Unit Dispatch

The legacy mode of operation of an HVAC unit for temperature control is typically through thermostat type controls, i.e., setting the control signal $u(k)$ according to the simple 'relay with hysteresis' control law [7,17,18]:

$$u(k) = \begin{cases} 1, & T(k) \leq T_R(k) - \Delta; \\ u(k-1), & T_R(k) - \Delta \leq \Delta T(k) \leq T_R(k) + \Delta; \\ 0, & T(k) \geq T_R(k) + \Delta; \end{cases}, \tag{3}$$

where $T_R(k)$ represents the temperature setpoint at time step k, and the deadzone of the relay (during which the output is held at its last value) is given by 2Δ. Such a mode of operation is simple to implement; in the following, it is wished to extend the basic relay-based approach to situations in which pre-heating (cooling) can be optimally managed for explicit DR events. For this, a rolling-horizon non-linear MPC employing a finite discrete input set is developed [20]. The optimization problem to be solved at each discrete-time step k is the minimization of the following multi-stage quadratic cost function:

Minimize:

$$J(k) = \sum_{i=0}^{M} (T(k+i+d) - T_R(k+i+d))^2 + \lambda \sum_{i=0}^{M-1} \Delta u(k+i)^2 \tag{4}$$

with respect to:

$$u(k+i), \quad 0 \le i \le M-1; \tag{5}$$

subject to:

$$T(k+i+d+1) = a \cdot T(k+i+d) + (1-a) \cdot (u(k+i) \cdot K_T + T_A(k+i+d)), \quad 0 \le i \le M-1; \tag{6}$$

$$u(k+i) \in \{0,1\}, \quad 0 \le i \le M-1; \tag{7}$$

$$U_L \le \sum_{i=0}^{M-1} (u(k+i) \cdot w(k+i)) \le U_U, \tag{8}$$

where $T(k)$ represents the zone temperature at time slot k, $T_R(k)$ represents the zone temperature reference (setpoint) at time slot k and $T_A(k)$ represents the ambient temperature at time slot k. The binary decision variable $u(k)$ represent the control input at time slot k and Δ is the difference operator such that $\Delta u(k) = u(k) - u(k-1)$. The scalar term λ is used to penalize changes in the control input and can be considered a 'move suppression' or regularization term. Integer $d \ge 1$ represents the time delay and $w(k) \in \{0, 1\}$ are indicator variables for defining a demand response window, such that DR is active during time slot k if $w(k)$ is '1' and not active otherwise. The integer M represents the length of the prediction horizon, and integers U_U and U_L represent upper and lower bounds on the allowed input activity during a defined DR window. Henceforth in this paper, only an upper bound on control activity (e.g., for handling STOR-style DR events) is considered, with the understanding that adding the lower bound (e.g., for DTU-style DR events) follows directly from the described methods. As with other predictive control problems, an appropriate length for the prediction horizon would be the number of samples required to capture the open-loop setting time of the process [21]. However, with reference to Figure 2, this should be extended somewhat to allow for preparatory control actions to also take place. For the thermal model, Equation (1), basic systems theory gives the settling time as 4.6τ, giving a suggested horizon length $M = \lceil (4.6\tau + W)/T_s \rceil$, where W is the length of time required for preparatory control. For example, a 30-min time constant and preparatory window W of one hour with time step of 90 s gives a horizon length M of 132 steps.

The first component of the objective function in Equation (4) is the sum of squared errors between the actual building temperature and its setpoint. A quadratic penalty term is appropriate as it approximately captures the relationship between temperature error and thermal comfort (see, e.g., [22]). During working hours, this will be typically 22.0 °C in winter and 24.5 °C in summer [22]. The second component is the sum of squared control moves weighted by the scalar term λ, which is used to suppress excessive switching of the controls. The indicator weights $w(k)$ and bound U_U are used in combination to penalize electricity consumption during a specific defined window of the prediction horizon (corresponding to DR events). DR events are enabled for a particular stage by setting the indicator variable $w(k)$ equal to '1'. As with all rolling horizon predictive control schemes, once the optimization has been solved for the current time step, only the first control (corresponding to $u(k)$) is applied. At time step $k + 1$, the process is repeated with repeated measurement information.

It is quite straightforward to see that for all but prediction horizons of one or two steps in length, and in the absence of any DR activity over the horizon (implying no constraint bounds in Equation (8)), then the control law resulting from the solution of Equations (4)–(7) is a switching surface dependent upon two factors. The first is the predicted difference between the reference temperature $T_R(k+d)$ and actual temperature $T(k+d)$ and the second is the previous state of the control input, $u(k-d-1)$. In this situation, the control defaults to a predictive relay-based controller with deadzone $\Delta \approx \sqrt{\lambda(k)}$. In the presence of upcoming DR events, however, the effective switching surface can change considerably to provide optimal pre-heating (or cooling). In addition, the solution of the optimization problem contains useful predicted quantities regarding upcoming DR events. In particular, since the future control signals can only be '1' or '0', the predicted electricity consumption of the HVAC unit during any upcoming DR event—denoted as *URC(k)*—is easily derived from either inspection of the solution vector, or from the left hand side (l.h.s.) of constraint (8), as follows:

$$URC(k) = K_E \cdot T_s \cdot \sum_{i=0}^{M-1} (u(k+i) \cdot w(k+i)). \tag{9}$$

For upcoming DR events, the integer U_U on the right hand side (r.h.s.) of constraint (8) can be used to limit *URC(k)* and specify a load curtailment/reduced consumption from the HVAC unit. These observations will be exploited in the sequel, during which the DR allocation mechanism will be presented. The optimization problem of Equations (4)–(8) is easily formulated as a mixed-integer quadratic program (MIQP), or alternatively as an approximate mixed-integer linear program (MILP) [16,21]. In order to improve the run-time efficiency of the approach (and enable bounded worst-case execution times), the DP method is chosen. DP (see Bellman, [23]) is a computational method for solving optimal control problems with separable additive performance indices. It is based on the recursive application of Bellman's 'Principle of Optimality' [23,24]:

> *'An optimal policy has the property that whatever the initial state and the initial decision are, the remaining decisions must form an optimal policy with regard to the state resulting from the first decision.'*

The mathematical form of this idea can be expressed as a backwards sequence featuring the solution of simpler optimization problems at each stage. Continuing backwards from the end stage N to the current stage k—and applying the principle of optimality at each stage—will result in the following recurrence relations for discrete DP [24]:

$$\begin{aligned} &J_N(x_N) = g_N(x_N); \\ &k = N-1, N-2, ..., 1, 0: \\ &J_k(x_k) = \min_{u_k \in U_k(x_k)} \{g_k(x_k, u_k) + J_{k+1}(f_k(x_k, u_k))\}; \end{aligned} \tag{10}$$

where $J_k(x_k)$ is the cost of entering stage k with state x_k, $g_k(x_k)$ is the cost for entering stage k with state x_k, $U_k(x_k)$ is the set of allowed controls for the input u_k when the state x_k is entered at stage k, and $f_k(x_k, u_k)$ is a function which maps the state x_k onto state x_{k+1} when control u_k is applied at step k. In discrete DP (DDP), the state vector is mapped onto a grid of size S and the controls onto a grid of size U. By iterating through the recursion and trying all admissible control values at each admissible set of state values, a vastly reduced search space is explored when compared to a pure brute-force search; at the end of the minimization, a solution grid is obtained and the optimal control is obtained from the position in the grid corresponding to the current state. In the current context, the state variable is the temperature $T(k)$ plus the previous applied control $u(k-1)$. The admissible controls are the current control $u(k)$, having two possibilities ($U = 2$) at each stage in the absence of DR, but some possibilities may not be admissible if they invalidate constraint (8) when DR is active. In the approach taken in this paper, constraint (8) is translated into a penalty function inserted into the objective; with an iterative tightening of an applied weight upon constraint violation (a maximum of 10 iterations is employed). The temperature is mapped into a discrete grid of over a suitable working range, e.g., 10–30 °C; the control is already discrete in nature. The transition function $f_k(x_k, u_k)$ is given by the linear Equation (6). In the case that the HVAC dynamics are actually non-linear, then this can easily be captured by an appropriate choice of transition function. During the recursion, the minimal cost function is stored for each admissible state along with the corresponding partial sum of the l.h.s of constraint (8). After the recursion, the value of *UDR(k)* is readily computable using Equation (9) and the final value of this partial sum. In many situations, temperature sensors either have 10-bit resolution—or can easily be cast or truncated into this range—giving a full state size $S = 2^{11}$. As the run-time complexity of the DDP algorithm is given by *O(M.U.S)*, the algorithm runs efficiently even for a relatively large prediction horizon M, which may be needed for best results. During the backwards recursion, only the current and next stage costs and the partial l.h.s. of constraint (8) are actually required, reducing the memory requirement to $O(U.S)$. As sampling time requirements of approximately one minute are sufficient in many instances (due to the comparatively slow thermal dynamics), then the algorithm is clearly suitable for deployment without undue problems in a small/low-cost embedded system.

3.2. Generation of Predicted Baseline and Reduced Consumption

As discussed earlier, the effective practical implementation of a DR scheme requires the generation of both an accurate and reliable predicted baseline and an accurate and reliable predicted reduced consumption [25]. As shown in Figure 3, the baseline represents a counterfactual prediction (either forecast or backcast, generated on-line or off-line) of electricity consumption for targeted assets during the time-period corresponding to a DR event, under the assumption that no corrective DR action will be/had been taken. As also shown in Figure 3, the reduced consumption represents a prediction (forecast, generated on-line) of electricity consumption for targeted assets during the time-period corresponding to a DR event, under the assumption that corrective DR action will be taken. Note that although the observed reduced consumption in Figure 3 is always lower than predicted, this is not necessarily the case generally, and the accuracy of prediction depends heavily upon the models and methods employed.

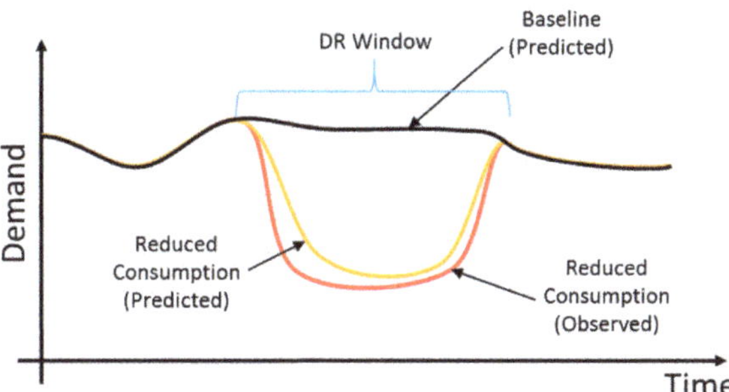

Figure 3. Concept of predicted baseline, predicted reduced consumption and reduced consumption.

Both the predicted baseline and predicted reduced consumption are also dependent upon many factors including occupancy, ambient temperature and weather conditions, and it has proved a challenging and ongoing task to select suitable methods for their implementation [25]. The baseline consumption is needed to gauge performance of controlled assets during a DR event and for post-event financial settlement, where it is compared to the actual measured consumption. Therefore, the baseline should be accurate as possible. In the context of the research presented here, it is required to generate a reliable baseline across the rolling horizon for an HVAC unit in an on-line fashion. The methodology proposed is illustrated in Figure 4.

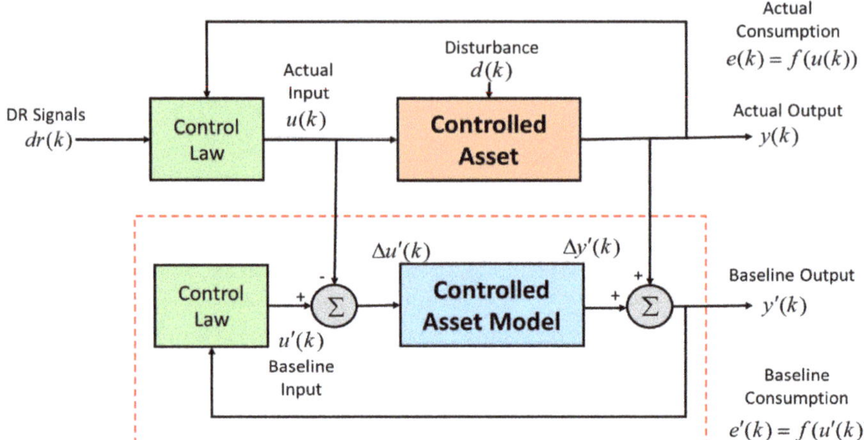

Figure 4. Methodology for online baseline generation.

As can be seen in Figure 4, the controlled asset is driven by input $u(k)$ and produces output $y(k)$, and consumption $E(k)$ is derived as a function of the input signal $f(u(k))$. The controlled asset is also driven by a disturbance sequence $d(k)$ which impacts the output $y(k)$; this sequence may consist of partially known (measured) values along with unmeasured or stochastic values. The asset input signal $u(k)$ is generated by closed-loop feedback controls, which are also driven by external DR signals represented by $r(k)$. The baseline is generated on-line (in real-time) by the components contained within the dashed red box; the asset model represents a discrete-time dynamic model of the controlled asset, while the asset model input signal $u'(k)$ is given by the same closed-loop feedback algorithm which drives the actual controlled asset. However, the DR signals $r(k)$ are suppressed from the control law in the baseline case to ensure the counterfactual sequence of inputs is generated. The baseline consumption sequence $E'(k)$ is derived as a function of the baseline input signal $f(u'(k))$. Since the disturbances $d(k)$ and plant-model mismatch can both impact upon the baseline accuracy, the counter-factual baseline output is generated on-line as follows. The actual nominal control input $u(k)$ is first subtracted from the baseline input $u'(k)$ to generate an input deviation signal $\Delta u'(k)$, which acts as input to the asset model to produce an output deviation signal $\Delta y'(k)$. The baseline output $y'(k)$ is then produced by adding the deviation signal $\Delta y'(k)$ to the measured output $y(k)$. This ensures that the effects of disturbances and plant-model mismatch on the actual output $y(k)$ are contained within the baseline output $y'(k)$, which also captures the impact of the baseline control sequence $u'(k)$ to ensure accuracy of the baseline. In the context of the current work, the controlled asset is the HVAC system, the controlled asset model is represented by Equation (2) with the measured part of the disturbance sequence being the ambient temperature, and the control law is the rolling-horizon non-linear MPC procedure given by Equations (4)–(8).

For the effective application of aggregated DR, it is required to generate both a predicted baseline and a predicted reduced consumption across the rolling horizon for each individual HVAC unit in an on-line fashion. At each timestep k, let the baseline consumption of the HVAC unit during an upcoming DR window be denoted as $UBL(k)$ and the reduced consumption of the HVAC unit during an upcoming DR window be given as $URC(k)$. Computation of this latter quantity as a by-product of the application of the DP algorithm was detailed in the previous Section. In order to compute the former quantity $UBL(k)$, the DP algorithm can be applied to solve the optimization problem using baseline input $u'(k)$ and output $y'(k)$ as input data and setting constraint (8) r.h.s. $U_U = \infty$. As discussed in the previous section, the control law resulting from the solution of Equations (4)–(8) in the absence of any DR signals is a predictive relay-based controller. Therefore to simplify the process of generating the baseline controls $u(k+i|k)$ and $y(k+i|k)$ (i.e., i-step predictions of the asset input and output, incorporating measured disturbances such as weather forecast), Equation (3) may be employed at each stage with deadzone $\Delta = \lambda^2/2$. Thus, both quantities can be computed in a straightforward procedure integrating both unit DR controls and unit baseline/reduced consumption predictions in an ITU-based edge device.

3.3. Coordinated Dispatch Scheme

Consider the arrangement of HVAC units (with their edge-based local dispatch optimizers) in the presence of a (cloud-based) aggregator at the network core, in the presence of a wider DR marketplace as shown in Figure 5. Assume that a flexible IP-based communications infrastructure is employed to interconnect the core and network edges (e.g., using Web Services, OpenADR protocol, etc.), and to provide loose synchronization of their clocks (e.g., using Network Time Protocol (NTP)).

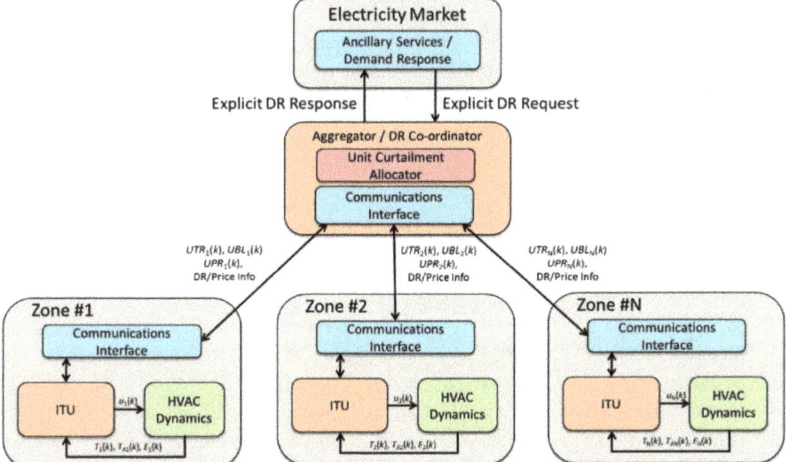

Figure 5. Arrangement of actors and signal exchanges in the coordinated dispatch scheme.

For the aggregated DR coordination scheme, the following approach is employed. In a multiple-zone scenario, let the predicted electricity consumption at time step k for an upcoming DR event involving HVAC unit j be given as $URC_j(k)$. Similarly, let the predicted baseline electricity consumption at time step k for an upcoming DR event involving HVAC unit j be given as $UBL_j(k)$. This quantity is computable as detailed in the last section. Since for each HVAC unit j and for every time step k it must hold that $UBL_j(k) \geq URC_j(k)$ (i.e., the predicted unit baseline consumption is not less than the unit DR consumption during a DR event), at step k the explicit unit predicted reduction in load—denoted as $UPR_j(k)$—for an upcoming DR event is given by: $UPR_j(k) = UBL_j(k) - URC_j(k)$. This quantity is easily computed in each ITU as detailed in the previous section. Then the predicted aggregated explicit reduction in load for an upcoming DR event at time step k—denoted $APR(k)$—for N participating HVAC units can be computed by the aggregator as:

$$APR(k) = \sum_{j=1}^{N} UPR_j(k). \tag{11}$$

Let the target aggregated electricity reduction consumption for an upcoming DR event be given as $ATR(k) \geq 0$, and the target reduction in load for individual HVAC unit j be given as $UTR_j(k)$. Since it has been assumed that the HVAC controls are binary in nature, in a given DR window the available load curtailment is limited between zero and the maximum available from the predicted baseline in discrete steps. As such, assume that each HVAC unit $j \in N$ offers an agreed price-schedule for load curtailment X_j, such that a specific load $L_{j,l}$ may be reduced for a price $p_{j,l}$ by selecting one of $l \in X_j$ different discrete price/curtailment options. Only one (or none) of the price/curtailment options can be selected for a given HVAC unit, for any given DR event. The objective for the DR coordinator is then to (i) at the start of a preparatory event, allocate individual HVAC unit load curtailments to meet the aggregate DR requirements; and (ii) should an HVAC unit opt-out or become unavailable—or environmental/pricing conditions otherwise change leading to an invalid initial allocation—reallocate curtailments to best suit the new conditions. The allocation/reallocation problem for N available units at time-step k can be written as a variant of a standard integer knapsack problem, as follows:

Minimize:

$$\sum_{j=1}^{N} \sum_{l \in X_j} p_{j,l}\, x_{j,l} \tag{12}$$

subject to:

$$\sum_{j=1}^{N} UTR_j(k) \geq ATR(k) \tag{13}$$

$$UTR_j(k) = \sum_{l \in X_j} L_{j,l}\, x_{j,l},\ 1 \leq j \leq N; \tag{14}$$

$$UTR_j(k) \leq UBL_j(k),\ 1 \leq j \leq N; \tag{15}$$

$$\sum_{l \in X_j} x_{j,l} \leq 1,\ 1 \leq j \leq N; \tag{16}$$

$$x_{j,l} \in \{0,1\}, \quad 1 \leq j \leq N; l \in X_j. \tag{17}$$

where $x_{j,l} \in \{0, 1\}$ are binary variables that indicate whether load reduction level $L_{j,l}$ is active for HVAC unit j. Equation (12) defined the main objective, to minimize DR costs while meeting the target reduction in demand (constraint (13)). Constraint (14) allocates load reduction level to individual unit targets, while constraint (15) ensures that individual unit load reductions are not greater than the individual unit baselines for the upcoming event. Constraint (16) enforces mutual exclusion in the choice of load reductions to individual units based upon the available price/curtailment options for that unit. Note that the presence of this constraint allows that any costs for activating a particular HVAC unit for DR purposes can be added into the corresponding costs for discrete load curtailment. The presence of the constraints also suggests that the problem can be efficiently solved using standard DP techniques, using a slight variant of the standard knapsack DP approach [26]. The time and space complexity of this approach is $O(N.D.X)$, where N is the number of participating units, D is the level of load curtailment requested and X is the largest cardinality of curtailment choices among the participating units [23,24]. Alternatively, efficient Branch-And-Bound techniques are known and can be applied [26]. Should the problem defined by Equations (12)–(17) prove infeasible, this will be due a violation of constraint (13), indicating that there is not enough capacity to achieve the aggregated DR value. In this case, either further units should be brought into the DR scenario, or the aggregator should report that it might not be able to deliver the required target to the market.

4. Evaluation

In this section, the design and results obtained from both a computational study and experimental test are reported in order to illustrate the proposed approach and to provide validation of its effectiveness in providing explicit DR in an aggregated environment.

4.1. Computational Study

For the computational study, five HVAC units are considered. The thermal dynamic model chosen for each unit represented a typical heating/cooling zone with thermal time constant ≈20 min, sized with a 200 kW direct expansion air conditioner heat pump. The dispatch algorithm was configured to have a setpoint of 22 °C and deadzone Δ = 1 °C. An ambient temperature profile was generated from recorded winter seasonal data in the UK in each simulation run, with the units configured for heating purposes. The five HVAC units offer DR reductions as follows: Four of the units (A–D) offer 0–100 kWh in 3.3 kWh steps @ 0.25€/kWh, and 103.3 kWh to 160 kWh in 3.3 kWh steps @ 0.35€/kWh EUR/kWh. The final unit (E) offers 0–160 kWh in 3.3 kWh steps @ 0.2€/kWh. The example considers the case of an aggregate load curtailment (STOR request) of 500 kWh being required for an event lasting one hour.

Application of the modified knapsack DP algorithm (Equations (12)–(17)) yields the simple minimum-cost solution as 160 kWh from HVAC unit E, 40 kWh from HVAC unit D and 100 kWh each from units A–C, giving the required 500 kWh at a total cost of €117. Four simulation runs were thus considered for an HVAC unit in the presence of the STOR event: (i) no DR activated (baseline); (ii) 40 kWh reduction through DR; (iii) 100 kWh reduction through DR; and (iv) 160 kWh reduction through DR. The responses obtained are given in Figures 6–9. Each figure indicates zone temperature in °C (red trace), HVAC on/off status (black trace) and temperature setpoint in °C (dashed blue trace) plotted versus time, in minutes. The figures also indicate the location of the DR window on the time axis. The DR window commences after 380 min for a duration of 60 min; the notification of the upcoming DR event occurs after 140 min.

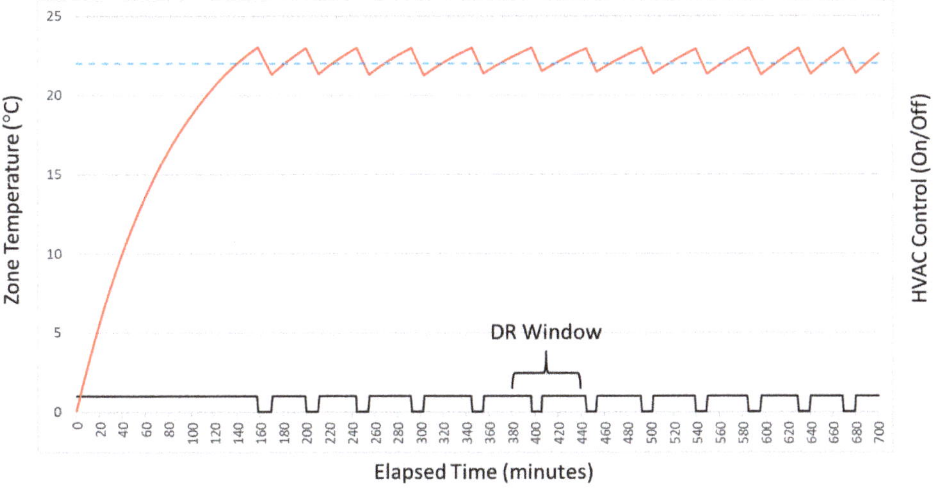

Figure 6. Baseline STOR scenario (red trace, temperature; black trace, HVAC control).

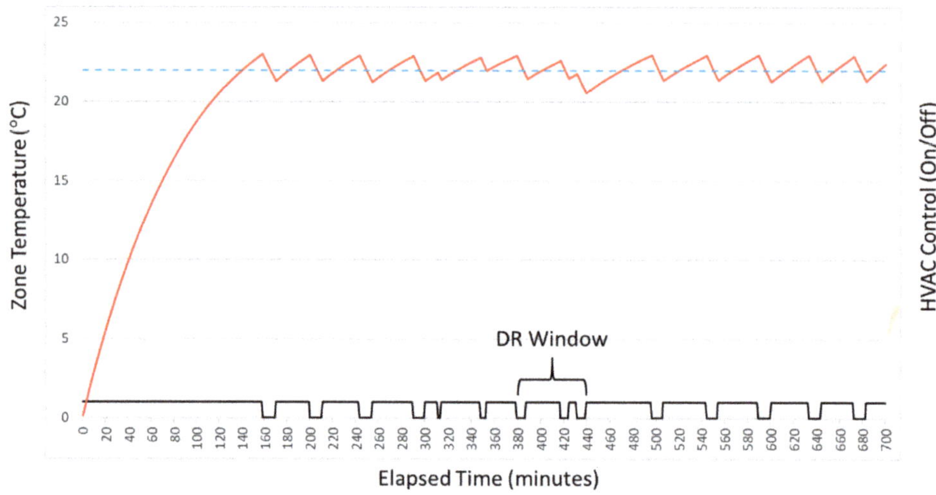

Figure 7. The 40 kWh reduction STOR scenario (red trace, temperature; black trace, HVAC control).

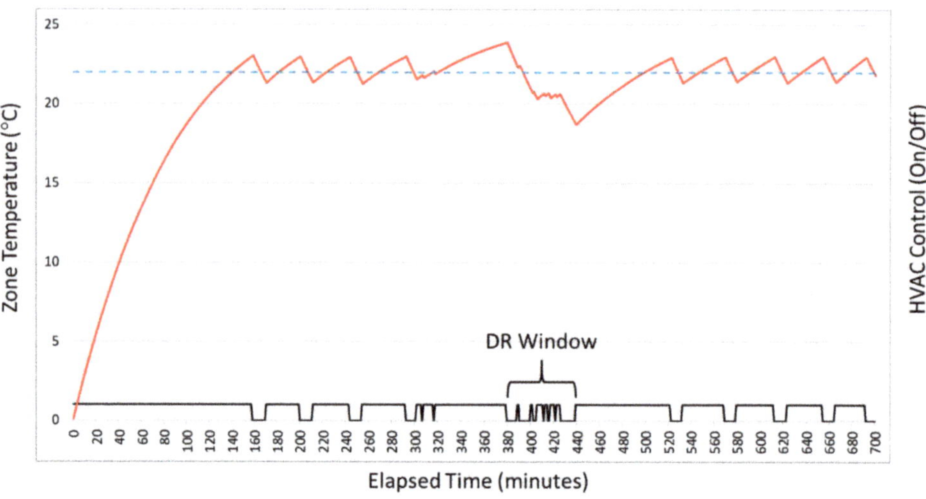

Figure 8. The 100 kWh reduction STOR scenario (red trace, temperature; black trace, HVAC control).

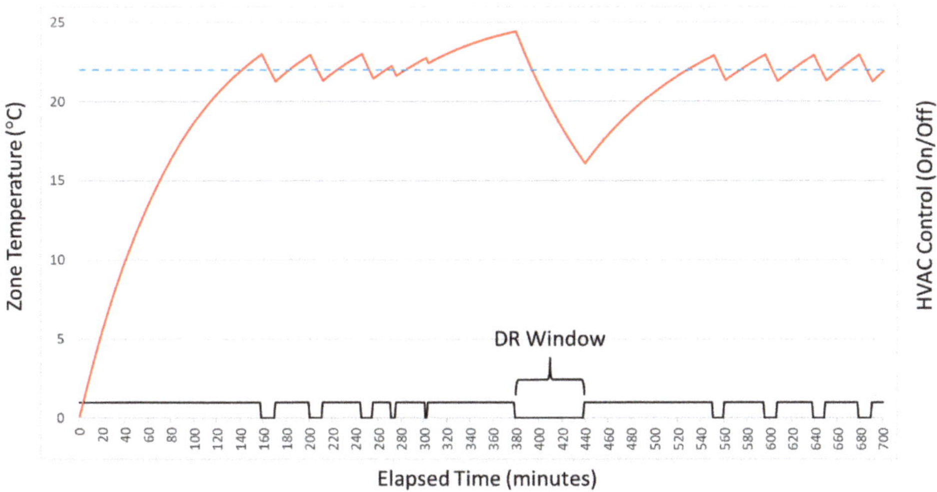

Figure 9. The 160 kWh reduction STOR scenario (red trace, temperature; black trace, HVAC control).

In the baseline case, with zero DR activity, after initially heating to the setpoint the temperature cycles ±1 °C around the setpoint, and continues to do so during the DR window, with the HVAC unit operating as normal. After initially achieving setpoint, the maximum, minimum and average temperatures recorded were 23.0, 21.2 and 22.2 °C respectively. In this case, 160 kWh is consumed during the DR window. In the 40 kWh STOR case, after initially heating to the setpoint the temperature again starts to cycle ±1 °C around the setpoint; but after approximately 290 min, deviations from the regular cycle can be observed as the unit prepares for DR activity during the DR window. After initially achieving setpoint, the maximum, minimum and average temperatures recorded were 23.0, 20.6 and 22.1 °C respectively. In this case, 120 kWh is consumed during the DR window. Focusing now upon the 100 kWh STOR case, after initially heating to the setpoint the temperature again starts to cycle ±1 °C around the setpoint. After approximately 290 min, deviations from the regular cycle can be observed as the unit prepares for DR activity during the DR window, during which 60 kWh is consumed. A distinct period of pre-heating can be observed prior to this window. After initially achieving setpoint, the maximum, minimum and average temperatures recorded were 23.9, 18.7 and 22.0 °C respectively. Finally, concerning the 160 kWh STOR case, the unit again starts to cycle the temperature ±1 °C around the setpoint after initially heating. A pronounced period of pre-heating begins at the 260-min mark as the unit prepares for DR activity during the DR window. In this case, zero kWh is consumed during the DR window, i.e., the unit remains in the 'off' state. After initially achieving setpoint, the maximum, minimum and average temperatures recorded were 24.5, 16.1 and 21.7 °C respectively.

As can be observed in Figures 8 and 9, the proposed dispatch strategy is effective in producing the DR actions required in order to surgically reduce consumption during DR events. Although this has some negative impact upon the zone temperature, this impact is minimized through the optimization which has been developed. Accurate baselining and optimal activation of the unit both prior to and during the DR window itself are evident. Figure 10 plots the electricity consumption (in kWh) for all five HVAC units over the duration of the simulation time in both the DR and no DR (baseline) cases. The aggregated consumption for the no-DR case during the DR window was found to be 833.3 kWh, while the aggregated consumption for the DR case during DR window was found to be 333.3 kWh. The commanded reduction of 500 kWh has clearly been achieved, and an accurate baseline provided for settlement purposes. Figures 6–10 also clearly indicate that consumption has been shifted away from the DR window to occur both before and after the window itself.

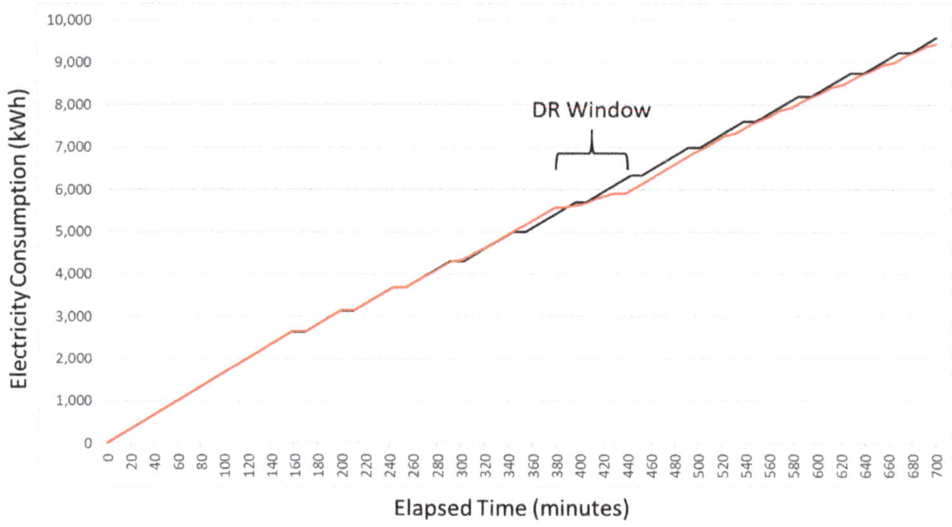

Figure 10. Electricity consumption in kWh (red trace, DR case; black trace, baseline case).

4.2. Experimental Evaluation

For the experimental evaluation, testing was performed at an experimental site on the Teesside University campus in Middlesbrough, UK, within the context of the Demand Response in Blocks of Buildings (DR-BoB) Horizon 2020 project. The experiment was undertaken on four HVAC/AHU (air handling units) providing both heating and cooling services to a large building housing office spaces and lecture rooms. Figure 11 shows a photograph of the locations on the rooftop at the demonstration site building. The units were configured to operate in cooling mode, with circulating water setpoint set at 5.25 °C with deadzone $\Delta = 1.25$ °C. The cooled water feeds into individual room heat exchangers to provide air conditioning and cooling in the office spaces and lecture rooms.

To provide remote override and operation of the HVAC controls, and provide an interface to measured temperatures, a BACNet connection to an additional programmable control device was employed. This device represented the ITU, and allowed the implementation of the dispatch controls. An internet connection to a cloud-based coordination application sending the simulated DR signals was also implemented over a secured public IP connection. The HVAC dynamics are non-linear in the HVAC units, being partially dependent upon the thermal demand. This demand is predictable and manifests as a time varying coefficient $a(k)$ in Equation (2). In addition, a small transmission zero is present and was included in the model, which provides a very good fit (≈90%) to historic recorded data. The experimental test was carried out on the afternoon of the 1st of September 2017, during which a DR event requiring a reduction of 200 kWh was simulated between the hours of 12:30 and 15:30. Results for the response of HVAC/AHU #1 are shown in Figures 12 and 13. Figure 12 displays the chiller return temperature (in °C) for the unit during the course of the day, with the actual temperature displayed in red and baseline temperature displayed in black. The ambient temperature outside the building is displayed in green and the setpoint in blue. The unit is activated at approximately 09:30 and the water temperature is brought towards the setpoint. The disturbance at approximately 10:00 is best attributed to a load change as internal heat exchangers are switched on as the building becomes occupied. Pre-cooling of water temperature begins at 12:00 prior the DR event itself. The electricity consumption of the unit is shown Figure 13. The red trace displays the actual (metered) consumption of the unit, while the black trace displays the baseline (counterfactual) consumption of the unit. Figures 12 and 13 show evidence of pre-cooling, and the generation of an accurate baseline indicating a post-event reduction in electricity consumption from the unit of 54.13 kWh. The overall aggregated reduced consumption obtained from HVAC units #2, #3 and #4 were at similar levels, meeting the aggregated target of 200 kWh. Overall, these results provide a practical, experimental and real-world validation of the techniques described in this research.

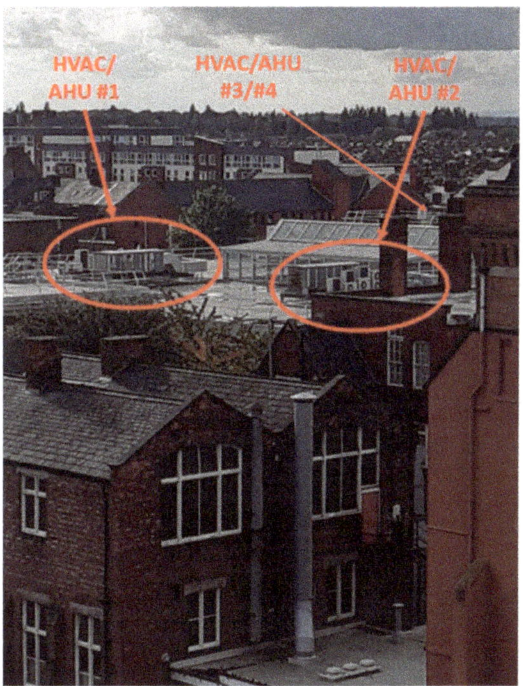

Figure 11. Photograph indicating positioning of the four HVAC units at the demonstration site.

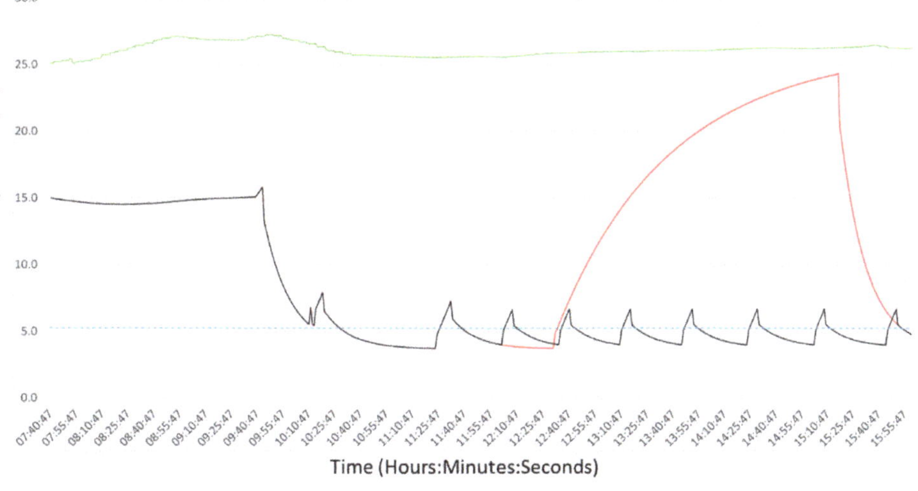

Figure 12. Recorded temperatures in °C (red trace, DR case; black trace, baseline case; green trace, ambient temperature).

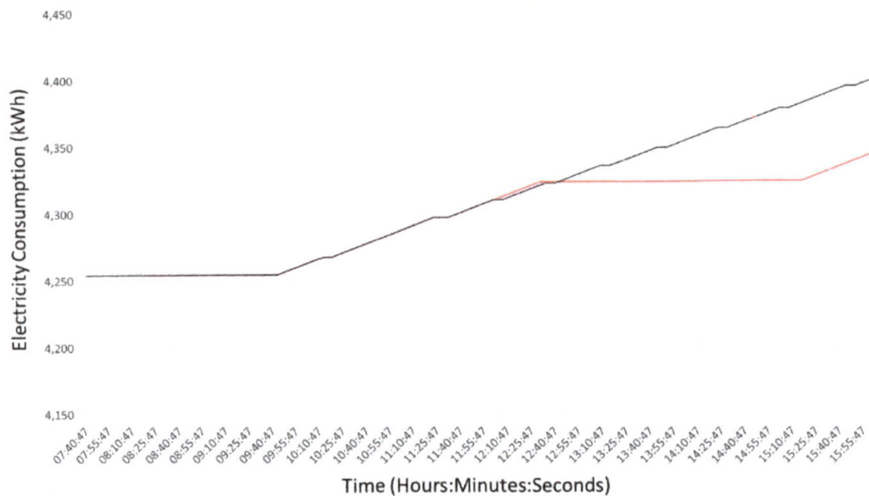

Figure 13. Electricity consumption in kWh (red trace, DR case; black trace, baseline case).

5. Conclusions

This paper argues that since energy consumption is estimated to be about 40% accountable to buildings, sustainable demand side management is one possibility to mitigate the dependence of buildings on the electrical grid. As such, a novel approach to coordinated demand response (DR) for aggregated HVAC is presented to support surgical (explicit) DR in buildings and blocks or groups of buildings, which are possibly spread across multiple geographic locations. This approach presents a large-scale decentralized optimization framework that leverages concepts of Industry 4.0, incorporating distributed DP solutions for individual unit dispatch with a centralized DP allocation procedure. In addition, an accurate baselining procedure to enable effective calculation of reduced consumption and post-event settlement at the individual unit level has been proposed. The approach has been validated using calibrated simulation studies and experimental tests, and it is demonstrated that surgical DR can be effected by a number of coordinated units. Future work includes further representative testing of the proposed approach and its commercialization. In addition, future work will also investigate the integration of other forms of DR assets (e.g., combined heat and power (CHP) plant [27] and smart home appliances [28] under rolling-horizon techno-economic optimization) into the aggregated DR framework.

Author Contributions: Conceptualization, M.S. and R.C.; methodology, M.S., S.R. and R.C.; software, M.S.; validation, M.S., S.R.; formal analysis, M.S.; investigation, M.S. and T.C.; resources, N.D.; data curation, M.S. and S.R.; writing—original draft preparation, M.S.; writing—review and editing, M.S., S.R., R.C., T.C. and N.D.; visualization, M.S. and S.R.; supervision, M.S.; project administration, S.R. and N.D.; funding acquisition, M.S., R.C., T.C. and N.D.

Funding: The work presented in this paper was carried out as part of the Demand Response in Blocks of Buildings (DR-BOB) project (01/03/16–28/02/19) which is co-funded by the EU's Horizon 2020 framework programme for research and innovation under grant agreement No. 696114. The authors wish to acknowledge the European Commission for their support, the efforts of the project partners, and the contributions of all those involved in DR-BOB.

Acknowledgments: An early version of the HVAC dispatch algorithm described in Section 3.1 of this paper was presented at the IAPE 2019 conference in Oxford, UK [29]. The authors acknowledge the support provided by John Broderick (University of Manchester, UK), Chris Ogwumike (Teesside University, UK) and Vladimir Vukovic (Teesside University, UK) in the project from which this paper has arisen.

Conflicts of Interest: The authors declare no conflict of interest. The funders had no role in the design of the study; in the collection, analyses, or interpretation of data; in the writing of the manuscript, or in the decision to publish the results.

References

1. Isaksson, A.J.; Harjunkoski, L.; Sand, G. The impact of digitalization on the future of control and operations. *Comput. Chem. Eng.* **2018**, *114*, 122–129. [CrossRef]
2. Mahdavinejad, M.S.; Rezvan, M.; Barekatain, M.; Adibi, P.; Barnaghi, P.; Sheth, A.P. Machine learning for internet of things data analysis: A survey. *Digit. Commun. Netw.* **2018**, *4*, 161–175. [CrossRef]
3. Perez-Lombard, L.; Ortiz, J.; Pout, C. A review on buildings energy consumption information. *Energy Build.* **2008**, *40*, 394–398. [CrossRef]
4. Crosbie, T.; Short, M.; Dawood, M.; Charlesworth, R. Demand response in blocks of buildings: Opportunities and requirements. *J. Entrep. Sustain. Issues* **2017**, *4*, 271–281. [CrossRef]
5. Arteconi, A.; Polonara, F. Demand side management in refrigeration applications. *Int. J. Heat Technol.* **2017**, *35*, S58–S63. [CrossRef]
6. Proffitt, E. *Profiting from Demand Side Response*; Technical Publication; The Major Energy Users' Council in Association with National Grid: London, UK, 2016.
7. Shi, Q.; Li, F.; Hu, Q.; Wang, Z. Dynamic demand control for system frequency regulation: Concept review, algorithm comparison, and future vision. *Electr. Power Syst. Res.* **2018**, *154*, 75–87. [CrossRef]
8. Luo, F.; Ranzi, G.; Dong, Z. Rolling horizon optimization for real-time operation of thermostatically controlled load aggregator. *J. Mod. Power Syst. Clean Energy* **2017**, *5*, 947–958. [CrossRef]
9. Williams, S.; Short, M.; Crosbie, T. On the use of thermal inertia in building stock to leverage decentralised demand side frequency regulation services. *Appl. Therm. Eng.* **2018**, *133*, 97–106. [CrossRef]
10. Ramanathan, B.; Vittal, V. A framework for evaluation of advanced direct load control with minimum disruption. *IEEE Trans. Power Syst.* **2008**, *23*, 1681–1688. [CrossRef]
11. Hao, H.; Sanandaji, B.; Poolla, K.; Vincent, T.L. Aggregate flexibility of thermostatically controlled loads. *IEEE Trans. Power Syst.* **2015**, *30*, 189–198. [CrossRef]
12. Vrettos, E.; Anderson, G. Combined load frequency control and active distribution network management with thermostatically controlled loads. In Proceedings of the IEEE International Conference on Smart Grid Communications (SmartGridComm), Vancouver, BC, Canada, 21–24 October 2013.
13. Mathieu, J.L.; Koch, S.; Callaway, D.S. State estimation and control of electric loads to manage real-time energy imbalance. *IEEE Trans. Power Syst.* **2013**, *28*, 430–440. [CrossRef]
14. Wang, H.; Meng, K.; Luo, F.; Dong, Z.Y.; Verbič, G.; Xu, Z.; Wong, K.P. Demand response through smart home energy management using thermal inertia. In Proceedings of the Australia Universities Power Engineering Conference (AUPEC 2013), Hobart, TAS, Australia, 29 September–3 October 2013.
15. Luo, F.; Zhao, J.; Wang, H.; Wang, H.; Tong, X.; Chen, Y.; Dong, Z.Y. Direct load control by distributed imperialist competitive algorithm. *J. Mod. Power Syst. Clean Energy* **2014**, *2*, 385–395. [CrossRef]
16. Drgona, J.; Klauco, M.; Kvasnica, M. MPC-based reference governors for thermostatically controlled residential buildings. In Proceedings of the IEEE 54th Annual Conference on Decision and Control (CDC 2015), Osaka, Japan, 15–18 December 2015.
17. Callaway, D.S. Tapping the energy storage potential in electric loads to deliver load following and regulation, with application to wind energy. *Energy Convers. Manag.* **2009**, *50*, 1389–1400. [CrossRef]
18. Zhou, L.; Li, Y.; Wang, W.; Hu, X. Provision of supplementary load frequency control via aggregation of air conditioning loads. *Energies* **2015**, *8*, 14098–14117. [CrossRef]
19. Afram, A.; Janabi-Sharifi, A. Review of modelling methods for HVAC systems. *Appl. Therm. Eng.* **2014**, *67*, 507–519. [CrossRef]
20. Mathews, E.; Richards, P.; Lombard, C. A first-order thermal model for building design. *Energy Build.* **1994**, *21*, 133–145. [CrossRef]
21. Camacho, E.F.; Bordons, C. *Model Predictive Control*, 2nd ed.; Springer: Berlin, Germany, 2004.
22. Djongyang, N.; Tchinda, R.; Njomo, D. Thermal comfort: A review paper. *Renew. Sustain. Energy Rev.* **2010**, *14*, 2626–2640. [CrossRef]
23. Bellman, R.E. *Dynamic Programming*; Princeton University Press: Princeton, NJ, USA, 1957.
24. Bertsekas, D.P. *Dynamic Programming and Optimal Control*; Athena Scientific: Belmont, MA, USA, 2005.
25. Chelmis, C.; Aman, S.; Saeed, M.R.; Frincu, M.; Prasanna, V.K. Estimating reduced consumption for dynamic demand response. In Proceedings of the AAAI Workshop on Computational Sustainability, Austin, TX, USA, 1 April 2015.

26. Martello, S.; Toth, P. *Knapsack Problems: Algorithms and Computer Implementations*; John Wiley and Sons: Hoboken, NJ, USA, 1990.
27. Short, M.; Crosbie, T.; Dawood, M.; Dawood, N. Load forecasting and dispatch optimization for decentralized co-generation plant with dual energy storage. *Appl. Energy* **2017**, *186*, 304–320. [CrossRef]
28. Ogwumike, C.; Short, M.; Abugchem, F. Heuristic optimization of consumer electricity costs using a generic cost model. *Energies* **2016**, *9*, 6. [CrossRef]
29. Short, M. Optimal HVAC dispatch for demand response: A dynamic programming approach. In Proceedings of the International Conference on Innovative Applied Energy (IAPE'19), Oxford, UK, 15 March 2019.

Article

Parameter Selection for the Virtual Oscillator Control Applied to Microgrids

Daniel Alves Costa [1,2,*], Leonardo A. B. Tôrres [1], Sidelmo Magalhães Silva [1], Alberto De Conti [1] and Danilo Iglesias Brandao [1]

[1] Graduate Program in Electrical Engineering, Universidade Federal de Minas Gerais, Av. Antônio Carlos 6627, Belo Horizonte, MG 31270-901, Brazil; leotorres@ufmg.br (L.A.B.T.); sidelmo@eee.ufmg.br (S.M.S.); conti@cpdee.ufmg.br (A.D.C.); dibrandao@ufmg.br (D.I.B.)

[2] Department of Mechatronics Engineering, Centro Federal de Educação Tecnologica de Minas Gerais, R. Alvares de Azevedo 400, Divinópolis, MG 35503-822, Brazil

* Correspondence: daniel.a.c@cefetmg.br

Abstract: Virtual Oscillator Control (VOC) is a promising technique that allows several inverters connected to a microgrid to naturally synchronize, without communication. However, the selection of the VOC parameters often require iterative or optimization procedures that render its practical use not straightforward. In this paper, this problem is overcome with the proposition of a novel methodology for determining the dead-zone type VOC parameters based on the describing function method. The methodology consists of a set of analytical equations that use as input data few basic electrical system parameters from the converter and from the microgrid, namely, the operating voltage and frequency ranges, besides rated power. The proposed set of equations is used to calculate the parameters required to control an inverter in voltage mode. The validity of the proposed approach is demonstrated in experiments that encompass different situations such as pre-synchronization, connection, and disconnection of a second inverter from a microgrid.

Keywords: Virtual Oscillator Control; parameter tuning; voltage-mode inverter; microgrid

Citation: Costa, D A.; Torres, L.A.B.; Silva, S.M.; De Conti, A.; Brandao, D.I. Parameter Selection for the Virtual Oscillator Control Applied to Microgrids. *Energies* **2021**, *14*, 1818. https://doi.org/10.3390/en14071818

Academic Editosr: Michael Short, Tracey Crosbie and Maher Al-Greer

Received: 24 February 2021
Accepted: 22 March 2021
Published: 24 March 2021

1. Introduction

Decentralized energy systems are becoming a promising alternative to the traditionally centralized model. This approach reduces the need for investments in transmission infrastructures, optimizes line losses, and assists in reactive compensation and voltage regulation [1,2]. At the same time, interest in using renewable energy for the energy matrix has increased considerably in recent years [3]. The use of photovoltaic panels, wind turbines, and other types of clean energy, besides the reduction of environmental impacts, has already become economically viable, especially in isolated places [4].

Observing this trend in the energy market, it is imperative to develop efficient and robust control strategies for power inverters operating in parallel, connected to the distribution system, or operating isolated from the grid. This type of structure is the basis for Parallel Connected Uninterruptible Power Supplies (PCUPS) and Micro Grid-based Electrical Power Systems (MGEPS) [5,6]. Important challenges in this architecture are related to the reduction or elimination of direct communication among inverters while ensuring stability and synchronization between generating units regardless of loads, regulating frequency and voltage, and providing load distribution according to the power of each generating device [7].

Parallel inverters synchronization methods can be divided into centralized, master/slave, distributed, and without communication (fully decentralized) [8]. This paper deals with the approach to control parallel inverters in a microgrid system without communication. The main advantage of this class of techniques is to avoid a common mode failure for the power distribution network associated with possible communication channel faults, thus improving the system overall reliability [9]. Usually the price to pay is the difficulty in distributing the power demand, since the load that each device must assume is unknown after unanticipated disconnections of faulted inverters.

Among methods of parallelism without communication, one of the most widely used is the Voltage and Frequency Droop Method (VFDM), or just Droop Control [10]. One of the first works on VFDM for converter control is presented in Chandorkar et al. [11], where transmission lines are considered inductive. VFDM then emulates the behavior of a synchronous generator controller, imposing an inverse relationship between frequency and active supplied power, while the exchanged reactive power is varied by voltage amplitude control. When the line cannot be considered purely inductive, the relationship between these variables needs to be modified accordingly [12]. In addition, when the line impedances to the load of each inverter are different, the method, as initially proposed, fails to properly distribute power between the generating units [12]. An approach used to mitigate the impacts of the two issues mentioned above is the use of a virtual impedance [13]. This impedance is designed to contribute to load distribution and can be chosen in such way that the equivalent impedance is inductive. Virtual impedances may even vary over time to accommodate changes in the electrical system [13].

Smooth operation of VFDM inverters is highly dependent on the way active and reactive power are estimated. Changing the underlying computation can improve the performance of a given implementation or lead to instability, as each estimation method adds different amounts of delay, usually associated with low-pass filters, to the control loop in each inverter. A simple way to obtain these power variables is presented in Andrade et al. [14].

In addition, previously developed synchronization methods that rely on estimated active and reactive power may exhibit slow convergence to the target power balance condition and usually need specific apparatus for the pre-synchronization and the power estimation procedures, such as phase-locked loops (PLLs). Other types of synchronization methods, e.g., those based on rotating (synchronous) reference frames, rely directly on PLLs to implement their strategies. In Wu et al. [15], it is shown the effects of PLLs on the converter stability, particularly how the interaction of PLLs with the grid/regulators can lead to instability.

These drawbacks have motivated the development of a new approach, the Virtual Oscillator Control (VOC), initially proposed in Torres et al. [16,17], with mathematical proofs for guaranteeing global asymptotic stability of the synchronization condition for symmetric electrical networks. To validate the theoretical developments, experimental results for the parallel operation of two inverters were presented, together with a pre-synchronization strategy to reduce the amplitude of transients at the time of connection of a new device in the electrical network [18]. In Johnson et al. [7,19], similar theoretical analysis and new experimental results were reported, and the term VOC was likely used for the first time.

Inverters controlled using the VOC technique usually have shorter settling times after load changes in comparison with other techniques. This is mainly because they do not depend on active and reactive power estimates commonly employed in other approaches. VOC is conceived considering time-domain variables [17], whereas conventional techniques are often based on phasors, taken as steady-state quantities or approximate mean values [20]. The steady state behavior of an inverter using VOC is similar to an inverter using VFDM [20,21]. Thus, it would be possible to use in the same network inverters controlled by both techniques. However, no work to date has explored this possibility. In Johnson et al. [21], the convergence time comparison is made between VOC and VFDM. It is possible to notice the superiority of the first method in the considered cases.

The first VOC studies have focused only on isolated microgrids, like the one shown in Figure 1. In more recent publications [22–24], a dispatchable VOC (dVOC) strategy was proposed. In this new approach, it is possible to set desired reference values for power injection that satisfy the power flow equations (steady-state condition). Nevertheless, the studied cases focus on isolated grids. Although Minghui Lu et al. [25] highlights that the VOC presented is grid-connected capable, neither stability nor synchronization studies in this context were presented. Furthermore, the Unified Virtual Oscillator Control (uVOC) is proposed in Awal et al. [26], which is built on the theoretical results of dVOC. The authors claim that uVOC can operate as Grid-Forming and Grid-Following with built-in fault protection, which allows a smooth transition between connected/isolated operation and protection of hardware against short circuits. The VOC can be used in conjunction with the active and reactive injection control to obtain dispatchable characteristics [27], this has advantage over the methods dVOC and uVOC because it is independent on the X/R ratio of the line. Surprisingly, the VOC method has also been applied in direct current (DC) systems [28,29]. In Lin at al. [28], a current mode oscillator structure is used to control distributed active filters to attenuate the ripple in the DC line. On the other hand, in Duarte et al. [29] the VOC voltage output, after a coordinate transformation, is used to control DC/DC converters connected to a DC microgrid.

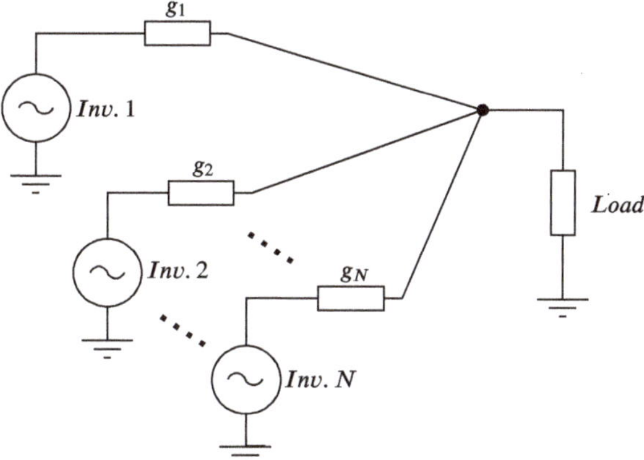

Figure 1. Interconnected system with N inverters.

In most works about VOC, the underlying mathematical developments and methods used to choose appropriate parameters can render the widespread usage of this technique somewhat difficult. This paper proposes a novel strategy to determine VOC design parameters for application on parallel inverters connected to a microgrid (Section 2.2). Different from the work in Johnson et al. [21], this paper deals with dead-zone (or saturation) type oscillators, for which the parameter calculation method presented there is not applicable. In addition, the use of a saturation function instead of the cubic function, considered in that work, can potentially generate less third harmonic distortion for nominal load operation, if the parameters are chosen following our tuning methodology, by avoiding the saturation limits in this condition. Finally, the use of a piecewise-linear function (the saturation or dead-zone function, depending on the point-of-view) facilitates the discrete-time numerical implementation of the VOC strategy on Digital Signal Processors (DSPs) by the use of *exact* discretization methods for linear and time-invariant systems. The development presented in this paper could be used directly in applications such as in Ali et al. [27], Duarte et al. [29] and, with some modifications, in Lin at al. [28], with the benefits described above.

The proposed strategy consists of a set of analytical equations that use as input data only basic electrical parameters from the converter and from the microgrid, namely, the rated power and voltage and frequency ranges. It has the advantage over existing procedures to avoid the use of optimization, numerical solutions or graphical analyzes, and therefore significantly simplifies the VOC tuning process. In addition, a simple pre-synchronization strategy is presented (Section 3) to ensure smooth connection transients. Experimental results are presented in Section 4 to validate the proposed approach for a single-phase system. Other works has exploited the use of this type of controller on three-phase converters. The ideas presented in Johnson et al. [30] and Rosse et al. [31], concerning three-phase implementations, are compatible with the method presented here.

2. An Overview of The VOC Method

The concept of the VOC method is to emulate the dynamics of an oscillator through a power inverter. In the case of cooperation to feed the same load, this oscillator must have the property of naturally synchronizing with a unknown but finite number of several other power oscillators, that is, by construction the output voltages of all oscillators converge to the same quasi-sinusoidal oscillatory pattern, with low harmonic distortion, as a natural result of being connected to the same microgrid [17]. One of the possible oscillatory systems capable of presenting this property is shown in Figure 2, consisting of a RLC filter and a current source controlled by a nonlinear voltage function, ϕ. The dynamic equations of this circuit are integrated by an appropriate numerical method, implemented in a DSP, which generates the voltage reference, $v_{o_{ref}}$, in that case equal to v_{osc}. This means that the components shown in Figure 2 are virtual ones, the reason why this method has been called Virtual Oscillator Control.

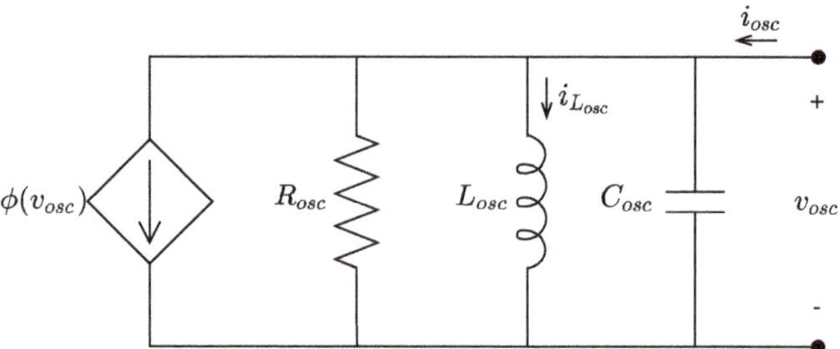

Figure 2. Oscillator example, compatible with Virtual Oscillator Control (VOC) method and used in this work.

Considering the circuit in Figure 2, and applying Kirchhoff's laws, one has that

$$\begin{aligned} \dot{\vec{x}} &= \begin{bmatrix} 0 & 1/L_{osc} \\ -1/C_{osc} & -1/(R_{osc}C_{osc}) \end{bmatrix} \vec{x} + \begin{bmatrix} 0 \\ 1/C_{osc} \end{bmatrix} u, \\ y &= [0\ 1]\,\vec{x}, \end{aligned} \tag{1}$$

where $\vec{x} = [i_L \ \ v_{osc}]^T$, $u = i_{osc} - \phi(v_{osc})$, and i_{osc} is defined as the current flowing into the inverter. The controlled current source in Figure 2 behaves as a nonlinear resistor with piecewise constant non-positive resistance values, and it is able to inject current into the circuit. For limited amplitude oscillations, the ϕ function must also be bounded. The following saturation-like function is used:

$$\phi(v_{osc}) = \begin{cases} -\alpha v_{osc}, & \text{if} \quad |v_{osc}| < \lambda\,, \\ -\alpha\lambda, & \text{if} \quad v_{osc} \geq \lambda\,, \\ \alpha\lambda, & \text{if} \quad v_{osc} \leq -\lambda\,. \end{cases} \tag{2}$$

where $\lambda > 0$ and $\alpha > 0$ are constant parameters to be designed.

Now, consider $N \geq 1$ inverters controlled using this technique, with identical parameters, connected in parallel as shown in Figure 1. The theoretical justification for the synchronization of a unknown but finite number of power oscillators in a symmetric electrical network is related to the *incremental* passivity of the oscillators, together with the incremental passivity of the electrical network, in the sense briefly presented in Appendix A. Using this concept it was shown that a sufficient condition for synchronization, among others, is that

$$\alpha < \eta + \alpha_p, \tag{3}$$

where η is the passivity characteristic of the electrical network and α_p is the passivity characteristic of the linear time-invariant (LTI) system (1) (see Section 2.1). In this case, all units in Figure 1 will synchronize naturally, sharing equally the load [17]. It is important to emphasize that inequality (3) is only a sufficient condition. In other words, even if the system violates this condition, it can still achieve synchronization in some cases. In Torres et al. [17], although the connection impedances $g_1, g_2, \ldots, g_N$ were considered identical, the value of g_j or the load characteristic (resistive/inductive) does not require modifications to the method structure, as opposed to VFDM.

2.1. Passivity Constraints on the Virtual Oscillator Design

Considering the LTI system presented in Equation (1), it is possible to define an energy storage function ([32] Definition 6.3) and compute its time-derivative as

$$\begin{aligned} S^{\text{OSP}} &= \frac{L_{osc} i_{L_{osc}}^2}{2} + \frac{C_{osc} v_{osc}^2}{2} = \frac{L_{osc} x_1^2}{2} + \frac{C_{osc} x_2^2}{2}, \\ \dot{S}^{\text{OSP}} &= L_{osc} x_1 \dot{x}_1 + C_{osc} x_2 \dot{x}_2 = u x_2 - x_2^2 / C_{osc}. \end{aligned}$$

As $x_2 = y$, one has that

$$\dot{S}^{\text{OSP}} = -y\rho(y) + uy, \tag{4}$$

and it can be concluded that the LTI system Equation (1) is Output Strictly Passive (OSP) ([32] Definition 6.3). Although the *RLC* filter is passive, the nonlinear oscillator should not be, otherwise oscillations would not exist. The analysis of this condition can be performed by substituting $u = i_{osc} - \phi(y)$ in Equation (4), and recognizing that the system input is now $\tilde{u} = i_{osc}$,

$$y\tilde{u} = \dot{S}^{\text{OSP}} + \frac{y^2}{R_{osc}} + y\phi(y). \tag{5}$$

However, $\phi(y)$ belongs to sector $[-\alpha, 0]$, so $-\alpha y^2 \le y\phi(y) \le 0$ and Equation (5) can be rewritten as

$$y\tilde{u} \ge \dot{S}^{\text{OSP}} + y^2\left(\frac{1}{R_{osc}} - \alpha\right). \tag{6}$$

Thus, the system will be OSP ([32] Definition 6.1) if

$$\alpha < \frac{1}{R_{osc}} \tag{7}$$

Moreover, making $\tilde{u}(t) = 0$ and $y(t) = 0, \forall t \ge 0$, in Eqation (1) leads to the conclusion that $x_2(t) = y(t) \equiv 0 \Rightarrow \dot{x}_2(t) = 0, \forall t \ge 0$, and therefore $x_1(t) \equiv 0$. This means that the system is zero-state observable ([32] Definition 6.5) and, from ([32] Lemma 6.7), the virtual oscillator, with $i_{osc} = 0$, would be globally asymptotically stable, which is an undesirable condition. Therefore, one must have

$$\alpha \ge \alpha_p^* > \alpha_p := \frac{1}{R_{osc}}, \tag{8}$$

where α_p^* represents a safety margin for sustained oscillations when $i_{osc} \ne 0$. So, the condition expressed in Equation (8) is a necessary one. This will be explored in the next section.

2.2. Practical VOC Parameters Selection

In Johnson et al. [21], the average oscillator models are used to obtain parameters for a Van der Pol oscillator with cubic function, considering a quasi-stationary operation condition. However, the described procedure can be laborious and somewhat non-intuitive. The use of cubic function nonlinearity has some disadvantages, as third harmonic generation (see Section 4.1). Besides, as reported in that work, an adaptation was required to eliminate an algebraic loop resulted from the use of approximate integration by means of the trapezoidal rule applied to discretize the smooth nonlinear differential equations. In order to overcome these drawbacks, a new approach for Voltage-Controlled Voltage Source Inverters (VCVSIs) control design, using VOC, is presented in this section. As described in Section 2, an oscillator with saturation nonlinearity will be used. The parameters are selected to reduce harmonic generation. As the saturation is a piecewise linear function, this allows the use of exact discretization of the linear time-invariant systems that are used in the DSP numerical implementation.

From Section 2, the necessary parameters to apply the VOC method are R_{osc}, L_{osc}, C_{osc}, α, and λ. The parameters L_{osc} and C_{osc} are associated with the frequency of v_{osc}, while the amplitude is determined by the set of parameters R_{osc}, α and λ. Furthermore, it is desirable to satisfy inequality Equation (3) such that α must be chosen in order to guarantee both sustained oscillations and synchronization of inverters. From Equation (1), it is possible to infer that the current i_{osc} also influences the amplitude of the oscillations. In the event that $|i_{osc}| \ll |\phi(v_{osc})|$, the load current will have little impact on voltage regulation. However, the more insensitive to the load current, the slower the synchronization process will become.

The describing function method is applied to the oscillator in Figure 2. For a single-input single-output (SISO) nonlinear system, represented by a linear system with a nonlinear function feedback with $i_{osc} = 0$, there might be a periodic solution with a frequency and amplitude close to ω and a, if it is possible to solve the equation

$$G(j\omega)\Phi(a) + 1 = 0\,, \tag{9}$$

where $G(j\omega)$ is the frequency response of the LTI system Equation (1) and $\Phi(a)$ is the describing function obtained for the nonlinear function ϕ ([32] Section 7.2). The expression in Equation (9) is known as the harmonic balance equation. For memoryless, time-invariant and odd functions, $\Phi(a)$ is real and can be calculated as

$$\Phi(a) = \frac{2}{\pi a}\int_0^{\pi} \phi(a\sin\theta)\sin\theta\, d\theta\,. \tag{10}$$

For the addressed problem, ϕ corresponds to the saturation function, Equation (2), with $\alpha > 0$. In this case, the solution of Equation (10) is given by

$$\Phi(a) = \begin{cases} -\alpha, & \text{if } 0 \le a \le \lambda\,, \\ -\frac{2\alpha}{\pi}\left[\arcsin(\kappa) + \kappa\sqrt{1-\kappa^2}\,\right], & \text{if } a > \lambda\,, \end{cases} \tag{11}$$

where $\kappa = \lambda/a$. By inspection we have

$$-\alpha \le \Phi(a) < 0. \tag{12}$$

As $\Phi(a)$ is real, we can split Equation (9) in two equations and solve separately as

$$1 + \Phi(a)\Re\{G(j\omega)\} = 0, \tag{13}$$

$$\Im\{G(j\omega)\} = 0. \tag{14}$$

The transfer function for the system in Equation (1) is

$$G(s) = \frac{s}{s^2 + s\frac{1}{R_{osc}C_{osc}} + \frac{1}{L_{osc}C_{osc}}}. \tag{15}$$

Making $s = \jmath\omega$, we have

$$G(\jmath\omega) = \frac{\omega^2 L_{osc}^2 / R_{osc} + \jmath L_{osc}\omega(1 - L_{osc}C_{osc}\omega^2)}{\omega^2 L_{osc}^2 / R_{osc}^2 + (1 - L_{osc}C_{osc}\omega^2)^2}. \tag{16}$$

Solving Equation (14) for ω it is found that the oscillation of the system, if it exists, must have an approximate frequency of

$$\omega = \sqrt{\frac{1}{L_{osc}C_{osc}}}. \tag{17}$$

This result was already expected, as this is the central frequency of the RLC bandpass filter shown in Figure 2. Solving now Equation (13) using Equation (17) we have

$$\Phi(a) = -1/R_{osc}. \tag{18}$$

In addition, as $R_{osc} > 0$ we can replace Equation (18) in Equation (12), thus

$$\alpha \geq \frac{1}{R_{osc}}. \tag{19}$$

Therefore, if Equation (19) is met, according to the describing function method, the system represented by Equation (1), with feedback ϕ and with $i_{osc} = 0$ (disconnected from the electrical network), has a limit-cycle with amplitude a and frequency ω.

Although the describing function method is based on approximations and does not guarantee the existence of oscillations [32], the condition expressed in Equation (19) is consistent with that obtained in Equation (8). In addition, the method is useful in determining an approximate relation between the parameters R_{osc}, α, λ, and the steady-state amplitude a. On the other hand, it also gives an approximate relation between the steady-state frequency ω and the parameters L_{osc} and C_{osc}.

In the previous analysis, i_{osc} was considered to be zero. However, this condition is not realistic, as the main objective of the inverter is to supply power to the grid. To partially circumvent this issue, it is possible to imagine that a load connected to the oscillator is actually part of it, changing the values of the internal impedances. In the special case where the load is a pure resistance, R_L, the resulting equivalent resistance, R_{osc}^*, of the new oscillator would be the computed from the parallel connection between R_{osc} and R_L. This means that regardless of the value of R_L, the new value of the resistance of the oscillator would be less than the original value. Therefore, the load seen by the inverter when connected to the network has a direct impact on the oscillation condition of Equation (19). For a given value of α, it is possible for the system to become asymptotically stable when connected to a load with a sufficiently small resistance value as pointed out in Equation (8). This will be associated with the maximum allowed active power output, or rated power. Furthermore, as already discussed, R_{osc} has an impact on the steady state amplitude, a, and R_{osc}^* must be used to define the minimum oscillation amplitude. Similarly, looking at Equation (17), inductive and capacitive loads will directly impact the ultimate system's oscillation frequency.

To formalize the previous discussion, consider that the N inverters controlled by the VOC technique shown in Figure 1 are synchronized. Using the fact that these oscillators are incrementally passive, it can be inferred that all states between oscillators are equal in steady state. In this condition, the microgrid can be represented as shown in Figure 3.

An equivalent circuit for a single inverter that is mathematically equivalent to the one in Figure 3 is shown in Figure 4.

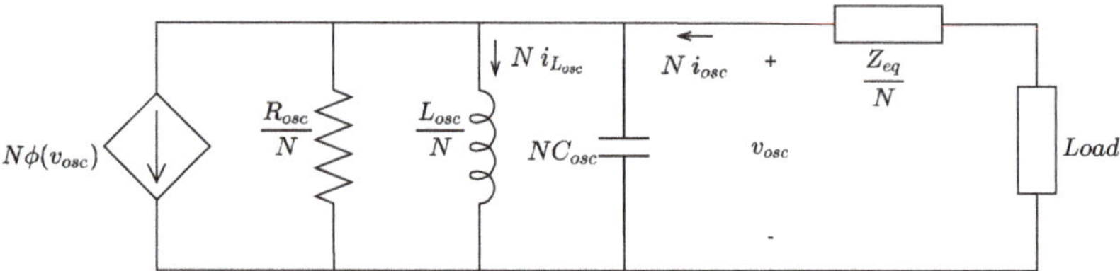

Figure 3. Microgrid schematic representation after synchronization.

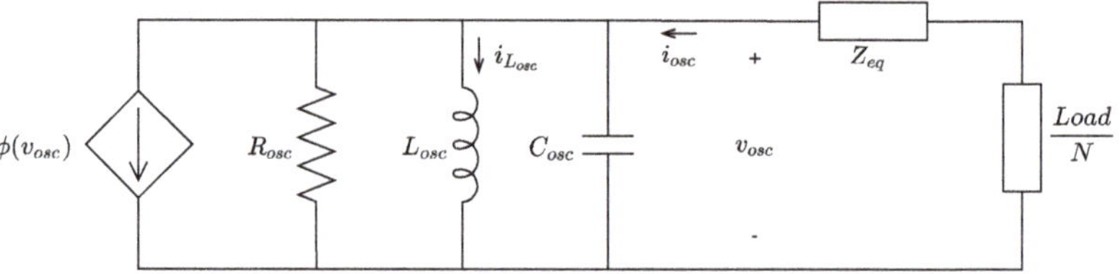

Figure 4. Operating condition of one inverter after synchronization.

Neglecting the harmonics that can be generated by the load or by ϕ and considering that the system operates in steady state, that is, that phasor analysis can be applied, it is possible to decompose $Z_T = Z_{eq} + Z_{Load}/N$ in a parallel RL or RC circuit in order to keep I_{osc} unchanged. In this case, $Z_T = R_T//jX_T$, thus $R_T = -V_{osc}/\Re\{I_{osc}\}$ and $X_T = -V_{osc}/\Im\{I_{osc}\}$. In addition, the power absorbed by the grid is given by $S = -V_{osc}\bar{I}_{osc}$ and $S = -V_{osc}\Re\{I_{osc}\} + jV_{osc}\Im\{I_{osc}\} = P + jQ$. Therefore, it is possible to determine the values of R_T and X_T as

$$\begin{aligned} R_T &= V_{osc}^2/P, \\ X_T &= -V_{osc}^2/Q. \end{aligned} \tag{20}$$

Considering what has been presented so far, it is possible to describe an algorithm for choosing the VOC parameters. The input data are the minimum and maximum voltage values ($V_{\max}$ and $V_{\min}$), nominal frequency and maximum frequency deviation (f_n and Δf [Hz]), and inverter rated power (P_n [W] and Q_n [VAr]).

First, we will define α. The lowest bound is given by Equation (8), but for the new equivalent oscillator we must use $R_{osc}^* = R_{osc}//R_T$ and $\alpha_p^* = 1/R_{osc}^*$. When the inverter delivers nominal power, the oscillator voltage will be minimum. From Equation (20), $R_T = V_{\min}^2/P_n$. Considering the synchronization condition expressed in Equation (3) one has that

$$\frac{1}{R_{osc}} + \frac{P_n}{V_{\min}^2} \leq \alpha < \eta + \frac{1}{R_{osc}}. \tag{21}$$

For simplicity, we choose α as

$$\alpha = \frac{1}{R_{osc}} + \frac{P_n}{V_{\min}^2} = \alpha_p^*. \tag{22}$$

This is a good choice because there is no need to deal with the grid structure or impedance to calculate η. It is true that this does not guarantee synchronization, at least not without knowing the passivity characteristic η of the electrical network. This represents a compromise between a simplified design procedure, which is the main purpose in this work, and the computation of a property that depends on the exact knowledge of the electrical network parameters but that would guarantee the satisfaction of the sufficient, and therefore strong, condition (3) for inverters synchronization. More information on the synchronization criteria and η calculation are found in Torres et al. [17].

Using again the minimum voltage condition together with Equation (11) and Equation (18), for the value of $\Phi(V_{\min}\sqrt{2})$ to be equal to $-1/R^*_{osc}$, λ must be greater than or equal to $V_{\min}\sqrt{2}$ in Equation (11). The maximum voltage condition will happen for the no-load condition. This time $\Phi(V_{\max}\sqrt{2})$ must be equal to $1/R_{osc}$. Looking at the second part in Equation (11) we note that κ must be less than one, otherwise this equation has no solution. Thus λ must be less than $V_{\max}\sqrt{2}$. Again, for simplicity we choose

$$\lambda = V_{\min}\sqrt{2}. \tag{23}$$

Using the minimum value allowed for λ will maximize R_{osc}, which helps filtering high frequencies created by ϕ. Moreover, for nominal load operation, the peak values of $|v_{osc}|$ will be close to λ then $\phi(v_{osc}) \approx -\alpha v_{osc}$, and practically no harmonic is generated by the oscillator.

After defining λ, and using the maximum voltage condition, it is possible to write the second part in Equation (11) as

$$\frac{\alpha}{\gamma} = \frac{1}{R_{osc}}, \tag{24}$$

where

$$\begin{aligned} \gamma &= (\pi/2)[\arcsin(\kappa) + \kappa\sqrt{1-\kappa^2}]^{-1}, \\ \kappa &= V_{\min}/V_{\max}. \end{aligned} \tag{25}$$

Using now Equation (22) and Equation (24), it is possible to find

$$\begin{aligned} \alpha &= \frac{P_n}{V_{\min}^2}\left(\frac{\gamma}{\gamma-1}\right) \\ R_{osc} &= \frac{V_{\min}^2}{P_n}(\gamma-1). \end{aligned} \tag{26}$$

The reactive power delivered by the converter can be associated with an inductive or capacitive current. If X_T is inductive, the new inductance L^*_{osc} will be less than L_{osc} and the system frequency will be greater than the nominal frequency f_n; in this case, $f_{\max} := f_n + \Delta f$. On the other hand, with X_T capacitive, C^*_{osc} will be greater than C_{osc} and the frequency will be less than the nominal frequency; in this case, $f_{\min} := f_n - \Delta f$. From circuit theory, it is known that

$$L^*_{osc} = \frac{L_{osc}L_T}{L_{osc}+L_T}, \tag{27}$$

$$C^*_{osc} = C_{osc} + C_T. \tag{28}$$

From Equation (17),

$$\frac{f_{\max}^2}{f_n^2} = \frac{L_{osc}}{L^*_{osc}}, \quad \frac{f_{\min}^2}{f_n^2} = \frac{C_{osc}}{C^*_{osc}}.$$

Applying these relations to Equations (27), (28) and (20), remembering that $X_L = \omega L$ and $X_C = -1/(\omega C)$, it is possible to write

$$L'_{osc} = \frac{1}{2\pi f_{\max}}\frac{f_{\max}^2 - f_n^2}{f_n^2}\frac{V_{\min}^2}{|Q_n|}, \tag{29}$$

$$C^a_{osc} = \frac{1}{2\pi f_{\min}}\frac{f_{\min}^2}{f_n^2 - f_{\min}^2}\frac{Q_n}{V_n^2} = \frac{1}{2\pi}\frac{f_{\min}}{f_n^2 - f_{\min}^2}\frac{|Q_n|}{V_{\min}^2}. \tag{30}$$

As Equation (17) must also be satisfied with $\omega = 2\pi f_n$, it is not possible to satisfy Equations (29) and (30) simultaneously. However, C_{osc} must be maximized in order to minimize the filter pass band, given by

$$\beta = \frac{1}{R_{osc} C_{osc}} \,. \tag{31}$$

The lower the value of β, the lower is the voltage harmonic distortion due to the nonlinear function ϕ. C_{osc} is calculated again starting from Equation (17) using Equation (29). The result is given by

$$C_{osc}^{b} = \frac{1}{2\pi} \frac{f_{\max}}{f_{\max}^2 - f_n^2} \frac{|Q_n|}{V_{\min}^2} \,. \tag{32}$$

C_{osc} must be chosen from $\max\left\{C_{osc}^{a},\, C_{osc}^{b}\right\}$. Comparing Equations (30) and (32) it can be concluded that $C_{osc}^{b} > C_{osc}^{a}\ \forall f_n > 0$, so (32) should be used. Once C_{osc} has been determined, the value of L_{osc} can be calculated using

$$L_{osc} = \frac{1}{4\pi^2 f_n^2 C_{osc}} \,. \tag{33}$$

We can summarize the presented method as follows:

1. First, define the input parameters $V_{\min}$, $V_{\max}$, f_n, Δf, P_n, and Q_n. Acquire the parameters from the inverter and from the microgrid (the inverter rated power and grid voltage and frequency). Define the aspects of power quality for your application. What is the minimal and maximum voltage amplitude and frequency allowed on your system? A good starting point is to use $V_{\min} = 0.95V_n$ and $V_{\max} = 1.05V_n$ and $\Delta f = 0.5$ Hz. Besides, some regulation or standard can be used to drive this choice [33].
2. Use $V_{\min}$ to calculate λ with Equation (23).
3. Use $\kappa = V_{\min}/V_{\max}$ to calculate γ with Equation (25).
4. Use γ from previous step, P_n, and $V_{\min}$ to calculate α and R_{osc} with Equation (26).
5. Use $f_{\max} = f_n + \Delta f$, Q_n and V_{min} to calculate C_{osc} with Equation (32).
6. Use C_{osc} from previous step and f_n to calculate L_{osc} with Equation (33).

In Section 4, an example is presented with the actual parameters extracted from an implemented test setup.

3. A Pre-Synchronization Strategy

Although it is shown that devices operating with VOC will achieve synchronization asymptotically regardless of the initial condition, the insertion of a new inverter in the network may cause unwanted transients [17]. With that in mind, it is important to perform the pre-synchronization of a device before connection. As shown in the next section, pre-synchronization allows the oscillator output voltage to be close to the line voltage before closing the connection switch. The voltage at the end of the line, however, is out of phase with the voltage of the inverters already in operation due to the connection impendance. Only after closing the switch final synchronization will take place.

The pre-synchronization method that is used in this work was inspired by Johnson et al. [7]. When a VOC inverter, with parameters obtained as described in Section 2.2, is inserted into the network, even if with large amplitude and phase differences, it quickly adapts and synchronizes with the rest of the system. The pre-synchronization process takes advantage of this feature. Thus, it is only required to emulate the converter operation before the switch is effectively closed. For this purpose, the circuit shown in Figure 5 is emulated in the DSP that controls the converter.

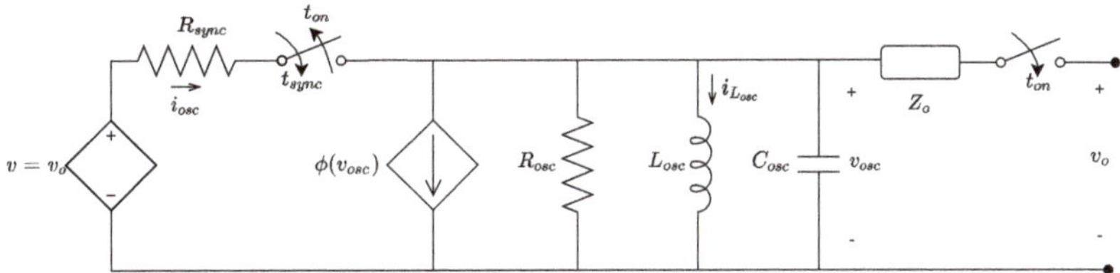

Figure 5. VOC schematic with pre-synchronization circuit.

As seen in the figure, a virtual source with a series resistor is used. The virtual source is a mirror of the voltage measured at the inverter terminals, where it will be connected. Therefore, the procedure consists of making the output current during the pre-synchronization period equal to

$$i_{osc} = \frac{v_o - v_{osc}}{R_{sync}}, \quad for\ t_{sync} < t < t_{on}. \tag{34}$$

The virtual resistor, R_{sync}, must have a resistance value of the order of the equivalent connection resistance. If this parameter is unknown, $R_{sync} = (V_{\min}^2/P_n)/100$ can be used. The obtained value has little influence on the pre-synchronization result as long as the procedure described in Section 2.2 is followed. In any case, actual synchronization will follow after connection.

4. Results

4.1. Simulation Results

The main purpose of this section is to compare the dead-zone VOC, and design procedure introduced in this work, with the methodology presented in Johnson et al. [21]. It will be shown that less harmonic distortion, especially third harmonic, can be obtained by following the tuning methodology presented here, which relies on the use of a piecewise-linear nonlinear function, in comparison to the approach relying on a cubic nonlinear function. In addition, the reduction of harmonic content is achieved without harming dynamic behavior. In fact, it is possible to notice a reduction in synchronization time between devices that use this new approach. The simulation platform MATLAB ® [34] was used along with Simulink ® to obtain the results presented in this section.

First, it is important to understand the source of the harmonics related to both methods. The cubic function employed in Johnson et al. [21], considering a static application, will always generate the same relative third harmonic content because

$$\sin(\theta)^3 = \frac{1}{4}[3\sin(\theta) + \sin(3\theta)]. \tag{35}$$

and the ratio of the third harmonic amplitude to the amplitude of the fundamental component is always the same. Let $\delta_{3:1}$ be the relation between the third harmonic and the fundamental amplitude regarding only the static behavior of the nonlinear function, for the cubic function $\delta_{3:1} = 1/3 \approx 33\%$. For a saturation function the harmonic generation depends on how much an input sine wave is distorted, and this is regulated by the λ parameter. Remembering, the saturation function is defined as

$$\phi(v_{osc}) = \begin{cases} -\alpha v_{osc}, & \text{if} \quad |v_{osc}| < \lambda\,, \\ -\alpha\lambda, & \text{if} \quad v_{osc} \geq \lambda\,, \\ \alpha\lambda, & \text{if} \quad v_{osc} \leq -\lambda\,. \end{cases} \tag{36}$$

The value assigned to λ from Section 2.2 is $V_{\min}\sqrt{2}$. The term $\sqrt{2}$ is used because λ is compared with v_{osc}, which is an instantaneous voltage. The minimum voltage is reached with nominal load, so in this condition $\phi(v_{osc}) \approx -\alpha\, v_{osc}$, and practically no harmonic distortion is generated by the oscillator. The worst case is the "no load" condition. In this scenario, the output voltage will be $V_{\max}$ and the nonlinear function will cut the sine wave above $V_{\min}\sqrt{2}$. These voltages, $V_{\min}$ and $V_{\max}$, are normally chosen to be close to the nominal voltage, thus only a small piece of the sine wave is removed. For saturation function the $\delta_{3:1} = \Phi_3/\Phi$, where $\Phi(a)$ is defined by Equation (11), and $\Phi_3(a)$ is defined as

$$\Phi_3(a) = \begin{cases} 0, & \text{if } 0 \leq a \leq \lambda\,, \\ \frac{4\alpha}{\pi a}\left[a\kappa^3 \cos\zeta + \frac{\lambda}{3}\cos 3\zeta\right], & \text{if } a > \lambda\,, \end{cases} \tag{37}$$

with $\zeta = \arcsin(\lambda/a)$. Using the parameters from Tables 1 and 2, the worst-case distortion rate for saturation function will be $\delta_{3:1} = 3.07\%$, that is, more than ten times better than the cubic function.

However, all harmonics generated by the nonlinear function are attenuated by the RLC filter in the oscillator. Finally, the total harmonic distortion is obtained from the combination between harmonic source and filter attenuation in a feedback system—the oscillator. Therefore, let $\delta_{3:1}^{ef}$ be the effective relation between the third harmonic and the fundamental amplitude. For a more detailed discussion we will present some simulation results. The simulation is configured such that there is only one oscillator connected, or not, to a load. The objective is to compare the harmonic distortion and frequency deviation obtained by using oscillators designed using our Proposed Method (PM) with oscillators designed using the Reference Method (RM) published in Johnson et al. [21]. The input parameters are taken from the RM, and they are replicated in Table 1.

Table 1. Input parameters from Reference Method (RM) [21].

Parameter	Value	Description
$V_{\max}$	126 V	Max. voltage
$V_{\min}$	114 V	Min. voltage
P_n	750 W	Inv. nominal power
Q_n	750 VAr	Allowed reactive power
f_n	60 Hz	Grid nominal frequency
Δf	0.5 Hz	Allowed freq. var.
$t_{rise}^{\max}$	0.2 s	Rise time
$\delta_{3:1}^{\max}$	2 %	Max. third-to-first harmonic ratio

Using these parameters on both selection guides generates the output parameters presented in Tables 2 and 3. The last two parameters from Table 1 are used only in the RM.

Table 2. Output parameters Proposed Method (PM).

Parameter	Value	Description
λ	161.220 V	Saturation limit
α	1.659 S	Saturation gain
R_{osc}	624.26 mΩ	Oscillator capacitance
C_{osc}	9.223 mF	Oscillator capacitance
L_{osc}	762.9 µH	Oscillator inductance

Table 3. Output parameters RM.

Parameter	Value	Description
κ_v	126 V/V	Voltage-scaling factor
κ_i	0.152 A/A	Current-scaling factor
σ	6.093 S	Conductance
α	4.062 A/V^3	Coefficient of cubic current source
C_{osc}	175.908 mF	Oscillator capacitance
L_{osc}	39.999 μH	Oscillator inductance

Note that the values obtained for the C_{osc} are quite different, and the value for the RM is approximately 19 times higher than for the PM. The larger capacitor in the RM is used to guarantee proper frequency regulation ([21] Equation (47)). With C_{osc} value, the effective third-to-first harmonic ratio can be calculated, $\delta_{3:1}^{ef} = 1.14\%$ for the RM ([21] Equation (41)). After the parameter definition, the two types of oscillators were implemented on simulation, on continuous and discrete-time domains. The integration step time for continuous-time simulation is $t_{solv} = (1/(48\,\text{kHz}))$ and $T_{sam} = (1/(24\,\text{kHz}))$ for the sample time in the discrete-time case. The simulation run for 100 s after voltage stabilization, and then the FFT algorithm was applied over the steady state portion of the data.

In the following, three simulation scenarios are presented, first with no load, second with nominal RL load, and third with nominal RC load. The results are respectively shown in Tables 4–6, where the acronyms "c." and "d." stand for continuous and discrete, respectively; "f" is the fundamental frequency; h_1 is the fundamental amplitude; and h_3 is the third harmonic amplitude.

Table 4. No load simulation.

Description	f [Hz]	h_1 [V]	h_3 [V]	$\delta_{3:1}^{ef}$ [%]
RM c.	59.97	177.8	2	1.12
RM d.	59.99	176.3	1.86	1.06
PM c.	59.99	176.1	0.88	0.5
PM d.	59.99	176.1	0.88	0.5

Table 5. Nominal RL load.

Description	f [Hz]	h_1 [V]	h_3 [V]	$\delta_{3:1}^{ef}$ [%]
RM c.	60.48	160.0	1.380	0.86
RM d.	60.5	160.3	1.240	0.77
PM c.	60.5	161.0	0.000	0.00
PM d.	60.5	162.1	0.015	0.01

Table 6. Nominal RC load.

Description	f [Hz]	h_1 [V]	h_3 [V]	$\delta_{3:1}^{ef}$ [%]
RM c.	59.48	159.4	1.380	0.90
RM d.	59.5	160.2	1.240	0.91
PM c.	59.5	161.6	0.010	0.01
PM d.	59.5	160.2	0.003	0.00

Note that in all cases the PM was better than the RM in terms of $\delta_{3:1}^{ef}$. In addition, among the investigated operating conditions, the RM had a small $\delta_{3:1}^{ef}$ change, unlike the PM. This was expected, because the cubic function always generates the same $\delta_{3:1} = 1/3$, no matter the voltage amplitude. The $\delta^{ef}_{3:1}$ calculated using Equation (41) from [21], is in accordance with the results. Moreover, the RM has a little, but noticeable, error on frequency regulation, and they are different for continuous and discrete implementation. To verify $\delta^{ef}_{3:1}$ changing behavior related to load condition, another simulation was performed, now with 50% of nominal RL load, and the result is presented in Table 7. There seems to be an almost linear relationship between $\delta^{ef}_{3:1}$ and the load percentage when the PM is applied.

Table 7. RL load 50% of nominal, (*) using C_{osc} and L_{osc} from RM.

Description	f [Hz]	h_1 [V]	h_3 [V]	$\delta^{ef}_{3:1}$ [%]
RM c.	60.23	170.0	1.629	0.96
RM d.	60.25	170.0	1.763	1.04
PM c.	60.25	170.7	0.463	0.27
PM d.	60.25	170.8	0.468	0.27
PM c.*	60.01	171,5	0.011	0.01
PM d.*	60.01	171,0	0.011	0.01

The last two lines in Table 7 are related to the PM when C_{osc} and L_{osc} parameters from the RM are used. These values give a view of the impact of the oscillator structure on frequency regulation and harmonics generation. However, the higher the capacitance value and the lower the inductance, the more energy is stored in the oscillator. This will potentially impact the synchronization procedure. If there is a difference between the states of each oscillator, more energy needs to be dissipated in the network, and more time could be required for the synchronization. Even so, in the PM one has a considerable margin to increase the capacitance value when compared to the RM if some other performance target needs to be met. This gives space for future improvements in this designing guide.

In the sequence, a comparison between the PM and RM synchronization process will be presented. For the following simulations, two identical oscillators, with parameters from Tables 2 and 3, are connected to an RL load, with 50% of nominal power, through two identical lines with $R_{line} = 1\ \Omega$ and $L_{line} = 2$ mH, as shown in Figure 1. The oscillators are constructed in discrete-time domain and the connection network in continuous-time domain, with $t_{solv} = (1/(48\,\text{kHz}))$ and $T_{sam} = (1/(24\,\text{kHz}))$. Initially, only one oscillator is connected to the load, and after some time the second oscillator is added to the system. For the results presented in Figures 6 and 7, the pre-synchronization procedure is not used. Instead, the initial conditions of each oscillator are calculated to generate a lag of 1° between then. When the clock reaches 10 ms the second oscillator is connected to the load through the RL line. As can be seen the oscillators behave similarly for the PM and RM.

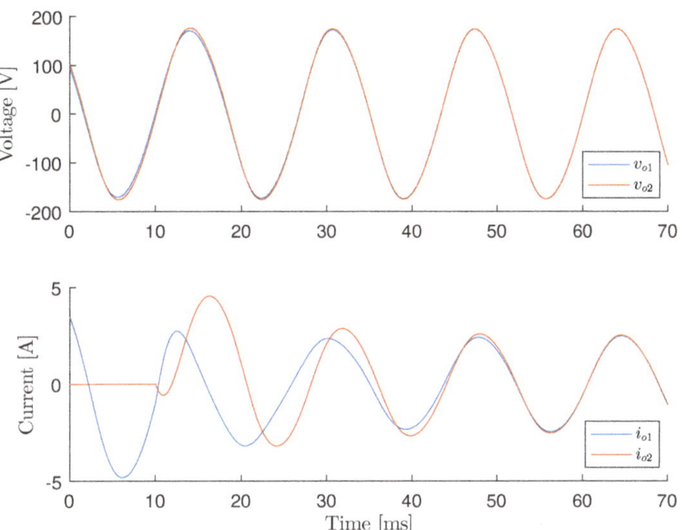

Figure 6. Synchronization of two oscillators designed with RM without pre-synchronization.

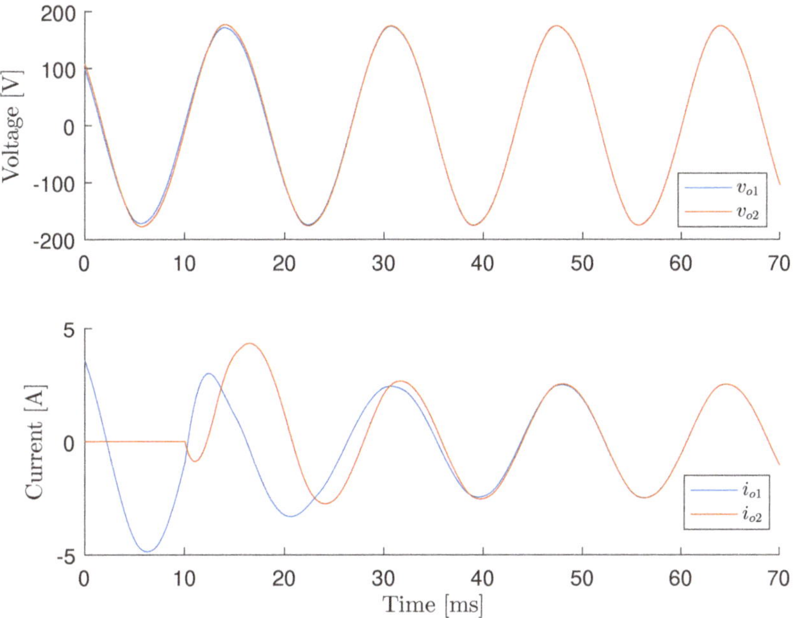

Figure 7. Synchronization of two oscillators designed with PM without pre-synchronization.

For a more detailed analysis, the graph of Figure 8 was constructed, where the difference between the output voltages and output currents for each method is compared.

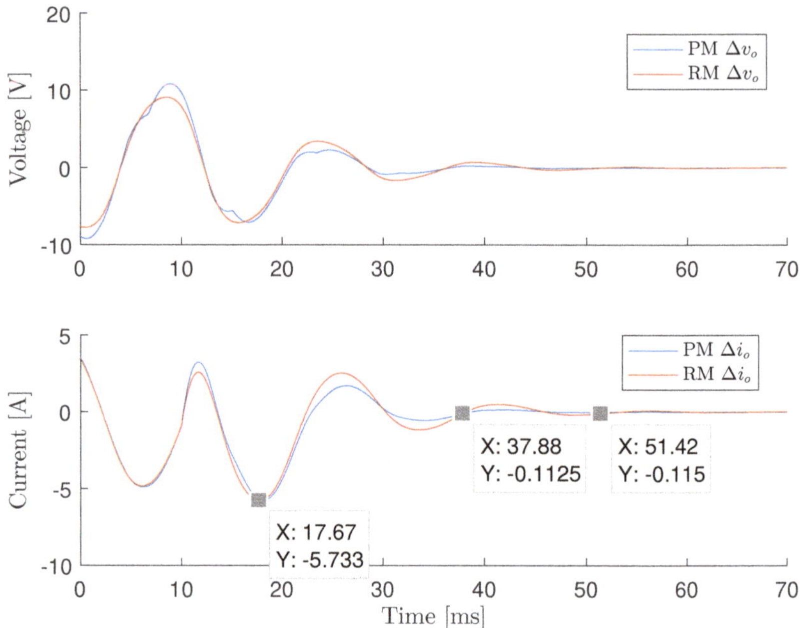

Figure 8. Difference between output voltage and output current of each oscillator without pre-synchronization.

As can be seen in Figure 8, the PM reaches synchronization between oscillators faster than RM. Defining the settling time, t_{set}, as the time it takes for a signal to fall below 2% of its peak value, for PM $t_{set} \approx 27.9$ ms and for RM $t_{set} \approx 41.4$ ms, considering Δi_o signal.

In order to check the total time of an oscillator on the network for entering into service, the next simulation also takes into account the pre-synchronization process following the technique presented in Section 3 with $R_{sync} = (V_{\min}^2/P_n)/100 = 173.3$ mΩ. The results are shown in Figures 9–11. The initial conditions of each oscillator are calculated to generate a lag of 90° between then. When the clock reaches 5 ms, the pre-synchronization process starts in the second oscillator, and at 30 ms mark it is connected to the load through the RL line. The pre-synchronization quickly changes the output voltage of the second oscillator for both methods, as seen in Figures 9 and 10. Besides, the output current of oscillator increases smoothly after connection.

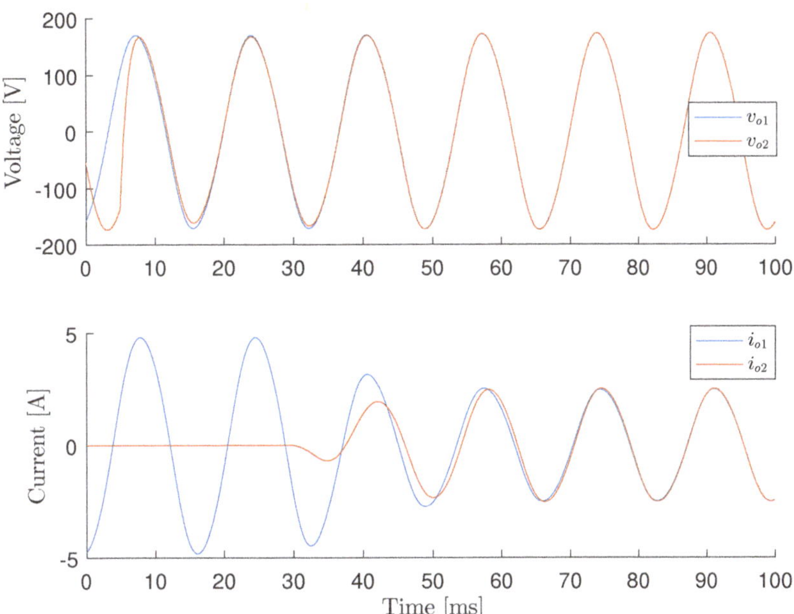

Figure 9. Synchronization of two oscillators designed with RM with pre-synchronization.

From Figure 11 it is possible to measure the settling time when the pre-synchronization is applied. In this case, this value will give an overview of the pre-synchronization and synchronization efficacy related with the VOC method applied. As in the previous case, the synchronization process for the PM is faster than in the RM. The values are $t_{set} \approx 26.4$ ms for PM and $t_{set} \approx 42.5$ ms for RM. For a quick comparison, Table 8 was constructed with the most important results presented in this section.

Table 8. Results Summary

Description	PM	RM	Units
Worst $\delta_{3:1}^{ef}$	0.5	1.12	%
Best $\delta_{3:1}^{ef}$	0	0.77	%
No pre-sync. t_{set}	27.9	41.4	ms
With pre-sync. t_{set}	26.4	42.5	ms

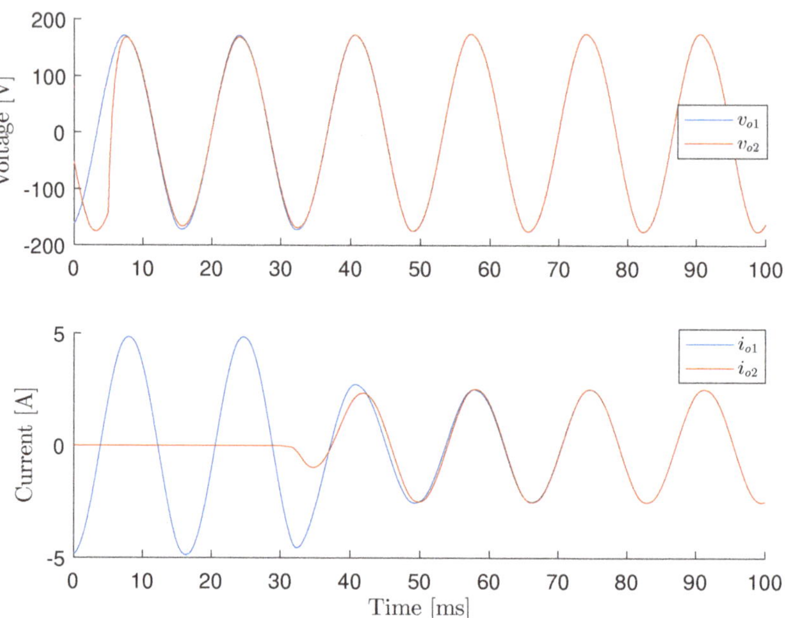

Figure 10. Synchronization of two oscillators designed with PM with pre-synchronization.

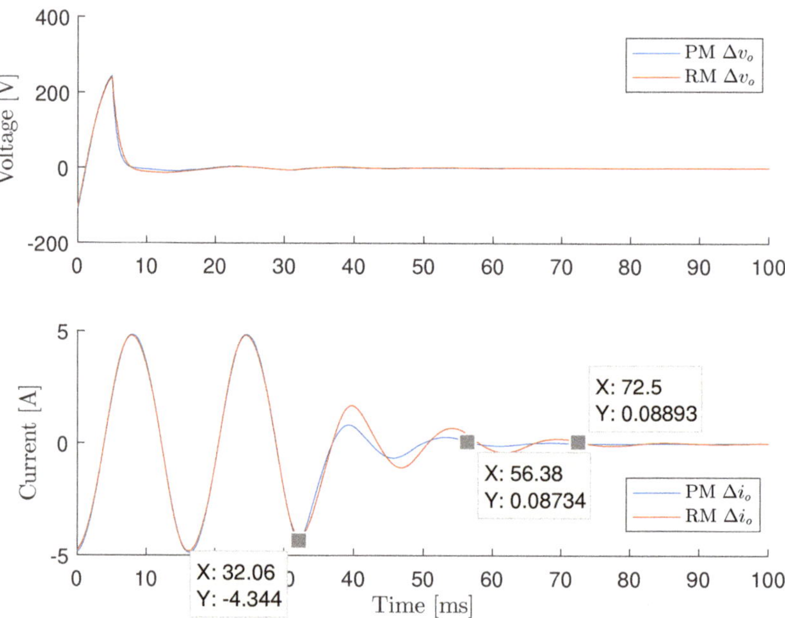

Figure 11. Difference between output voltage and output current of each oscillator with pre-synchronization.

4.2. Experimental Results

The VOC method was embedded in a single-phase commercial PV inverter [35] by firmware changing. The simplified schematic for the experiment is shown in Figure 12 and the assembly can be seen in Figure 13. The inverter output is connected to a coupling transformer, with a 1:1 ratio, that provides galvanic isolation between the input and output. The switch s_3 is used only for pre-synchronization strategy validation. When s_3 is closed, Inverter 2 automatically detects terminal voltage and starts the pre-synchronization procedure. Currents were measured with Hall effect current probes, while voltages were measured with differential probes. Data were stored in a four-channel digital scope, without any filtering. For reference, all relevant parameters are listed in Table 9. The two inverters have the same part numbers and are based on a H-brige IGBT (Insulated-gate Bipolar Transistor) topology with hard switching. The inductor and capacitor filters have 10% and 20% tolerances, respectively.

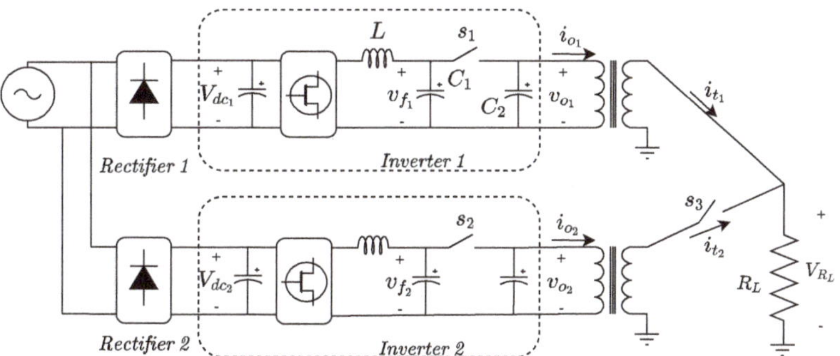

Figure 12. Schematic of the laboratory-scale prototype for testing the VOC.

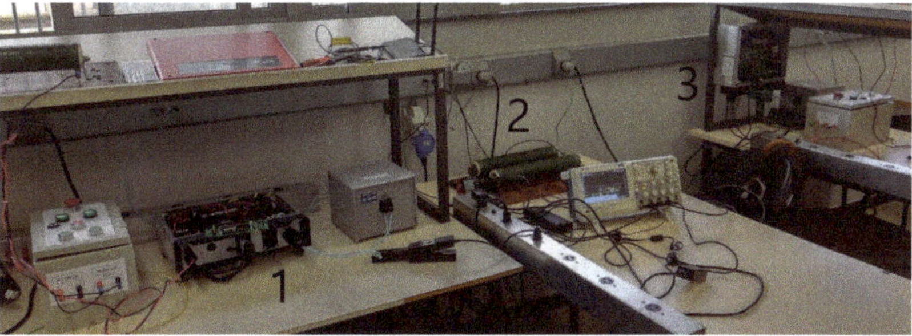

Figure 13. Picture of the inverter assembly for testing the VOC algorithm: (1) Inverter n°. 1, (2) Set of resistive loads, and (3) Inverter n°. 2.

The commands are made with triangular carrier unipolar PWM ($f_{pwm} = f_s$). The duty cycle is defined as

$$d = v_{osc}/V_{dc}. \tag{38}$$

All the calculations were made with fixed point math. To simplify implementation, the per unit system was used. Thus, before using the equations in Section 2.2, the input data must be converted using voltage and power bases. The used VOC parameters are presented in Table 9.

Table 9. Parameters of the experimental setup.

Parameter	Value	Description
V_n	127 V	Grid nominal voltage
f_n	60 Hz	Grid nominal frequency
L	2 mH	Filter inductance
R_{fl}	1 Ω	Filter inductor resistance
$C_1 = C_2$	3.3 µF	Filter capacitance
R_{fc}	10 mΩ	Filter capacitor ESR (60 Hz)
R_{w1}	919 mΩ	T. winding resistance
X_{w1}	737 mΩ	T. winding reactance
R_{m1}	2.83 kΩ	T. magnetizing resistance
X_{m1}	625 Ω	T. magnetizing reactance
R_{w2}	906 mΩ	T. winding resistance
X_{w2}	655 mΩ	T. winding reactance
R_{m2}	3.21 kΩ	T. magnetizing resistance
X_{m2}	863 Ω	T. magnetizing reactance
R_L	25 Ω	Load Resistance
V_b	200 V	Voltage base (PU)
P_b	4 kW	Power base (PU)
$V_{\max}$	1.05 V_n	Max. voltage
$V_{\min}$	0.95 V_n	Min. voltage
Δf	0.3 Hz	Allowed freq. var.
P_n	1.5 kW	Inv. nominal power
Q_n	300 VAr	Allowed reactive power
λ	0.853 V	VOC ϕ limit
α	29.634 S	VOC ϕ gain
R_{osc}	34.961 mΩ	VOC resistance
C_{osc}	109.473 mF	VOC capacitance
L_{osc}	64.273 µH	VOC inductance
f_s	24 kHz	Sample frequency

The values $V_{\max}$ and $V_{\min}$ are chosen symmetrically to V_n as recommended on Section 2.2. The rated power of this inverter is $P_n = 1.5$ kW. As the experiments will be performed with a resistive load, the Q_n values are chosen only to overcome the reactance from the transformers. A good starting point for Δf is 0.5 Hz, but that limit could vary regarding some standard or equipment needs. We are choosing this parameter more tightly, $\Delta f = 0.3$ Hz, because we will deliver little reactive power. For the same Q_n value, a more restricted Δf will increase C_{osc}, that normally leads to a longer synchronization time, regarding the same system. However, a bigger capacitance will help to reject more harmonics, because it reduces β.

The test was conducted as follows. Initially, Inverter 1 was started. After stabilization, it was connected to the microgrid by closing switch s_1. Then, switch s_3 was closed. The pre-synchronization procedure started automatically as soon as the rms value of v_{o2} exceeded 80 V. Subsequently, switch s_2 was closed to put inverter 2 into operation. After operating for a short period, switch s_2 was opened again so that Inverter 1 took over the entire load again. Results related to each of these steps are presented next. It must be noted that both inverters have the same firmware.

The first results are shown in Figure 14. This is related to pre-synchronization process of Inverter 2. As it can be seen, after a small transient the voltage in the inverter filter, v_{f2} , which is an image of VOC output voltage, replicates v_{o2}, except for a small lag. As will be seen below, this procedure ensures that the input transients of the inverters are well controlled.

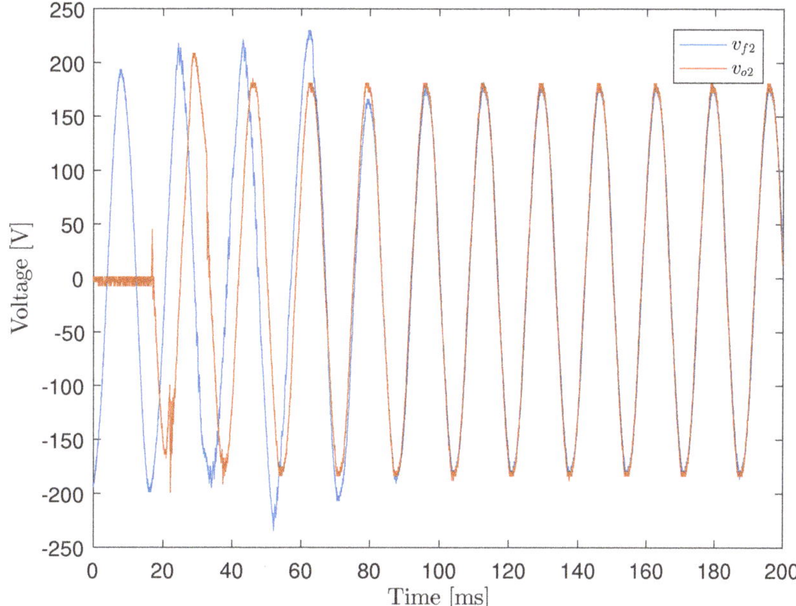

Figure 14. Pre-synchronization process for Inverter 2.

Figure 15 shows the moment when the second inverter is connected to the microgrid. The measured voltages are v_{o1} and v_{o2}, as well as currents i_{o1} and i_{o2} (see Figure 12).

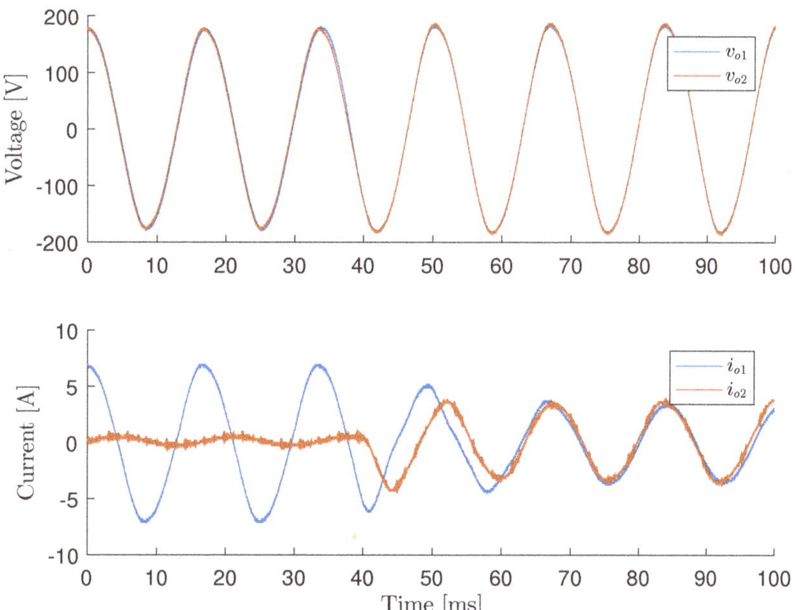

Figure 15. Connection of the second inverter in the microgrid.

As seen in Figure 15, current i_{o2} quickly synchronizes with i_{o1}. In addition, a soft transient is achieved due to the pre-synchronization procedure. The difference between i_{o1}

and i_{o2} is due to difference in transformers impedance. The more noticeable noise presented in i_{o2} is related to the higher sensitivity of the current probe 2. This noise comes mainly from common-mode electromagnetic interference generated by the inverter switching.

After some time in operation, switch s_2 was opened. The result is shown in Figure 16.

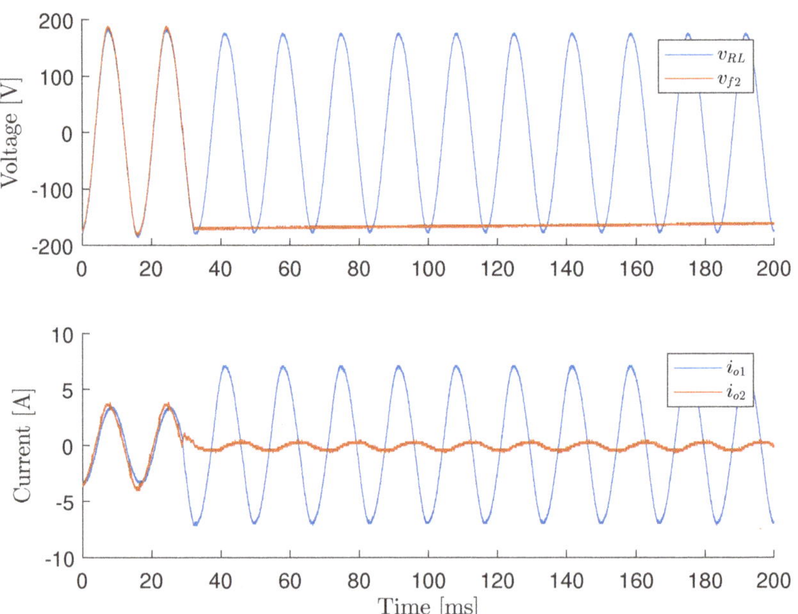

Figure 16. Disconnection of the second inverter from microgrid.

The measured voltages are v_{f2} and v_{RL}. Once requested to disconnect, the inverter goes to a state where the PWM is deactivated, so the filter voltage after this event will tend to zero. The capacitor discharge, compared to the network frequency, is slow. This ends up generating the "continuous" signal seen in the graph of Figure 16. Another point to be highlighted is the reduction of the voltage in the load after the Inverter 2 is interrupted. This reduction is due to both the increase in losses in Inverter 1 and transformer and the reduction in the reference voltage generated by the virtual oscillator. Although the values of $V_{\min}$ and $V_{\max}$ were chosen symmetrically in relation to V_n, this is not mandatory. The only restriction is that $V_{\max} > V_{\min}$. In this way the value of $V_{\min}$ can be chosen to compensate for possible losses in the inverter filter, if this is relevant in a specific application.

5. Conclusions

In this paper, a new method is proposed for determining the design parameters required for virtual oscillator control. In the the proposed method, all VCO parameters are directly computed with simple expressions that use as input data information that is readily available from the inverter and from the microgrid. The proposed method was implemented in a pair of converters that were set to operate as a microgrid in the laboratory. The obtained results demonstrate the validity of the method by investigating both the pre-synchronization process and the cooperative operation of both inverters in different operating conditions. The pre-synchronization ensures that the transients resulting from the insertion of a second inverter on an operating microcrid are smooth. This is essential for commercial applications, reducing the downtime of the device. It is also shown that the disconnection of inverter 2 does not create any trouble for the operation of inverter 1, which ultimately assumes all power that is delivered to the loads after disconnection. In future work, other scenarios need to be investigated, as well as pre-synchronization and cooperative operation. In addition, as presented in the results, the proposed method has potential advantages over the technique established in the literature. Following the design guide presented here, the dead-zone type VOC generates less third harmonic distortion and needs less time to synchronize when connected to a network.

Author Contributions: Conceptualization, D.A.C. and L.A.B.T.; methodology, D.A.C.; Resources D.I.B. and S.M.S.; Software, D.A.C.; validation, D.A.C., S.M.S., and D.I.B.; writing—original draft preparation, D.A.C., L.A.B.T., S.M.S., A.D.C., and D.I.B.; writing—review and editing, D.A.C., A.D.C., L.A.B.T., and S.M.S. All authors have read and agreed to the published version of the manuscript.

Funding: This research was funded by the Brazilian government agencies CAPES, FAPEMIG, and CNPQ.

Institutional Review Board Statement: Not applicable.

Informed Consent Statement: Not applicable.

Acknowledgments: The authors thank the Graduate Program in Electrical Engineering (PPGEE-UFMG) for supporting this research.

Conflicts of Interest: The authors declare no conflicts of interest.

Appendix A. Incremental Passivity

The notion of *incremental* passivity is closely related to the usual concept of passivity representing relations among inputs, outputs, and storage functions depending on system states [32]. The incremental notion is instead associated with *differences* between inputs, outputs, and trajectories in state space. According to the work in [16], a generic dynamical system represented by

$$\begin{aligned} \dot{\vec{x}} &= f(\vec{x}, \vec{u}), \\ \vec{y} &= h(\vec{x}, \vec{u}), \end{aligned} \tag{A1}$$

with inputs $\vec{u} \in \mathbb{R}^m$, outputs $\vec{y} \in \mathbb{R}^m$, and states $\vec{x} \in \mathbb{R}^n$, is said to be Incrementally Input Strictly Passive (IISP) (resp. Incrementally Output Strictly Passive (IOSP)) if there exists a positive definite incremental storage function $S_\Delta^{\mathrm{ISP}}(\Delta\vec{x})$ and passivity characteristic $\eta > 0$ (resp. $S_\Delta^{\mathrm{OSP}}(\Delta\vec{x})$ and passivity characteristic $\alpha_p > 0$) such that the following respective inequalities hold, $\forall t \geq 0$:

$$\begin{aligned} \dot{S}_\Delta^{\mathrm{ISP}}(\Delta\vec{x}) \leq & -\eta \left\| \vec{u}^1(t) - \vec{u}^2(t) \right\|^2 \\ & + \left[\vec{u}^1(t) - \vec{u}^2(t) \right]^\top \left[\vec{y}^1(t) - \vec{y}^2(t) \right], \\ \dot{S}_\Delta^{\mathrm{OSP}}(\Delta\vec{x}) \leq & -\alpha_p \left\| \vec{y}^1(t) - \vec{y}^2(t) \right\|^2 \\ & + \left[\vec{u}^1(t) - \vec{u}^2(t) \right]^\top \left[\vec{y}^1(t) - \vec{y}^2(t) \right], \end{aligned}$$

with $\Delta\vec{x} = \vec{x}(t; \vec{x}_0^1, \vec{u}^1) - \vec{x}(t; \vec{x}_0^2, \vec{u}^2)$ such that the time-derivatives in the left-hand side are taken along the difference between the solutions $\vec{x}(t; \vec{x}_0^1, \vec{u}^1)$ and $\vec{x}(t; \vec{x}_0^2, \vec{u}^2)$ to (A1) starting, respectively, at $\vec{x}(0) = \vec{x}_0^1$ with input $\vec{u}^1(t)$ and output $\vec{y}^1(t)$, and at $\vec{x}(0) = \vec{x}_0^2$ with input $\vec{u}^2(t)$ and output $\vec{y}^2(t)$. Note that passivity of LTI systems implies the corresponding incremental passivity property, but this is not true for nonlinear systems.

One way to prove that the VOC method applied to power oscillators connected to symmetric networks will lead to the synchronization of the oscillators is to show that the differences between state trajectories from different oscillators will vanish asymptotically because of their incremental passivity properties, and the fact that they are in feedback with the incrementally passive interconnection electrical network. This gives an overall incrementally strictly passive system. A more in depth discussion of this topic can be found in Torres et al. [16,17].

References

1. Chiradeja, P. Benefit of Distributed Generation: A Line Loss Reduction Analysis. In Proceedings of the IEEE/PES Transmission & Distribution, Conference & Exposition: Asia and Pacific, Dalian, China, 18 August 2005, doi:10.1109/tdc.2005.1546964.
2. Huang, B.B.; Xie, G.H.; Kong, W.Z.; Li, Q.H. Study on smart grid and key technology system to promote the development of distributed generation. In Proceeding sof the IEEE PES Innovative Smart Grid Technologies, Tianjin, China, 21–24 May 2012, doi:10.1109/isgt-asia.2012.6303265.
3. Ritchie, H.; Roser, M. Renewable Energy Charts. Available online: https://ourworldindata.org/renewable-energy (accessed on 15 October 2020).
4. Li, Z.; Zang, C.; Zeng, P.; Yu, H.; Li, S. Fully Distributed Hierarchical Control of Parallel Grid-Supporting Inverters in Islanded AC Microgrids. *IEEE Trans. Ind. Inform.* **2018**, *14*, 679–690, doi:10.1109/tii.2017.2749424.
5. Alberto, L.; Silva, F.; Bretas, N. Direct methods for transient stability analysis in power systems: state of art and future perspectives. In Proceedings of the IEEE Porto Power Tech Proceedings (Cat. No.01EX502), Porto, Portugal, 10–13 September 2001, doi:10.1109/ptc.2001.964750.
6. Borrega, M.; Marroyo, L.; Gonzalez, R.; Balda, J.; Agorreta, J.L. Modeling and Control of a Master–Slave PV Inverter With N-Paralleled Inverters and Three-Phase Three-Limb Inductors. *IEEE Trans. Power Electron.* **2013**, *28*, 2842–2855, doi:10.1109/tpel.2012.2220859.
7. Johnson, B.B.; Dhople, S.V.; Hamadeh, A.O.; Krein, P.T. Synchronization of Parallel Single-Phase Inverters With Virtual Oscillator Control. *IEEE Trans. Power Electron.* **2014**, *29*, 6124–6138, doi:10.1109/tpel.2013.2296292.
8. Shanxu, D.; Yu, M.; Jian, X.; Yong, K.; Jian, C. Parallel operation control technique of voltage source inverters in UPS. In Proceedings of the IEEE 1999 International Conference on Power Electronics and Drive Systems. PEDS 99 (Cat. No.99TH8475), Hong Kong, China, 27–29 July 1999, doi:10.1109/peds.1999.792823.
9. Patrick O'Connor, A.V.K. *Practical Reliability Engineering*; John Wiley and Sons Ltd.: Hoboken, NJ, USA, 2012.
10. Vaidya, S.; Somalwar, R.; Kadwane, S.G. Review of various control techniques for power sharing in microgrid. In Proceedings of the 2016 International Conference on Global Trends in Signal Processing, Information Computing and Communication (ICGTSPICC), Jalgaon, India, 22–24 December 2016, doi:10.1109/icgtspicc.2016.7955341.
11. Chandorkar, M.; Divan, D.; Adapa, R. Control of parallel connected inverters in stand-alone AC supply systems. In Proceedings of the Conference Record of the 1991 IEEE Industry Applications Society Annual Meeting, Dearborn, MI, USA, 28 September–4 October 1991, doi:10.1109/ias.1991.178359.
12. De Souza Azevedo, G.M. Controle e Operação de Conversores em Microrredes. Ph.D. Thesis, Univerdade Federal de Pernanbuco, Recife, Brazil, 2011.
13. Guerrero, J.; GarciadeVicuna, L.; Matas, J.; Castilla, M.; Miret, J. Output Impedance Design of Parallel-Connected UPS Inverters With Wireless Load-Sharing Control. *IEEE Trans. Ind. Electron.* **2005**, *52*, 1126–1135, doi:10.1109/tie.2005.851634.
14. Andrade, E.T.; Ribeiro, P.E.M.J.; Pinto, J.O.P.; Chen, C.L.; Lai, J.S.; Kees, N. A novel power calculation method for droop-control microgrid systems. In Proceedings of the 2012 Twenty-Seventh Annual IEEE Applied Power Electronics Conference and Exposition (APEC), Orlando, FL, USA, 5–9 February 2012, doi:10.1109/apec.2012.6166136.
15. Wu, H.; Wang, X. Design-Oriented Transient Stability Analysis of PLL-Synchronized Voltage-Source Converters. *IEEE Trans. Power Electron.* **2020**, *35*, 3573–3589, doi:10.1109/tpel.2019.2937942.
16. Torres, L.A.B.; Hespanha, J.P.; Moehlis, J. Power supply synchronization without communication. In Proceedings of the 2012 IEEE Power and Energy Society General Meeting, San Diego, CA, USA, 22–26 July 2012, doi:10.1109/pesgm.2012.6344905.
17. Torres, L.A.B.; Hespanha, J.P.; Moehlis, J. Synchronization of Identical Oscillators Coupled Through a Symmetric Network With Dynamics: A Constructive Approach With Applications to Parallel Operation of Inverters. *IEEE Trans. Autom. Control* **2015**, *60*, 3226–3241, doi:10.1109/tac.2015.2418400.
18. De Azevedo Ávila, M. Operação em Paralelo e sem Comunicação de Sistemas UPS: Uma abordagem baseada em Passividade. Master's Thesis, Universidade Federal de Minas Gerais, Belo Horizonte, Brazil, 2013.
19. Johnson, B.B.; Dhople, S.V.; Hamadeh, A.O.; Krein, P.T. Synchronization of Nonlinear Oscillators in an LTI Electrical Power Network. *IEEE Trans. Circuits Syst. I: Regul. Pap.* **2014**, *61*, 834–844, doi:10.1109/tcsi.2013.2284180.
20. Sinha, M.; Dorfler, F.; Johnson, B.B.; Dhople, S.V. Uncovering Droop Control Laws Embedded Within the Nonlinear Dynamics of Van der Pol Oscillators. *IEEE Trans. Control Netw. Syst.* **2017**, *4*, 347–358, doi:10.1109/tcns.2015.2503558.
21. Johnson, B.B.; Sinha, M.; Ainsworth, N.G.; Dorfler, F.; Dhople, S.V. Synthesizing Virtual Oscillators to Control Islanded Inverters. *IEEE Trans. Power Electron.* **2016**, *31*, 6002–6015, doi:10.1109/tpel.2015.2497217.
22. Seo, G.S.; Colombino, M.; Subotic, I.; Johnson, B.; Gros, D.; Dorfler, F. Dispatchable Virtual Oscillator Control for Decentralized Inverter-dominated Power Systems: Analysis and Experiments. In Proceeding sof the 2019 IEEE Applied Power Electronics Conference and Exposition (APEC), Anaheim, CA, USA, 17–21 March 2019, doi:10.1109/apec.2019.8722028.
23. Gros, D.; Colombino, M.; Brouillon, J.S.; Dorfler, F. The Effect of Transmission-Line Dynamics on Grid-Forming Dispatchable Virtual Oscillator Control. *IEEE Trans. Control Netw. Syst.* **2019**, *6*, 1148–1160, doi:10.1109/tcns.2019.2921347.
24. Colombino, M.; Groz, D.; Brouillon, J.S.; Dorfler, F. Global Phase and Magnitude Synchronization of Coupled Oscillators With Application to the Control of Grid-Forming Power Inverters. *IEEE Trans. Autom. Control* **2019**, *64*, 4496–4511, doi:10.1109/tac.2019.2898549.

25. Lu, M.; Dutta, S.; Purba, V.; Dhople, S.; Johnson, B. A Grid-compatible Virtual Oscillator Controller: Analysis and Design. In Proceedings of the 2019 IEEE Energy Conversion Congress and Exposition (ECCE), Baltimore, MD, USA, 29 September–3 October 2019, doi:10.1109/ecce.2019.8913128.
26. Awal, M.A.; Husain, I. Unified Virtual Oscillator Control for Grid-Forming and Grid-Following Converters. *IEEE J. Emerg. Sel. Top. Power Electron.* **2020**, 1–1, doi:10.1109/JESTPE.2020.3025748.
27. Ali, M.; Nurdin, H.I.; Fletcher, J. Dispatchable Virtual Oscillator Control for Single-Phase Islanded Inverters: Analysis and Experiments. *IEEE Transactions on Industrial Electronics* **2020**, *68*, 4812–4826, doi:10.1109/tie.2020.2991996.
28. Lin, J. Virtual Oscillator Control of Distributed Power Filters for Selective Ripple Attenuation in DC Systems. *IEEE Trans. Power Electron.* **2021**, *36*, 8552–8560, doi:10.1109/TPEL.2021.3054672.
29. Duarte, J.L.; Roes, M.; Xu, X.; Wang, F. Virtual Oscillator Control as Applied to DC Microgrids with Multiple Sources. In Proceedings of the 2020 IEEE 9th International Power Electronics and Motion Control Conference (IPEMC2020-ECCE Asia), Nanjing, China, 31 May–3 June 2020; pp. 2007–2011, doi:10.1109/IPEMC-ECCEAsia48364.2020.9368043.
30. Johnson, B.B.; Dhople, S.V.; Cale, J.L.; Hamadeh, A.O.; Krein, P.T. Oscillator-Based Inverter Control for Islanded Three-Phase Microgrids. *IEEE J. Photovoltaics* **2014**, *4*, 387–395, doi:10.1109/jphotov.2013.2280953.
31. Rosse, A.; Denis, R.; Zakhour, C. Control of parallel inverters using nonlinear oscillators with virtual output impedance. In Proceedings of the 2016 18th European Conference on Power Electronics and Applications (EPE16 ECCE Europe), Karlsruhe, Germany, 5–9 September 2016, doi:10.1109/epe.2016.7695112.
32. Khalil, H.K. *Nonlinear Systems;* Prentice Hall: Upper Saddle River, NJ, USA, 2002.
33. IEEE . *IEEE Standard for Interconnection and Interoperability of Distributed Energy Resources with Associated Electric Power Systems Interfaces;* IEEE Std 1547-2018 (Revision of IEEE Std 1547-2003); IEEE: Piscataway, NJ, USA, 2018; pp. 1–138, doi:10.1109/IEEESTD.2018.8332112.
34. *MATLAB Version 9.0.0.341360 (R2016a)*; The Mathworks, Inc.: Natick, MA, USA, 2016.
35. *Inverter Model: PHB 1500NS, Processing Unit: TMS320F28034*; PHB Eletronica Ltda.: São Paulo - SP, Brazil, 2018

Article

Two-Stage Optimal Microgrid Operation with a Risk-Based Hybrid Demand Response Program Considering Uncertainty

Ho-Sung Ryu and Mun-Kyeom Kim *

School of Energy System Engineering, Chung-Ang University, 84 Heukseok-ro, Dongjak-gu, Seoul 06974, Korea; ghtjd0580@cau.ac.kr

* Correspondence: mkim@cau.ac.kr; Tel.: +82-2-5271-5867

Received: 7 October 2020; Accepted: 13 November 2020; Published: 19 November 2020

Abstract: Owing to the increasing utilization of renewable energy resources, distributed energy resources (DERs) become inevitably uncertain, and microgrid operators have difficulty in operating the power systems because of this uncertainty. In this study, we propose a two-stage optimization approach with a hybrid demand response program (DRP) considering a risk index for microgrids (MGs) under uncertainty. The risk-based hybrid DRP is presented to reduce both operational costs and uncertainty effect using demand response elasticity. The problem is formulated as a two-stage optimization that considers not only the expected operation costs but also risk expense of uncertainty. To address the optimization problem, an improved multi-layer artificial bee colony (IML-ABC) is incorporated into the MG operation. The effectiveness of the proposed approach is demonstrated through a numerical analysis based on a typical low-voltage grid-connected MG. As a result, the proposed approach can reduce the operation costs which are taken into account uncertainty in MG. Therefore, the two-stage optimal operation considering uncertainty has been sufficiently helpful for microgrid operators (MGOs) to make risk-based decisions.

Keywords: two-stage optimization; risk-based hybrid demand response; uncertainties; conditional value at risk; improved multi-layer artificial bee colony algorithm

1. Introduction

In a smart-grid environment, distributed energy resources (DERs), such as renewable energy, have gained more attention than traditional power generation units owing to the increasing trend to address environmental concerns [1,2]. Some of the most widely used renewable energy resources are wind turbine (WT) and photovoltaic (PV) systems because of the feasibility of existing related technologies. However, these forms of DERs are dependent on fluctuations and the unpredictive nature of wind and solar resources. Since the difficulty in managing distributed uncertainties by conventional power systems, microgrids (MGs) that can usefully control DERs have been introduced as a new concept [3]. MGs manage a cluster of loads and DERs, operating as a control to offer power locally. MGOs should be able to confirm the reliability of systems considering uncertain risk indices. In this regard, the accurate evaluation of MGs is a challenging task due to the uncertainties inherent in renewable energy. Moreover, methods are needed to mitigate the effects of uncertainty when scheduling optimal operations [4]. In this situation, using flexible energy, such as demand response (DR), can provide the required demand control for the system and can be used reliably in a relatively short time under the same conditions [5]. Thus, the DR program (DRP) for power systems is expected to advance steadily [6]. This is because power system infrastructure is focused on both stability and economics, and the DRP is a flexible and inexpensive resource for operating a system. MGOs can use the DRP to reduce the peak load, save on the power reserve, and ensure power reliability. The operators can also encourage customers to use less power during periods when demand reduction is required owing to uncertainties. Thus, customers engage in contracts with MGOs to reduce demand when requested, and MGOs offer incentive costs to customers that reflect the amount of demand reduced, enabling contracts to be maintained.

Various approaches have been used recently to address uncertain problems inherent in MG operation. To minimize operational costs under a deterministic and probabilistic environment, a stochastic approach for MG operation using energy storage system (ESS) was proposed in Reference [7]. In Reference [8], optimal planning for interconnected MGs under uncertainty in large-scale distribution systems was presented to improve reliability and economics. Further, to improve system operation and management efficiency, a stochastic resource planning strategy for MGs was introduced in Reference [9] to optimally manage resources required for both the generation and demand sides. In Reference [10], both the power generation of each unit and the exchange with upstream networks were assessed through the optimal operation of MGs. Artificial intelligence algorithms were used to forecast wind speeds and optimal set volumes for DERs, and ESS capacity was determined based on forecasted data to optimize the total operating costs by Motevasel et al. [11]. Moreover, a probabilistic methodology of uncertainties caused by wind and solar generation and load consumption for estimating spinning reserve requirement was presented in Reference [12]. The DRP was considered to control the frequency of a smart MG with renewable generation, and mixed-integer linear programming was used to solve the proposed model for wind-power generation under uncertainty [13]. A stochastic planning approach was suggested to model the probabilistic behavior of wind and DRs in an energy market [14]. The authors in Reference [15] studied the effects of demand-side management of appliances on the reliability, loss, and voltage profiles of power systems, and customer comfort was also considered. Using a risk-based stochastic optimization framework, the minimum required ESS to secure the desired voltage stability margin for distribution systems was computed by Jalali et al. [16]. The authors in Reference [17] presented a flexible risk control strategy with an ESS to help remedy the removal of line overload in post-contingency situations. To minimize operating costs of MGs using the genetic algorithm, the optimal energy and power capacity of energy storage systems were determined [18]. The operational costs of MGOs were reduced by selling remaining energy at a high load level by Samadi et al. [19]. This study addressed a daily power prediction module to provide MGs with solar power output data for DER scheduling. However, the reserve for compensating wind and PV power fluctuations was not considered within the daily DER schedule. The authors in Reference [20] solved the optimal power flow through particle swarm optimization on a system of MGs with WT. To date, several important studies on realizing optimal operation for MG systems accompanied with uncertainties have been conducted [6–20]. However, the risk strategy of uncertainty was mostly not considered when computing the optimization problem. Although uncertainty is inevitable in MG operation, few studies have considered risk scenarios. The DRP can be used in various approaches to improve system reliability and reduce operational costs, but it has generally been used as a single solution. Therefore, it is necessary not only to operate MGs considering risk strategy but also to utilize the hybrid DRP according to the risk-averse tendency of MGOs.

In this study, we present a two-stage optimization model for optimal operation in grid-connected MG considering the DRP with a risk strategy. Load, PV, and WT are considered as uncertainty variables, and each unit is modeled through a certain probability distribution function (PDF). Monte Carlo simulation and k-means clustering are used implement the scenarios. The risk-based hybrid demand response program (RH-DRP) is proposed to reduce risk-based operational costs by determining optimal DR volume configuration in accordance with risk aversion parameter. In the two-stage optimization problem, the expected operational costs are calculated in first-stage and then, the risk-based operational costs are determined via the conditional value at risk (CVaR) index in second-stage. This is successfully addressed using the improved multi-layer artificial bee colony (IML-ABC), a modified conventional ABC algorithm that improves convergence speed.

The main contributions of this paper can be summarized as follows:

- The proposed two-stage optimization is developed to calculate the risk-based operation costs while considering uncertainty based on scenarios. It assists MGOs make decisions from more additional degree of freedom.
- The effectiveness of applying the RH-DRP is evaluated in terms of the total risk-based operation costs reduction, and the superiority of the RH-DRP is demonstrated via comparison analysis with different DR strategies.
- The IML-ABC shows better performance in searching for optimal solutions in comparison with the different optimal algorithms. Furthermore, it achieves simulation time reduction with the rapid convergence speed and can be applied for a variety of optimization problems.

The remainder of this paper is organized as follows. Section 2 introduces uncertainty modeling for WT, PV, and load. Section 3 shows the proposed RH-DRP strategy, and Section 4 provides formulations for the risk-based two-stage optimization problem. Section 5 presents the solution method used by the IML-ABC algorithm and summarizes the overall process of the optimal operation approach. Section 6 presents the numerical analysis results, and, finally, Section 7 concludes the paper.

2. Uncertainty Modeling for Microgrid Operation

2.1. Uncertainty Modeling

In this study, a probabilistic model is used to minimize operating costs in a microgrid considering uncertainty. Accurate prediction is not possible due to the stochastic behavior of wind and solar irradiance and it is always related to the uncertainty error of the plan for the next day. Thus, to consider more realistic compliance, we use probability density function (PDF) to model the behavior of wind, solar, and load to achieve optimal results considering uncertainties.

2.1.1. Wind Generation Modeling

The Weibull PDF has been regularly used to model wind speed at a forecasted time [21] and can be expressed as

$$PDF_w(v) = \left(\frac{d}{C}\right)\left(\frac{v}{C}\right)^{d-1} \exp\left[-\left(\frac{v}{C}\right)^{d}\right], \quad (1)$$

where d and C are the factors that characterize the Weibull PDF and determine the shape and scale, respectively, and v is the wind speed.

$$P_w(v) = \begin{cases} 0 & v_{wind} < v_{ci} \\ P_r \times \frac{(v_{wind}-v_{ci})}{(v_r-v_{ci})} & v_{ci} \le v_{wind} < v_r \\ P_r & v_r \le v_{wind} < v_{co} \\ 0 & v_{co} \le v_{wind} \end{cases}. \quad (2)$$

The output power of WT can be calculated using the WT power curve parameters: where P_r, v_{ci}, v_r, and v_{co} are the rated power, cut-in speed, rated speed, and cut-off speed of the WT, respectively.

2.1.2. Solar Generation Modeling

Solar generation output depends on the amount of sun irradiance. Forecasted solar power generation is commonly calculated using the beta PDF expressed as follows [22]:

$$PDF_B(si) = \begin{cases} \frac{\Gamma(a+b)}{\Gamma(a)\Gamma(b)} \times si^{(a-1)} \times (1-si)^{(b-1)} & 0 \le si \le 1,\ a \ge 0,\ b \ge 0 \\ 0 & otherwise \end{cases}, \quad (3)$$

$$a = \frac{\mu_s \times b}{1 - \mu_s}, \tag{4}$$

$$b = (1 - \mu_s) \times \left(\frac{\mu_s \times (1 + \mu_s)}{\sigma_s{}^2} - 1 \right). \tag{5}$$

Here, si is the amount of solar irradiance (kWh/m^2), and a and b are the beta PDF parameters that can estimate the mean (μ_s) and standard deviation (σ_s) according to the solar irradiance data, respectively.

Depending on the characteristics of the PV panels, solar irradiance can be converted to solar power by

$$P_{PV}(si) = \eta_{pv} \times A_{pv} \times si, \tag{6}$$

where $P_{pv}(si)$ represents the amount of PV output power for irradiance, η_{pv} is the efficiency of the PV panels, and A_{pv} is the total surface area of the PV panels.

2.1.3. Load Modeling

Uncertainty in load consumption caused by the stochastic behavior of power consumers can be modeled as a typical probability function with a normal PDF [23].

$$PDF_n(load) = \frac{1}{\sqrt{2\pi} \times \sigma_l} \exp\left(\frac{(P_{load} - \mu_l)^2}{2\sigma_l{}^2} \right), \tag{7}$$

where P_{load} is load demand, and μ_l and σ_l are the mean value and standard deviation of the load, respectively.

2.2. Scenario Generation and Reduction

In stochastic cases, the high occurrence probability has the most effect on decision makers. However, high probability situations do not always arise, whereas low probability situations may occur at any time [24]. If the low probability scenario has a critical effect, it should be considered in the decision making. The MGO would then be confronted with a problem with uncertainty, such as renewable power and load consumption in MG operation. Our study conducted scenario modeling using Monte Carlo simulations to analyze various scenarios, including the risk situation for all low probabilities with a critical effect. The generated scenarios for the load demand (ϕ_{Ln}) and power generation of WT (ϕ_{WTn}) and PV (ϕ_{PVn}) are expressed as follows:

$$\phi_{Ln} = (L_n, \psi_{Ln}), \quad \sum_1^N \psi_{Ln} = 1, \tag{8}$$

$$\phi_{WTn} = (WT_n, \psi_{WTn}), \quad \sum_1^N \psi_{WTn} = 1, \tag{9}$$

$$\phi_{PVn} = (PV_n, \psi_{PVn}), \quad \sum_1^N \psi_{PVn} = 1\,, \tag{10}$$

$$S_n = (\phi_{Ln}, \phi_{WTn}, \phi_{PVn}), \quad \sum_1^N \psi_L^n \times \psi_{WT}^n \times \psi_{PV}^n = 1, \tag{11}$$

where L^n, WT^n, and PV^n are the n_{th} scenario states of the load, wind power, and solar power, respectively; Ψ_{Ln}, Ψ_{WTn}, and Ψ_{PVn} are the probability in the n_{th} scenario of the load, wind power, and solar power, respectively; and S^n is the possible states and probability of the n_{th} scenario.

The scenarios for load, wind, and PV are generated based on Equations (8)–(10). Subsequently, the integrated scenarios are modeled through Equation (11).

To minimize the calculation time, in our work, k-means clustering method is utilized to divide the scenarios into groups so that the sum of the squared distances from the data objects in the group is minimal [25].

$$\arg\min_{S} \sum_{i=1}^{K} \sum_{x \in S_i} ||x - \mu_i||^2,\ S_{k-means} = \{S_1, S_2, \ldots, S_K\}, \tag{12}$$

where $S_{k\text{-}means}$ denotes the groups divided using k-means clustering, and μ_i is the mean of the points in S_i.

3. Demand Response Strategy

Load management by adjusting consumer behavior has been implemented as DRP, which is aimed at peak shaving over high demand periods in the MGs operation [26]. The DRP includes modifying the electricity consumption patterns and incentives to promote change, and these incentives are primarily used when market prices are high or when system reliability is decreased. Hence, a new hybrid type of DRP, which considers the economics and risks of uncertainty, is proposed in this study. The RH-DRP can improve economics through load shifting in a day-ahead market and then, risk effects be reduced in response to power shortages because of uncertainties in MGs.

3.1. Demand Response Elasticity

Elasticity is an economic measure that assesses the percentage of change in demand due to price fluctuations [27]. In terms of electricity consumption, this percentage value changes as power demand varies with the changes in electricity market prices. This expected value is negative as higher electricity prices possibly results in load reductions. The elasticity of demand for electricity is calculated as follows:

$$E(t) = \frac{\Delta D(t)/D(t)}{\Delta MP(t)/MP(t)}, \tag{13}$$

where $E(t)$ is DR elasticity, and $D(t)$ and $MP(t)$ are the power demand capacity and electricity market price in time t, respectively.

The two types of elasticity of demand are self-price elasticity and substitution elasticity. Self-price elasticity disregards period variation, but it considers the variation of consumption according to electricity price changes. Meanwhile, substitution elasticity is related to shifts in the power consumption of electricity within a day. Thus, it has a constraint that power capacity after demand response is equal to the initial power consumption.

3.2. Pattern of Each Power Consumer

The value of elasticity depends on the power consumption patterns of the consumer. Therefore, the electricity value can be categorized as follows: industrial, commercial, and residential. Industrial consumers generally have the largest power demand among the three categories. Industrial loads are closely related to running factories; thus, the possible capacities of industrial consumers to participate in the DR should be considered prior to contract signing with MGOs. Then, they would receive incentives that match their capacity to participate in the DRP, and this incentive should be worth more than the industrial consumers can gain from running a factory over a period. If the consumer cannot commit to the responsibility specified in the contract, they could be charged a penalty fee. The changed power demand capacity after the industrial DRP can be calculated using the following formula:

$$D_{new,i}(t) = E_i \times D_{old,i}(t) \times \frac{(MP(t) - MP_0 + \pi_i(t) - pen(t))}{MP(t)} + D_{old,i}(t), \tag{14}$$

where $D_{new,i}(t)$ and $D_{old,i}(t)$ are the power demand capacity after and before implementation of industrial DR, respectively; E_i, MP_0, $\pi_i(t)$, and $pen(t)$ are the elasticity default value, reference electricity market price, incentive price per kilowatt-hour, and penalty fee for the industrial consumer at time t, respectively.

Commercial and residential loads are smaller and more distributed than industrial loads. Their capacities are difficult to precontract prior to participation in a DRP; thus, the elasticity value is generally smaller than the elasticity value of industrial consumers. However, adjusting commercial and residential loads is easy as they do not suffer from economic damages (e.g., industrial consumers not being able to run their factories by participating in the DRP). Therefore, commercial and residential consumers are not obligated to pay a penalty fee and can appropriately participate in real-time transactions. The changed power demand capacity after commercial or residential DRP can be calculated using the formula as follows:

$$D_{new,cr}(t) = E_{cr} \times D_{old,cr}(t) \times \frac{(MP(t) - MP_0 + \pi_{cr}(t))}{MP(t)} + D_{old,cr}(t), \tag{15}$$

where $D_{new,cr}(t)$ and $D_{old,cr}(t)$ are the power demand capacity after and before implementation of commercial and residential DR, respectively; $\pi_i(t)$ is incentive price per kilowatt-hour for commercial and residential consumers.

3.3. Risk-Based Hybrid Demand Response Program

In this study, we propose an RH-DRP, which not only reduces MG operational costs through economic DR in a day-ahead market but also decreases risk costs due to uncertainty by applying a risk-based DR in real-time operations. The two necessary conditions of the RH-DRP are the following:

- The RH-DRP consists of economic and risk-based DR. Economic DR affects the economic improvement of MG operational scheduling and is applied via a contract with MGOs by considering the substitution elasticity in a day-ahead market. Meanwhile, risk-based DR reduces the risk costs of uncertainty and can be considered the self-price elasticity in real time without a precontract.
- It determines the load capacity that each consumer can offer to participate in the DR. Thus, MGOs can reasonably request DR from consumers considering by the stability and economics of the power system, and the consumers should respond immediately.

As the economic DR considers substitution elasticity, it has constraint. The amount after load shifting is the same as the initial load. In a real-time market, risk-based DR participants increase system stability and reduce the risk costs of MG operations. If MGOs fail to control power shortages caused by uncertainty, a critical problem occurs, such as blackouts. Therefore, the reliable operation of MGs should be ensured when using flexible resources, such as DR. In particular, MGOs should provide additional compensation to consumers participating in risk-based DR. The incentive prices depend on the consumer type and DR participation. The formula for each incentive price is expressed as follows:

$$Inc_{e,c}(t) = \begin{cases} \pi_{e,c}(t) \times \left[D_{old,c}(t) - D_{new,c}(t)\right] & D_{old,c} \geq D_{new,c} \\ 0 & D_{old,c} < D_{new,c} \end{cases}, \tag{16}$$

$$Inc_{r,c}(t) = [\pi_{r,c}(t) + ac(t)] \times D_{r,c}(t) \quad D_{r,c}(t) \geq 0\,, \tag{17}$$

where $Inc_{e,c}(t)$ and $Inc_{r,c}(t)$ are the incentive prices for the type of consumer c to participate in economic DR and risk-based DR at time t, respectively; $\pi_{e,c}(t)$ is the incentive price per kilowatt-hour for each consumer participating in economic DR at time t; $\pi_{r,c}(t)$, $ac(t)$, and $D_{r,c}(t)$ are risk-based DR incentive prices per kilowatt-hour for each consumer, the additional compensation for participating in risk-based DR, and the capacity to participate in risk-based DR, which is required owing to the uncertainties for each consumer, respectively.

4. Two-Stage Optimal Formulation

Figure 1 illustrates the proposed two-stage optimal operation process. In the first-stage, operational costs are calculated by considering the expected scenario in a day-ahead market. This stage does not consider a probabilistic scheme and is associated with the first optimal scheduling using economic DR by the MGO. The second-stage computes power shortage volume caused by uncertainty, and then, risk-based DR is utilized to minimize risk index determined through CVaR.

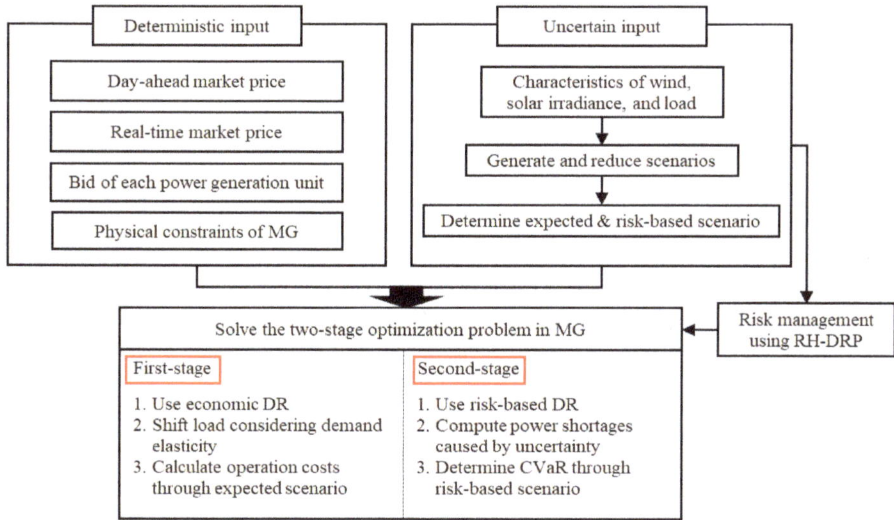

Figure 1. Schematic representation of two-stage optimal operation.

4.1. Objective Function

In the two-stage optimization model, the objective function for minimizing risk-based operational costs is as expressed follows:

$$obj.fun = Min\left(f_{first}(X) + f_{second}(X)\right), \tag{18}$$

where $f_{first}(X)$ and $f_{second}(X)$ are the first- and second-stage objective functions, respectively, and X is the decision variable vector.

4.1.1. First-Stage Objective Function

The first-stage of the objective function minimizes the operation costs of the expected scenario using the following formula:

$$Min\ \ f_{first}(X) = Min\sum_{t=1}^{T} Cost(t), \tag{19}$$

where *Cost*(*t*) is the operational costs for MGs in the day-ahead market at time *t*.

Operation costs are the sum of fuel costs for each generator, including start-up/shut-down costs, interactions between the utility and MG, and incentive prices of participating in economic DR. The detailed expression is as follows:

$$
\begin{aligned}
Cost(t) = &\sum_{i=1}^{N_G}\left[u_i(t)P_{Gi}(t)B_{Gi}(t) + S_{Gi}\left|u_i(t) - u_i(t-1)\right|\right] \\
&+\sum_{j=1}^{N_s}\left[u_j(t)P_{sj}(t)B_{sj}(t) + S_{sj}\left|u_j(t) - u_j(t-1)\right|\right] \\
&+\sum_{k=1}^{N_{pv}} P_{PVk}(t)B_{PVk}(t) + \sum_{l=1}^{N_w} P_{Wl}(t)B_{Wl}(t) \\
&+P_{Grid}(t)B_{Grid}(t) + \sum_{c=1}^{N_c} Inc_{e,c}(t),
\end{aligned}
\tag{20}
$$

where $P_{Gi}(t)$, $P_{sj}(t)$, $P_{PVk}(t)$, and $P_{Wl}(t)$ are the active power outputs of the ith diesel-generator (DG), jth storage device, kth PV panel, and lth wind generator at time t, respectively; $B_{Gi}(t)$, $B_{sj}(t)$, $B_{PVk}(t)$, and $B_{Wl}(t)$ are the bids of the generator, energy storage, solar energy, and wind energy at time t, respectively; S_{Gi} and S_{sj} are the start-up or shut-down costs for the ith generator and jth storage, respectively; $P_{Grid}(t)$ and $B_{Grid}(t)$ are the active power and bid price, which are bought and sold with the utility at time t, respectively; N_G, N_s, N_{pv}, N_w, and N_c are the total number of generators, storage devices, PV, wind generator units, and economic DR consumers, respectively.

When calculating operational costs, X is considered for each unit of output power, amount of load reduction of DR, and on/off mode in a day-ahead market, which can be calculated as follows:

$$X = [P_g, U_g]_{1\times 2nT}, \tag{21}$$

$$P_g = [P_{G1}, \ldots, P_{GN_G}, P_{s1}, \ldots, P_{sN_s}, P_{Grid}, P_{DR}], \tag{22}$$

$$U_g = [U_{G1}, \ldots, U_{GN_G}, U_{s1}, \ldots, U_{sN_s}, U_{Grid}, U_{DR}], \tag{23}$$

$$n = N_G + N_s + 2, \tag{24}$$

where P_g and U_g are the active power and state vector of all units during time t, respectively, and n and T are the numbers of decision variables and periods, respectively.

4.1.2. Second-Stage Objective Function

In stochastic optimal operation for MGs, operational costs vary according to each scenario, and some scenarios are expected to be very low or even have a negative effect. To consider such risk-based scenarios in MG operation, a risk management criterion is included in the mathematical formulations, where risk management implies reducing the negative effect of uncertainty. Several indices for risk management are standard deviation, shortfall probability, value-at-risk (VaR), CVaR, and so on [28]. To evaluate the risk costs of power shortages caused by uncertainties in wind, PV, and load, we used the α confidence level CVaR (α-CVaR). The α-CVaR is determined using the expected costs of (1 − α) × 100% for all worst-case scenarios. The mathematical formula is expressed as follows:

$$CVaR_\alpha(C_s) = VaR_\alpha + \frac{1}{1-\alpha}\sum_{s=1}^{Ns} \rho_s \times \varphi_s, \tag{25}$$

$$VaR_\alpha = \min\{\eta : P(C_s > \eta) > (1-\alpha)\},\ \forall\alpha \in (0,1), \tag{26}$$

$$\varphi_s \ge (C_s - VaR_\alpha(C_s)) \quad \varphi_s \ge 0,\ \ \forall s, \tag{27}$$

where $CVaR_\alpha$ and C_s are the risk costs calculated using the α-CVaR and the stochastic operational costs in scenario s, respectively; η is the smallest cost for a given confidence level α. VaR_α is the α confidence level VaR [29], as shown in Equation (26).

In risk-based MG operation, stochastic operational costs are considered power shortages due to uncertainties; they include the costs of purchasing real-time market and risk-based DR incentives, which are expressed by

$$C_s = \sum_{t=1}^{T} \left\{ Cost(t) + P_{rGrid}(t) B_{rGrid}(t) + \sum_{c=1}^{Nc} Inc_{r,c}(t) \right\}, \tag{28}$$

where $P_{rGrid}(t)$ and $B_{rGrid}(t)$ are the power and bid price purchased in the real-time market at time t, respectively.

Figure 2 depicts the procedure to calculate α-CVaR according to power shortages. The power shortage for each scenario can be computed based on the difference from the expected scenario in the day-ahead model; then, the α% worst scenarios are determined to calculate the α-CVaR.

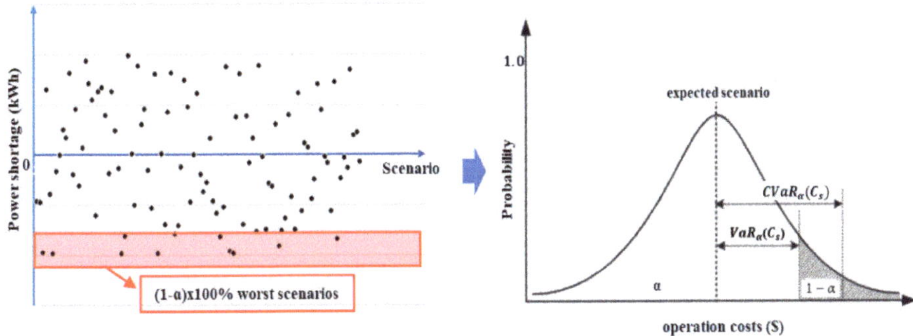

Figure 2. Determination of α confidence level conditional value at risk (α-CVaR) as a function of power shortages.

The second stage of the objective function is to minimize the α-CVaR in real time. The MGO can set parameter according to risk aversion tendency, and then the amount of power generation and risk-based DR capacity can be adjusted through this stage.

$$Min\ \ f_{second}(X) = Min \sum_{s=1}^{Ns} \beta \times CVaR_{\alpha}(C_{rt}^{s}), \tag{29}$$

where β is the risk aversion parameter, which can generally be adjusted from 0 to 1.

4.2. Constraints

4.2.1. Power Balance Constraints

The total power generation by each unit should satisfy and be equal to the power demand.

$$\sum_{i=1}^{N_G} P_{Gi}(t) + \sum_{j=1}^{N_s} P_{sj}(t) + \sum_{k=1}^{N_{pv}} P_{PVk}(t) + \sum_{l=1}^{N_w} P_{Wl}(t) + P_{Grid}(t) + P_{DR}(t) = \sum_{m=1}^{N_m} P_{Lm}(t), \tag{30}$$

$$P_{PVk}(t) = P_{PVk,\exp}(t) + \varepsilon_{PVk}(t), \tag{31}$$

$$P_{Wl}(t) = P_{Wl,\exp}(t) + \varepsilon_{Wl}(t), \tag{32}$$

$$P_{Lm}(t) = P_{Lm,\exp}(t) + \varepsilon_{Lm}(t). \tag{33}$$

Here, P_{DR}, P_{Lm}, and N_m are the total power capacity of the participating RH-DRP, amount of power in the m_{th} demand level, and total number of demands, respectively. Equations (31)–(33) represent the actual values of PV, WT, and load, which consist of the expected value and error.

4.2.2. Power Generation Constraints

The power output of each generator unit is limited by the lower and upper bounds.

$$P_{Gi,\min}(t) \le P_{Gi}(t) \le P_{Gi,\max}(t), \tag{34}$$

$$P_{PVk,\min}(t) \le P_{PVk}(t) \le P_{PVk,\max}(t), \tag{35}$$

$$P_{WI,\min}(t) \le P_{WI}(t) \le P_{WI,\max}(t), \tag{36}$$

$$P_{WI,\min}(t) \le P_{WI}(t) \le P_{WI,\max}(t). \tag{37}$$

Here, $P_{G,min}(t)$, $P_{PV,min}(t)$, $P_{WI,min}(t)$, and $P_{Grid,min}(t)$ are the lower bounds of the active power generated by DG, PV, WT, and the transaction with utility, respectively; $P_{G,max}(t)$, $P_{PV,max}(t)$, $P_{WI,max}(t)$, and $P_{Grid,max}(t)$ are the upper bounds of the active power of each unit at time t, respectively.

4.2.3. RH-DRP Constraints

The RH-DRP determines the maximum participating capacity under a contract with MGO, and the sum of each DR capacity is less than the maximum DR capacity.

$$D_{old,c}(t) - D_{new,c}(t) + D_{r,c}(t) \le D_{\max,c}(t), \tag{38}$$

where $D_{max,c}(t)$ is the maximum DR capacity of c at time t.

4.2.4. Energy Storage Constraints

Some limits on the charge and discharge capacities of ESS during each time interval exist, which can be calculated as follows:

$$W_{ess,t} = W_{ess,t-1} + \eta_{\text{charge}} P_{\text{charge}} \Delta t - \frac{1}{\eta_{\text{discharge}}} P_{\text{discharge}} \Delta t, \tag{39}$$

$$W_{ess,\min} \le W_{ess,t} \le W_{ess,\max}, \tag{40}$$

$$P_{\text{charge},t} \le P_{\text{charge,max}}\ , \tag{41}$$

$$P_{\text{discharge,t}} \le P_{\text{discharge,max}}, \tag{42}$$

where $W_{ess,t}$ is the amount of energy in the battery at time t, η_{charge} ($\eta_{discharge}$) is the efficiency of the ESS (dis)charging, P_{charge} ($P_{discharge}$) is the permitted capacity of the ESS (dis)charging during a Δt; $W_{ess,min}$ and $W_{ess,max}$ are the minimum and maximum limits on the amount of energy storage in ESS, respectively; $P_{charge,max}$ ($P_{discharge,max}$) is the maximum capacity of ESS (dis)charging during interval Δt.

5. Solution Method

5.1. Improved Multi-Level Artificial Bee Colony Algorithm

There are many optimization methods when solving an optimization problem. Among them, the ABC algorithm is widely used in various fields recently [30]. Especially, the ABC algorithm has been widely used to solve the MG operation problem and it has been modified to improve performance in recent years [31–33]. Generally, the ABC algorithm is based on the food foraging behavior of bees. A colony of bees is divided into three groups: employed bees, onlooker bees, and scout bees. Each type performs its respective tasks to find the most abundant resources. In population-based optimization techniques, such as the ABC algorithm, finding the first global minimum value is critical. Depending on the closeness of the first global minimum value to the final optimization value, the convergence speed and optimization accuracy be improved. Accordingly, the population and scope of exploration are crucial to the optimization algorithm. Here, the IML-ABC algorithm is proposed by adjusting the number of bees and food sources based on the number of iterations for finding faster and more accurate optimal solution. The IML-ABC algorithm consists of five levels (i.e., adjusting, initialization, employed bees, onlooker bees, and scout bees):

(i) Adjusting level: In the IML-ABC algorithm, the number of bees increases in the initial iteration and decreases according to the number of iterations to find the initial global minimum efficiently.

$$FN_n = \begin{cases} (1 + \sigma_{FN}) \times FN & 0 < n \leq \lambda \\ FN & \lambda < n \leq (iter_{\max} - \lambda) \\ (1 - \sigma_{FN}) \times FN & (iter_{\max} - \lambda) < n < iter_{\max} \end{cases}, n = \{1, 2, \ldots, iter_{\max}\}. \quad (43)$$

Here, FN_n and FN are the number of food sources in the nth iteration and mean number of food sources, respectively. In addition, σ_{FN}, λ, and $iter$ are the rate of food source adjustment, value of dividing the phase of the iteration, and number of iterations, respectively.

(ii) Initialization level: The IML-ABC algorithm makes a random initial population of food source positions. Each food source x_i (i = 1, 2, ... , SN) has a D-dimensional problem space and can be expressed as

$$x_{ij} = x_j^{\min} + rand[0, 1] \times (x_j^{\max} - x_j^{\min}), \quad (44)$$

where x_{ij} is the j_{th} decision variable of the i_{th} solution vector, and x_j^{min} and x_j^{max} represent the lower and upper limit values of the j components for the X_i vector, respectively.

(iii) Employed bees level: Each bee constantly explores to find a food source. When it finds the optimum solution, it selects a new and best position (v_{ij}) close to the reference position. In the IML-ABC algorithm, to determine a better global minimum solution, employed bees explore large ranges in initial iterations and search for the best solution in a narrow range as the number of iterations increases. This can be computed as follows:

$$v_{ij} = x_{ij} + \varphi_{ij}(x_{ij} - x_{kj}) \times (1 - R \times iter/iter_{\max}), \quad (45)$$

where v_{ij} represents the new position of the food source i for the jth component, φ_{ij} is a random number in the range [−1, 1], and R is the value of the exploration range reduction based on the number of iterations.

(iv) Onlooker bees' level: An onlooker bee finds new positions that are closed to the old position and searches for food sources according to the probability value associated with the corresponding food source. Then, the greedy method is applied; the probability value of the selected food source can be expressed as follows:

$$P_i = \frac{fit_i}{\sum_{i=1}^{SN} fit_i}, \quad (46)$$

$$fit_i = \begin{cases} 1/(1 + F_i), & F_i \geq 0 \\ 1 + |F_i|, & F_i < 0 \end{cases}, i = \{1, 2, \ldots, SN\}, \quad (47)$$

where P_i and fit_i are the probability value and fitness value of the ith food source evaluated by the ith employed bee, respectively; and F_i is the value of the objective function for the X_i solution.

(v) Scout bee level: If solution of X_i cannot be improved through the number of predetermined trials (*limit*), this solution is abandoned, and the corresponding employed bee is converted to a scout bee. This scout bee randomly attempts to find a new food source in the solution space to replace the abandoned solution using Equation (44). This procedure is repeated for several maximum cycles.

5.2. Improved Multi-Level Artificial Bee Colony Algorithm

Figure 3 summarizes the two-stage solution process for the optimal scheduling of MGs with uncertainties using RH-DRP. The procedure is performed sequentially as follows:

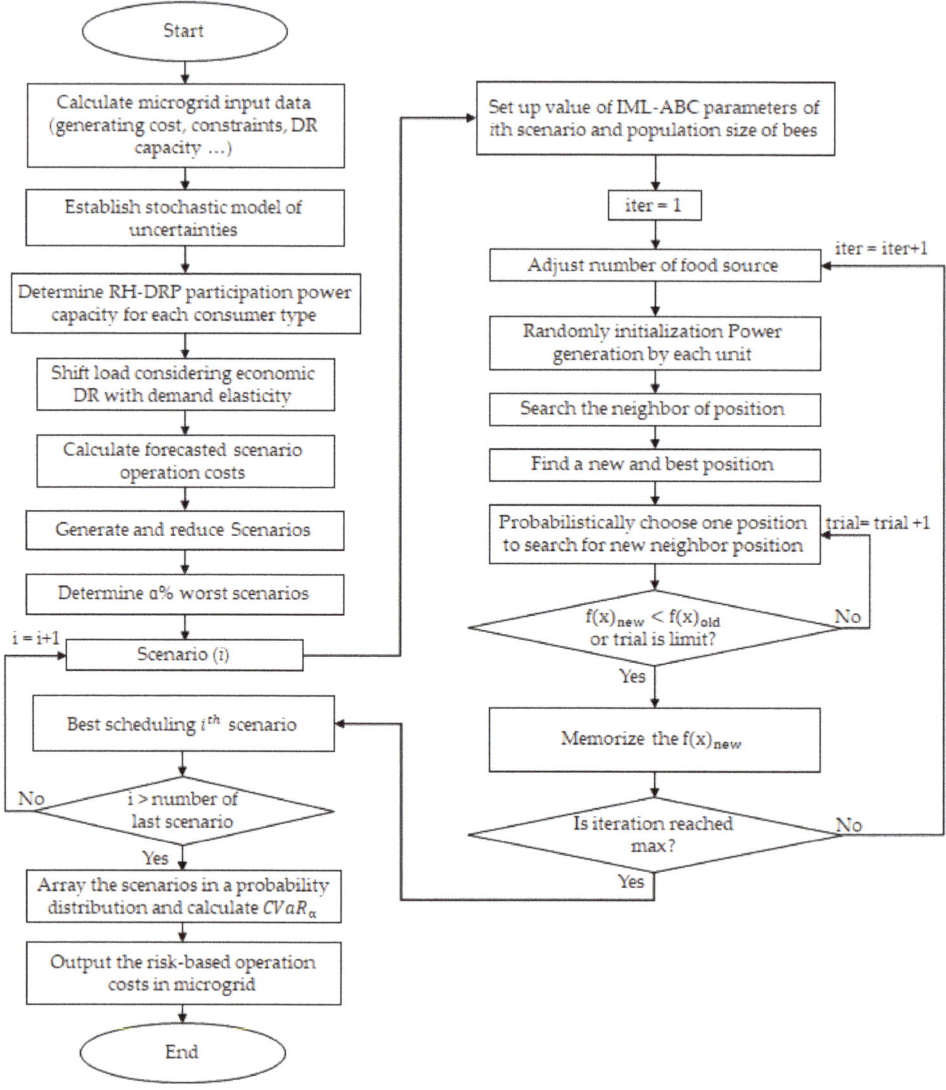

Figure 3. Overall process of the proposed optimal operation approach.

Calculate the MG input data based on fixed system information.

Step 1. Establish a stochastic model of uncertainties arising from renewable energies and load.

Step 2. Determine the RH-DRP participation power capacity for each consumer type.

Step 3. Shift load considering the economic DR with demand elasticity.

Step 4. Calculate expected scenario operational costs through first-stage objective function.

Step 5. Generate the scenarios using Monte Carlo simulation and reduce the scenarios using the k-means clustering technique.

Step 6. Determine $(1-\alpha)\times 100\%$ worst scenarios for calculating $CVaR_{\alpha}$.

Step 7. Start the IML-ABC algorithm loop for the *i*th scenario.

Step 8. Find the best scheduling for each scenario with the goal to minimize operation costs.

Step 9. Array the scenarios in a probability distribution and calculate $CVaR_{\alpha}$.

Step 10. Output the risk-based optimal operation costs in the MG.

6. Numerical Analysis

6.1. Input Data

The proposed two-stage risk-based optimal operation was examined on a low-voltage grid-connected MG (Figure 4). This MG system contained micro turbine (MT), fuel cell (FC), PV, WT, and ESS technologies. Table 1 depicts the unit data, including the bid, start-up/shut-down costs, and minimum/maximum power [21]. The total expected daily load was 1695 kWh, divided among industrial, commercial, and residential consumers, accounting for 50%, 33.33%, and 16.67% of the total initial demand, respectively [34]. Taking into account the increased load on uncertainty, we assumed that 20% of each group of consumers can participate in the DRP. Table 2 lists the elasticity value, incentive costs per power, and initial demand for each consumer [35].

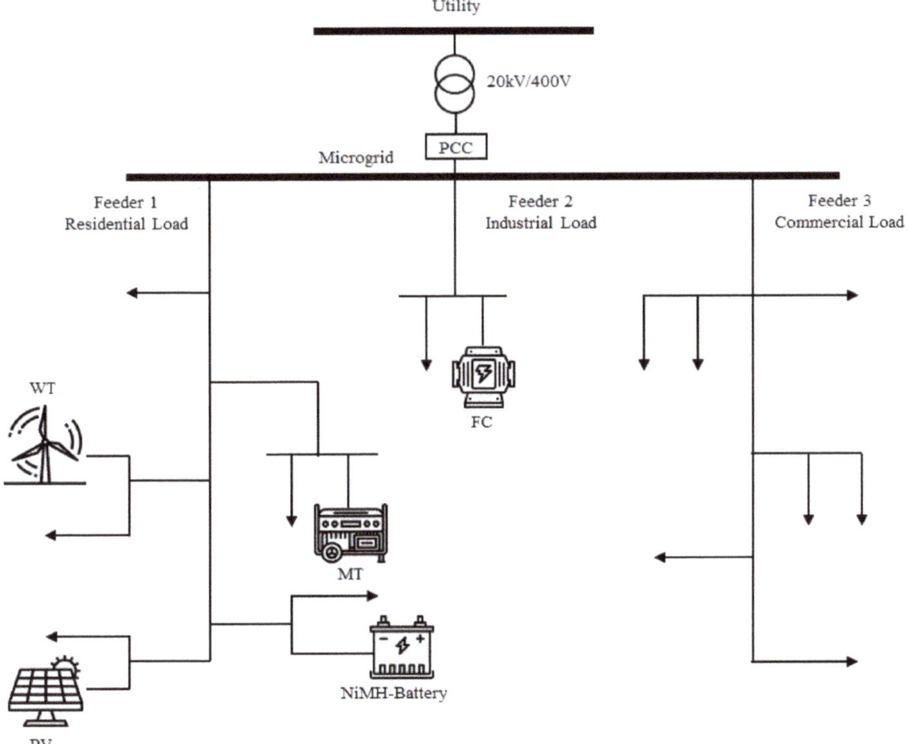

Figure 4. Diagram of a low-voltage grid-connected microgrid (MG).

Table 1. Limits and bids of installed distributed generator sources.

ID	Type	Min. Power (kW)	Max. Power (kW)	Bid ($/kWh)	Start-Up/Shut-Down Costs ($)
1	MT	6	30	0.457	0.96
2	FC	3	30	0.294	1.65
3	PV	0	25	2.584	0
4	WT	0	15	1.073	0
5	ESS	−30	30	0.38	0
6	Utility	−30	30	-	-

Table 2. Elasticity and incentive price for each consumer.

Type of Consumer	Elasticity	Incentive Costs per Power ($/kWh)	Initial Demand (kWh/day)
Industrial	−0.38	0.12	847.5
Commercial	−0.20	0.20	565
Residential	−0.14	0.18	252.5

Figure 5 shows the day-ahead market price for grid-connected MG operation. Here, the real-time market price is assumed to be 10% more expensive than the day-ahead market price, which is the APX hourly market price [36]. Table 3 presents the WT data and Weibull PDF factors. The solar panel was a 25 kW SOLAREX MSX (US) composed of 10 × 2.5 kW panels with 18.6% efficiency and 10 m^2 of total surface area [37]. In the ESS consisting of a nickel-metal hydride battery, charging and discharging efficiency was 95%, and the minimum and maximum amounts of storage were 5% and 100% of the battery capacity, respectively. Power generation units participate in the MG depending on their technical and economic features, and excess power is exchanged with the utility through the point of common coupling (PCC). Our operation model was performed with the proposed IML-ABC algorithm set to $FN = 200$ and $iter_{max} = 500$. All cases were simulated in MATLAB (R2019a) on a laptop computer with a 2.9 GHz Intel Core i5-9400 CPU and 16 GB RAM.

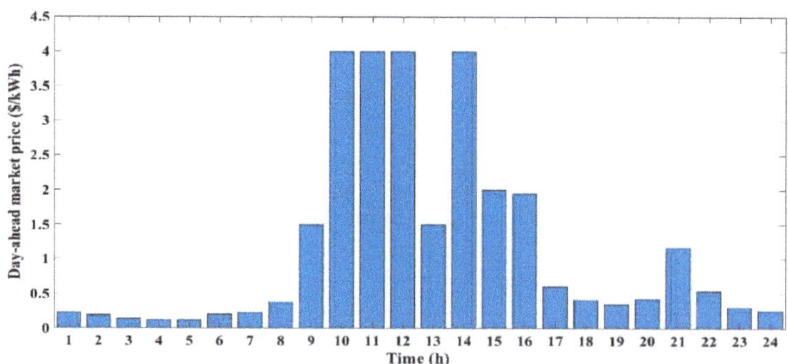

Figure 5. Day-ahead market price of a low-voltage MG.

Table 3. Wind turbine data and Weibull probability density function (PDF) factor.

Type	Parameter	Value
Wind turbine data	P_r (kW)	15
	V_r (m/s)	12.5
	V_{ci} (m/s)	3
	V_{co} (m/s)	25
Weibull PDF factor	*d*	2
	C	6.77

6.2. Proposed Optimization Results

In the first-stage, the MGO uses economic DR among the RH-DRP to optimize MG operation in a day-ahead market with expected scenario. This stage was simulated by considering the hourly average values of load, wind speed, and solar irradiance. Table 4 presents the expected power production of PV and WT for each hour. Here, it is assumed that the MGO must buy all the power produced by the PV/WT at each time of the day. In the second stage, risk costs arising from uncertainty are minimized by the MGO. Figure 6 show the generation scenarios and reduction results using the PDF for each unit. The 4000 scenarios were generated by considering the uncertainty in the MG via Monte Carlo simulations. Subsequently, 20 scenarios were selected via k-means clustering considering the cluster number estimation and simulation times. From the 20 generated scenarios, risk-based operational costs were determined by considering α-CVaR using the RH-DRP.

Table 4. Expected wind turbine (WT) and photovoltaic (PV) power.

Hour	PV (kW)	WT (kW)	Hour	PV (kW)	WT (kW)
1	0	1.7850	13	23.9000	3.9150
2	0	1.7850	14	21.0500	2.3700
3	0	1.7850	15	7.8750	1.7850
4	0	1.7850	16	4.2250	1.3050
5	0	1.7850	17	0.5500	1.7850
6	0	0.9150	18	0	1.7850
7	0	1.7850	19	0	1.3020
8	0.2000	1.3050	20	0	1.7850
9	3.7500	1.7850	21	0	1.3005
10	7.5250	3.0900	22	0	1.3005
11	10.4500	8.7750	23	0	0.9150
12	11.9500	10.4100	24	0	0.6150

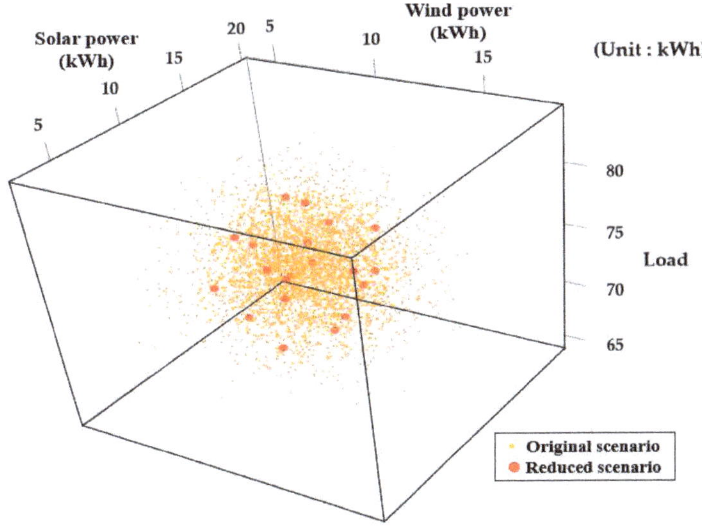

Figure 6. Scenario generation and reduction results.

Figure 7 illustrates the expected and risk-based scenario results of the daily load and power generation capacity of PV and WT. Here, the risk-based scenario is the average value comprising the 10% worst-case scenarios. The daily load increased by 162.69 kW, whereas the total power generated by PV and WT decreased by 9.62 kW and 5.25 kW, compared with the expected scenario, respectively.

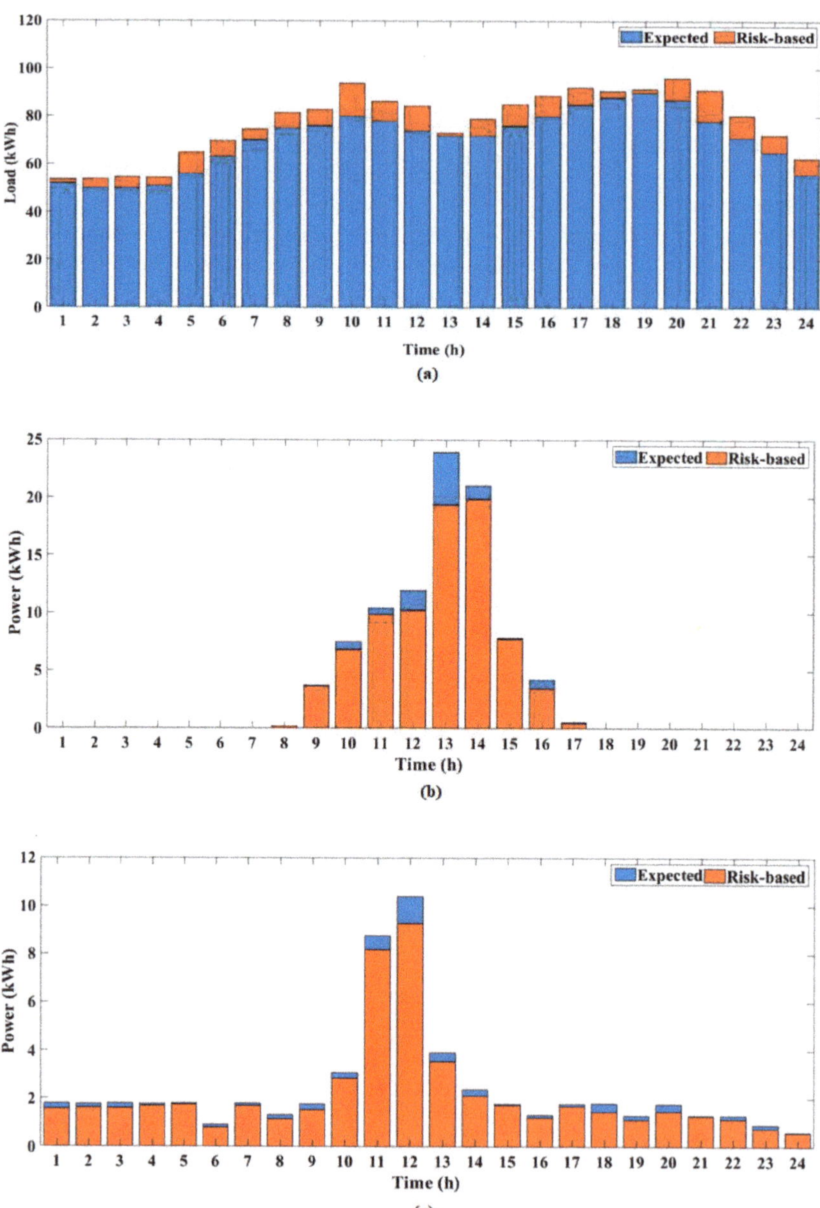

Figure 7. Expected and risk-based volumes. (**a**) Load, (**b**) photovoltaic (PV) generated, and (**c**) wind turbine (WT) generated.

Figure 8 shows the optimal results of expected operational costs and CVaR by changing the volume ratio of RH-DRP depending on the degree of risk aversion of MGO. We simulate when the risk aversion parameters are 0.1, 0.5, and 1 to consider various MGOs. MGO with β of 0.1 tends to ignore risk-based scenarios. MGO with β of 0.5 responds appropriately to risk-based scenarios, but do not attempt to avoid them entirely. MGO with β equal to 1 tries to completely avoid risk-based scenarios. When the volume of economic DR increases, the expected costs of MG operation decrease because the MGO can sell surplus power to the utility through load shifting at peak load times. However, risk-based DR capacity is not sufficient to address uncertainty problems; thus, the value of CVaR significantly increases as shown in Figure 8. Meanwhile, the high proportion of risk-based DR increases the expected costs and decreases the CVaR value. In addition, by increasing β, the MGO can decrease CVaR more effectively because they are more risk averse in MG operation.

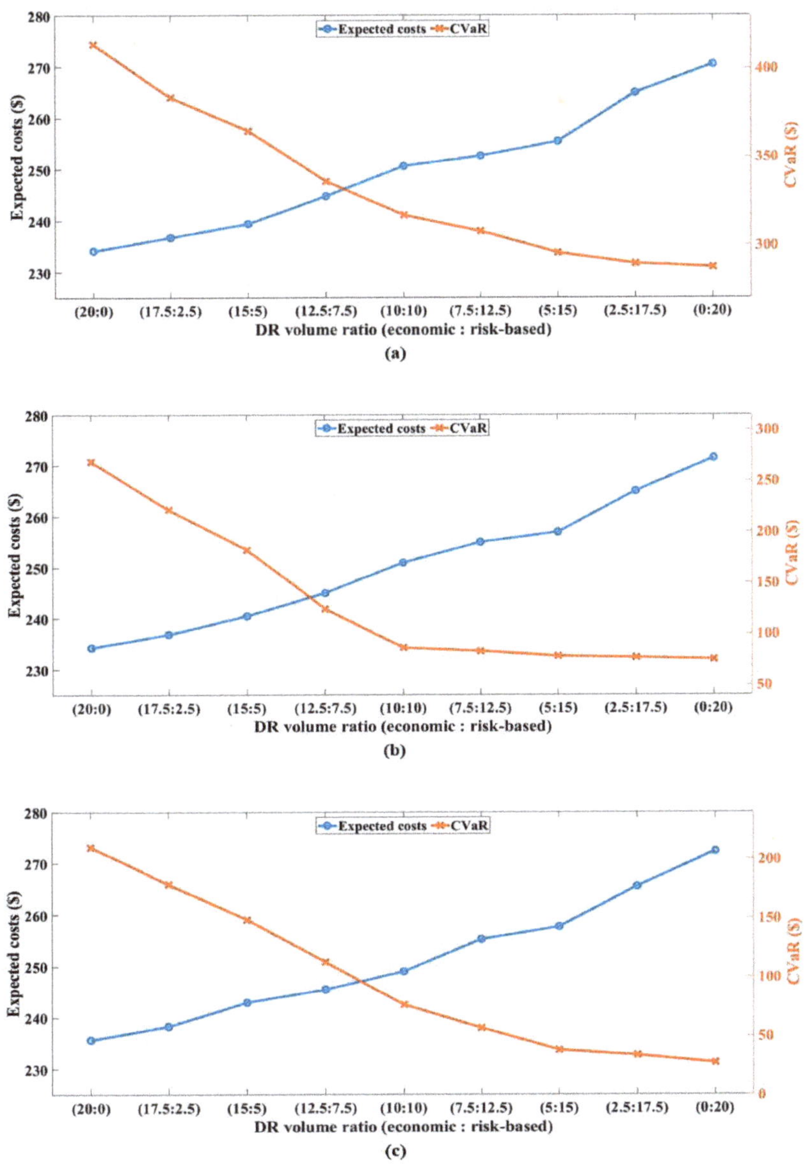

Figure 8. Expected operational costs and CVaR for each β. (**a**) $\beta = 0.1$, (**b**) $\beta = 0.5$, and (**c**) $\beta = 1$.

Figure 9 shows risk-based operational costs depending on risk aversion parameter β (0.1, 0.5, or 1). As shown in Figure 9, when β is 0.1, 0.5, and 1, the optimal solution is that the ratio of economic and risk-based DR is consist of 17.5:2.5, 10:10, and 5:15, respectively. As observed in these results, it should be noted that the MGO who can highly avoid risk should operate within the higher portion of the risk-based DR in RH-DRP. Table 5 summarizes the results of the risk-based optimal operation in MG. The lowest expected costs are determined when β is 0.1, but CVaR is very high, owing to insufficient management of risk problems. Meanwhile, the expected costs are slightly high when β is equal to 1, but the CVaR is significantly lower than in other cases. In other words, in a 10% worst scenario, in a low-voltage grid-connected MG operation, an MGO would incur operational costs of \$622.8660 ($\beta = 0.1$), \$338.3607 ($\beta = 0.5$), or \$293.6971 ($\beta = 1$). As can be seen from these results, RH-DRP is properly utilized according to the degree of risk aversion of MGO, and it is particularly effective in reducing risk costs for MGO that tend to avoid risk situation caused by uncertainty.

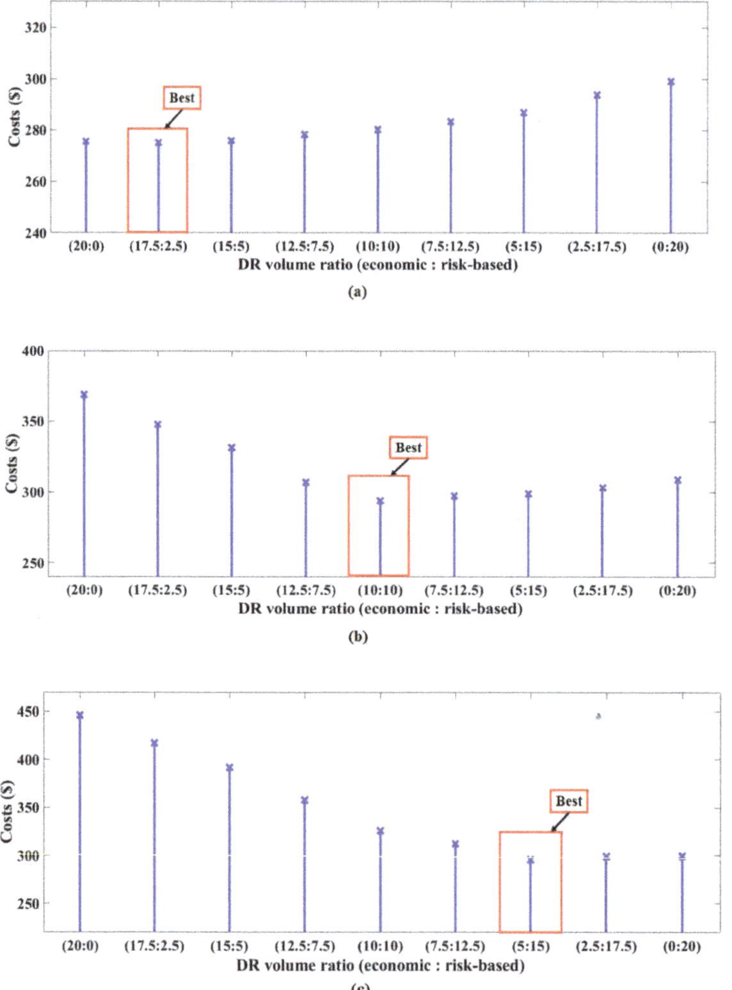

Figure 9. Risk-based operational cost as demand response (DR) volume ratio. (**a**) $\beta = 0.1$, (**b**) $\beta = 0.5$, and (**c**) $\beta = 1$.

Table 5. Risk-based optimal operation according to β.

Type	$\beta = 0.1$	$\beta = 0.5$	$\beta = 1$
Expected costs ($)	239.7650	251.9952	255.6907
CVaR ($)	383.1010	86.3655	38.0064
Operational costs ($) (10% worst scenario)	622.8660	338.3607	293.6971

Tables 6–8 present the optimal power scheduling in risk-based MG operation when β is equal to 0.1, 0.5, and 1, respectively. In early periods, the ESS charged and MT output are reduced due to cheap market prices. However, at peak load hours, the battery is fully discharged, and the MGO maximizes local generation to decrease operational costs to sell a considerable amount of energy to the main grid. Owing to low cost of power generation by the FC, the MGO then decides to use its maximum capacity for power supply. Based on the high operational cost of the MT, it is scheduled flexibly while considering the market price. Figure 10 shows RH-DRP participation in MG operation over time. The economic DR is primarily used from 9:00 to 17:00 to reduce the load in common. Through load reduction, the MGO can reduce operational costs by selling remaining power to the main grid when market prices are high. The risk-based DR responds to a power shortage caused by uncertainty and supports MG operation in terms of stability. In MG optimal operation, a larger β value confers a greater risk-based DR portion in RH-DRP. Therefore, it can be confirmed that RH-DRP is efficient for risk-based MG operations considering the risk aversion tendency.

Table 6. Optimal power scheduling ($\beta = 0.1$).

Hour	Units (kWh)					
	MT	FC	PV	WT	ESS	Utility
1	6.0715	30.0000	0.0000	1.7900	−6.8677	30.0000
2	6.0024	30.0000	0.0000	1.7800	−8.8357	30.0000
3	6.0446	30.0000	0.0000	1.7900	−8.8879	30.0000
4	6.0282	30.0000	0.0000	1.7800	−7.6826	30.0000
5	6.0493	30.0000	0.0000	1.7900	−1.8190	30.0000
6	6.0045	30.0000	0.0000	0.9200	7.3482	30.0000
7	6.0027	30.0000	0.0000	1.7900	14.3143	30.0000
8	6.0005	30.0000	0.2000	1.3100	27.4728	13.7642
9	29.997	30.0000	3.7500	1.7800	30.0000	−29.3026
10	29.0909	30.0000	7.5300	3.0900	30.0000	−30.0000
11	18.7481	30.0000	10.4500	8.7700	30.0000	−30.0000
12	12.1225	30.0000	11.9500	10.4100	30.0000	−30.0000
13	6.0163	30.0000	23.9000	3.9100	28.9134	−30.0000
14	10.3198	30.0000	20.0500	2.3700	30.0000	−30.0000
15	26.5753	30.0000	7.8700	1.7800	30.0000	−30.0000
16	29.9974	30.0000	4.2200	1.3100	30.0000	−25.8165
17	29.9980	30.0000	0.5500	1.7800	30.0000	−17.1583
18	6.0017	30.0000	0.0000	1.7900	30.0000	21.0241
19	6.0002	30.0000	0.0000	1.3000	28.7291	30.0000
20	6.0005	30.0000	0.0000	1.7800	30.0000	17.9934
21	29.9983	30.0000	0.0000	1.3000	30.0000	−23.3302
22	29.9999	30.0000	0.0000	1.3000	30.0000	−15.6415
23	6.0054	30.0000	0.0000	0.9200	4.7649	30.0000
24	6.0074	30.0000	0.0000	0.6200	−2.6961	30.0000

Table 7. Optimal power scheduling ($\beta = 0.5$).

Hour	Units (kWh)					
	MT	FC	PV	WT	ESS	Utility
1	6.0196	30.0000	0.0000	1.7900	−9.6310	30.0000
2	6.0204	30.0000	0.0000	1.7800	−11.8594	30.0000
3	6.0031	30.0000	0.0000	1.7900	−11.8298	30.0000
4	6.2472	30.0000	0.0000	1.7800	−10.9673	30.0000
5	6.0209	30.0000	0.0000	1.7900	−5.1570	30.0000
6	6.0083	30.0000	0.0000	0.9200	3.5573	30.0000
7	6.0050	30.0000	0.0000	1.7900	10.5224	30.0000
8	6.0014	30.0000	0.2000	1.3100	8.9664	30.0000
9	29.9970	30.0000	3.7500	1.7800	30.0000	−25.9567
10	29.9918	30.0000	7.5300	3.0900	30.0000	−27.3799
11	22.1811	30.0000	10.4500	8.7700	30.0000	−30.0000
12	15.3795	30.0000	11.9500	10.4100	30.0000	−30.0000
13	8.0987	30.0000	23.9000	3.9100	30.0000	−30.0000
14	13.4887	30.0000	20.0500	2.3700	30.0000	−30.0000
15	29.9203	30.0000	7.8700	1.7800	30.0000	−30.0000
16	29.9975	30.0000	4.2200	1.3100	30.0000	−22.2957
17	29.9998	30.0000	0.5500	1.7800	30.0000	−14.5210
18	6.0007	30.0000	0.0000	1.7900	30.0000	21.6680
19	6.0001	30.0000	0.0000	1.3000	25.6865	30.0000
20	6.0004	30.0000	0.0000	1.7800	30.0000	19.2081
21	29.9994	30.0000	0.0000	1.3000	30.0000	−19.8983
22	29.9998	30.0000	0.0000	1.3000	30.0000	−18.0410
23	6.0036	30.0000	0.0000	0.9200	2.5509	30.0000
24	6.0600	30.0000	0.0000	0.6200	−4.6747	30.0000

Table 8. Optimal power scheduling ($\beta = 1$).

Hour	Units (kWh)					
	MT	FC	PV	WT	ESS	Utility
1	6.4208	30.0000	0.0000	1.7900	−13.1397	30.0000
2	6.2111	30.0000	0.0000	1.7800	−15.0381	30.0000
3	6.0744	30.0000	0.0000	1.7900	−14.9114	30.0000
4	6.0550	30.0000	0.0000	1.7800	−13.8230	30.0000
5	6.0597	30.0000	0.0000	1.7900	−8.5424	30.0000
6	6.0133	30.0000	0.0000	0.9200	−0.2126	30.0000
7	6.0036	30.0000	0.0000	1.7900	6.3406	30.0000
8	6.0000	30.0000	0.2000	1.3100	30.0000	8.1371
9	29.9989	30.0000	3.7500	1.7800	30.0000	−22.7060
10	29.9963	30.0000	7.5300	3.0900	30.0000	−23.9606
11	25.5194	30.0000	10.4500	8.7700	30.0000	−30.0000
12	18.5466	30.0000	11.9500	10.4100	30.0000	−30.0000
13	11.1802	30.0000	23.9000	3.9100	30.0000	−30.0000
14	16.5702	30.0000	20.0500	2.3700	30.0000	−30.0000
15	29.9993	30.0000	7.8700	1.7800	30.0000	−26.8263
16	29.9983	30.0000	4.2200	1.3100	30.0000	−18.8726
17	29.9990	30.0000	0.5500	1.7800	30.0000	−10.8822
18	6.0004	30.0000	0.0000	1.7900	30.0000	20.9689
19	6.0002	30.0000	0.0000	1.3000	23.4763	30.0000
20	6.0002	30.0000	0.0000	1.7800	30.0000	19.9705
21	29.9992	30.0000	0.0000	1.3000	30.0000	−16.5598
22	29.9989	30.0000	0.0000	1.3000	30.0000	−19.6862
23	6.0179	30.0000	0.0000	0.9200	0.3184	30.0000
24	6.0214	30.0000	0.0000	0.6200	−7.3639	30.0000

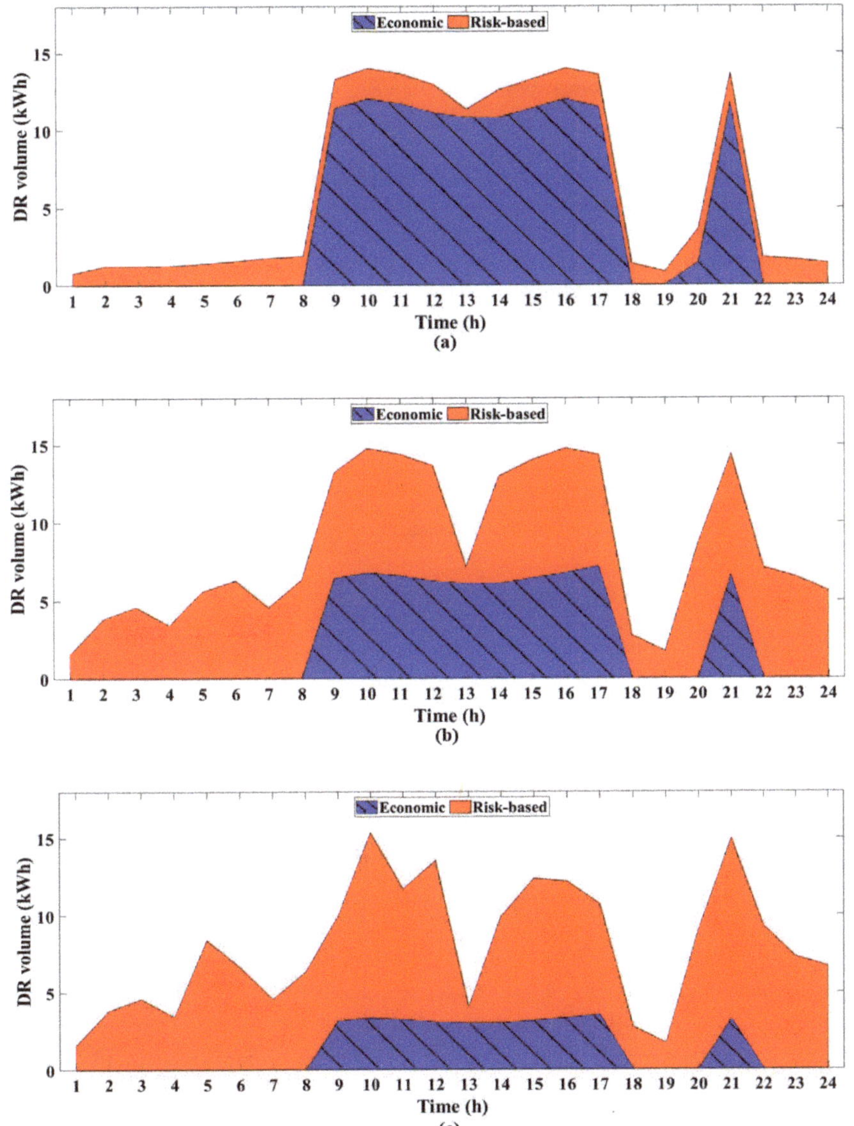

Figure 10. Optimal risk-based hybrid demand response program (RH-DRP) participation volume. (**a**) $\beta = 0.1$, (**b**) $\beta = 0.5$, and (**c**) $\beta = 1$.

6.3. Comparison of DR Analysis

In the comparison analysis, the effects of RH-DRP are verified by comparing with a different DR strategy for the typical low-voltage MGs operation. To evaluate these strategies, three cases are implemented as follows:

Case 1: MG operation model without DRP.
Case 2: MG operation model with prevalent economic DRP.
Case 3: MG operation model with RH-DRP.

All cases are simulated under the same constraints, but the DRP is different. In Case 1, risk-based costs are determined to purchase only real-time market prices without DRP. Case 2, which considers the prevalent economic DRP, is used to minimize operations costs but does not respond to the risk problem. Case 3 is applied with the proposed DRP strategy, which is the optimal risk-based operation in MG using RH-DRP.

Figure 11 indicates the risk-based operational costs and CVaR at various risk parameter for each case. Case 1 shows high CVaR and risk-based operational costs as peak load reduction and risk control constraints are not considered. As Case 2 utilizes all DRP capacity as economic DR, it shows lower operational costs than Case 1; however, the greater the β value to operate the MG, the higher the risk-based operating costs owing to the lack of risk management capabilities. For Case 3, the CVaR is generally lower that of the other cases because the volume ratio of RH-DRP is appropriately adjusted according to β and risk management is effectively performed; furthermore, owing to proper risk management, the risk-based operational cost does not increase significantly, even when MG is operated with a large value of β. Therefore, when MGOs operate MGs considering uncertainty problems, Case 3 promotes optimal economic and stable operation.

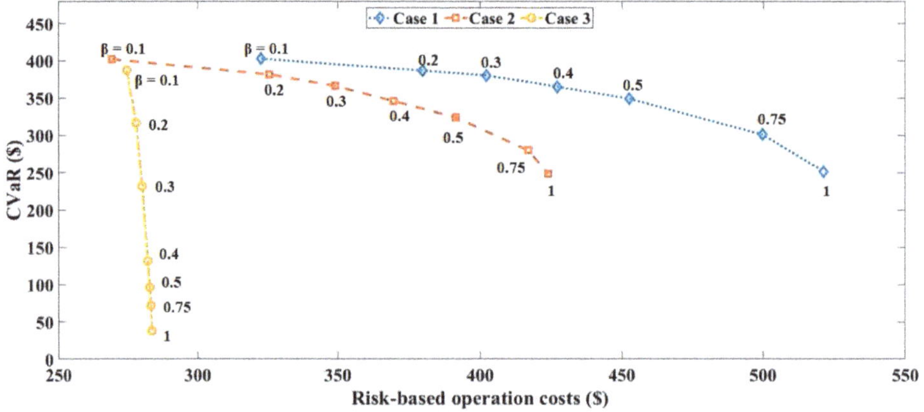

Figure 11. Optimal solution for each case.

6.4. Performance Test for the IML-ABC

To demonstrate the superiority of IML-ABC, performance tests were evaluated without DRP to fairly compare the IML-ABC with various algorithms. Table 9 presents optimal results for the forecasted operation costs or each algorithm. The simulation process for different algorithms was repeated 20 times, with the results of the best, worst, and average solutions shown. The IML-ABC algorithm reached an optimal solution compared with the other algorithms. The IML-ABC performs better than the conventional ABC algorithm in terms of simulation time. Figure 12 illustrates the convergence performances of ABC and IML-ABC. The IML-ABC algorithm found the optimal solution during the first iteration, and, when compared with the ABC algorithm, it improved the initial convergence rate. The IML-ABC and ABC reached less than a 0.1% error value compared for the final solution at 168 and 423 iterations, respectively. Therefore, these results verify that the simulation time necessary to find the optimal value is improved by reducing the number of unnecessary iterations.

Table 9. Comparison results for various algorithms.

Algorithm	Best Solution ($)	Worst Solution ($)	Average ($)	Mean Simulation Time (sec)
GA [38]	277.7444	304.5889	290.4321	-
PSO [38]	277.3237	303.3791	288.8761	-
AMPSO [38]	274.4317	274.7318	274.5643	-
θ-PSO [39]	275.3491	281.7841	277.6191	-
GSA [40]	275.5369	282.1743	277.8021	-
ABC	274.3801	274.9271	274.6536	35.28
IML-ABC	272.4761	272.6803	272.5782	29.11

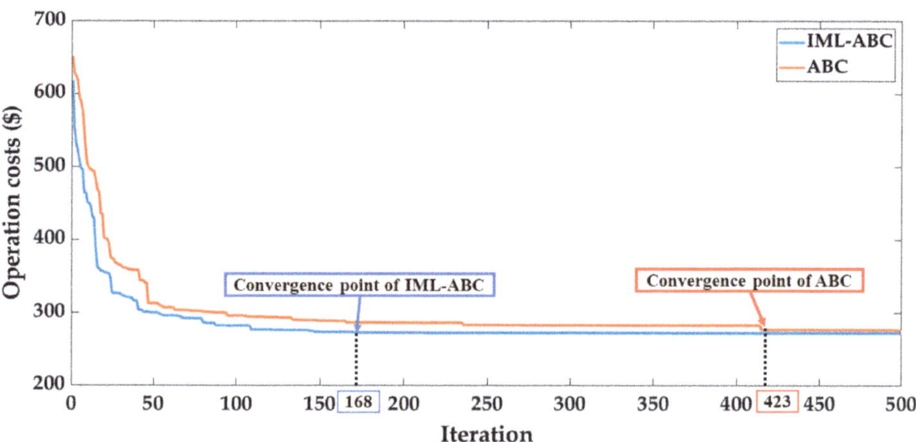

Figure 12. Convergence diagram using both artificial bee colony (ABC) and improved multi-layer (IML)-ABC algorithms.

7. Conclusions

In this study, we proposed two-stage optimal operation of MGs considering uncertainty problems with risk-based DR strategies. To account for uncertainty, WT, PV, and load were modeled as Weibull, Beta, and normal PDF, respectively. The proposed RH-DRP provides a hybrid solution using both economic and risk-based DR to reduce risk-based operational costs. The objective optimization problem consisted of two-stage optimization using the IML-ABC. In the first-stage, the expected operational costs were calculated by deterministic inputs; in the second stage, the risk-based operational costs considered CVaR. To verify the effectiveness of the proposed approach, simulation was conducted on a low-voltage grid-connected MG. The results demonstrate that reasonable operational costs can be determined under the risk aversion parameters compared to conventional optimization solutions. Moreover, our RH-DRP remarkably reduces risk-based operational costs via comparison with various DR strategies. The comparison with various other algorithms also confirmed the superiority of the proposed IML-ABC in identifying the optimal solution and reducing simulation times. Therefore, the proposed approach provides MGOs not only an additional degree of freedom in decision-making under uncertainty but also a solution to reduce risk-based operational costs. Our future work is under way to focus on not only reducing risk-based operating costs but also solving reliability about dynamic problems caused by uncertainty.

Author Contributions: H.-S.R. proposed the main idea of this paper, and M.-K.K. coordinated the proposed approach and thoroughly reviewed in the manuscript. All authors have read and agreed to the published version of the manuscript.

Funding: This research was supported by the Korea Electric Power Corporation (grant numbers: R18XA06-75). This research was also supported by Basic Science Research Program through the National Research Foundation of Korea (NRF) funded by the Ministry of Education (2020R1A2C1004743).

Conflicts of Interest: The authors declare no conflict of interest.

References

1. Ding, Z.; Hou, H.; Yu, G.; Hu, E.; Duan, L.; Zhao, J. Performance analysis of a wind-solar hybrid power generation system. *Energy Convers. Manag.* **2019**, *181*, 223–234. [CrossRef]
2. Millstein, D.; Wiser, R.; Bolinger, M.; Barbose, G. The climate and air-quality benefits of wind and solar power in the United States. *Nat. Energy* **2017**, *2*. [CrossRef]
3. Jo, K.-H.; Kim, M.-K. Stochastic Unit Commitment Based on Multi-Scenario Tree Method Considering Uncertainty. *Energies* **2018**, *11*, 740. [CrossRef]
4. Kim, M.K. Multi-objective optimization operation with corrective control actions for meshed AC/DC grids including multi-terminal VSC-HVDC. *Int. J. Electr. Power Energy Syst.* **2017**, *93*, 178–193. [CrossRef]
5. Mansouri, S.; Ahmarinejad, A.; Ansarian, M.; Javadi, M.; Catalao, J. Stochastic planning and operation of energy hubs considering demand response programs using Benders decomposition approach. *Int. J. Electr. Power Energy Syst.* **2020**, *120*, 106030. [CrossRef]
6. Kim, H.; Kim, M.-K. Multi-Objective Based Optimal Energy Management of Grid-Connected Microgrid Considering Advanced Demand Response. *Energies* **2019**, *12*, 4142. [CrossRef]
7. Sharma, S.; Bhattacharjee, S.; Bhattacharya, A. Probabilistic operation cost minimization of Micro-Grid. *Energy* **2018**, *148*, 1116–1139. [CrossRef]
8. Che, L.; Zhang, X.; Shahidehpour, M.; Alabdulwahab, A.; Abusorrah, A. Optimal interconnection planning of community microgrids with renewable energy y sources. *IEEE Trans. Smart Grid* **2017**, *8*, 1054–1063. [CrossRef]
9. Alipour, M.; Mohammadi-Ivatloo, B.; Zare, K. Stochastic Scheduling of Renewable and CHP-Based Microgrids. *IEEE Trans. Ind. Inform.* **2015**, *11*, 1049–1058. [CrossRef]
10. Niknam, T.; Golestaneh, F.; Malekpour, A. Probabilistic model of polymer exchange fuel cell power plants for hydrogen, thermal and electrical energy management. *J. Power Sources* **2013**, *229*, 285–298. [CrossRef]
11. Motevasel, M.; Seifi, A.R. Expert energy management of a micro-grid considering wind energy uncertainty. *Energy Convers. Manag.* **2014**, *83*, 58–72. [CrossRef]
12. Wang, M.Q.; Gooi, H.B. Spinning Reserve Estimation in Microgrids. *IEEE Trans. Power Syst.* **2011**, *26*, 1164–1174. [CrossRef]
13. Rezaei, N.; Kalantar, M. Smart microgrid hierarchical frequency control ancillary service provision based on virtual inertia concept: An integrated demand response and droop controlled distributed generation framework. *Energy Convers. Manag.* **2015**, *92*, 287–301. [CrossRef]
14. Albadi, M.; El-Saadany, E. A summary of demand response in electricity markets. *Electr. Power Syst. Res.* **2008**, *78*, 1989–1996. [CrossRef]
15. Aghajani, G.; Shayanfar, H.; Shayeghi, H. Demand side management in a smart micro-grid in the presence of renewable generation and demand response. *Energy* **2017**, *126*, 622–637. [CrossRef]
16. Jalali, A.; Aldeen, M. Risk-Based Stochastic Allocation of ESS to Ensure Voltage Stability Margin for Distribution Systems. *IEEE Trans. Power Syst.* **2018**, *34*, 1264–1277. [CrossRef]
17. Kim, S.; Lee, Y.; Kim, M.-K. Flexible risk control strategy based on multi-stage corrective action with energy storage system. *Int. J. Electr. Power Energy Syst.* **2019**, *110*, 679–695. [CrossRef]
18. Fossati, J.P.; Galarza, A.; Martín-Villate, A.; Fontán, L. A method for optimal sizing energy storage systems for microgrids. *Renew. Energy* **2015**, *77*, 539–549. [CrossRef]
19. Samadi, P.; Wong, V.W.; Schober, R. Load Scheduling and Power Trading in Systems With High Penetration of Renewable Energy Resources. *IEEE Trans. Smart Grid* **2016**, *7*, 1802–1812. [CrossRef]
20. Sortomme, E.; El-Sharkawi, M.A. Optimal Power Flow for a System of Microgrids with Controllable Loads and Battery Storage. In Proceedings of the 2009 IEEE/PES Power Systems Conference and Exposition, Seattle, WA, USA, 15–18 March 2009; pp. 1–5. [CrossRef]
21. Mohammadi, S.; Soleymani, S.; Mozafari, B. Scenario-based stochastic operation management of MicroGrid including Wind, Photovoltaic, Micro-Turbine, Fuel Cell and Energy Storage Devices. *Int. J. Electr. Power Energy Syst.* **2014**, *54*, 525–535. [CrossRef]

22. Mazidi, M.; Zakariazadeh, A.; Jadid, S.; Siano, P. Integrated scheduling of renewable generation and demand response programs in a microgrid. *Energy Convers. Manag.* **2014**, *86*, 1118–1127. [CrossRef]
23. Gazijahani, F.S.; Salehi, J. Optimal Bilevel Model for Stochastic Risk-Based Planning of Microgrids Under Uncertainty. *IEEE Trans. Ind. Informatics* **2017**, *14*, 3054–3064. [CrossRef]
24. Petrollese, M.; Valverde, L.; Cocco, D.; Cau, G.; Guerra, J. Real-time integration of optimal generation scheduling with MPC for the energy management of a renewable hydrogen-based microgrid. *Appl. Energy* **2016**, *166*, 96–106. [CrossRef]
25. Kanungo, T.; Mount, D.; Netanyahu, N.; Piatko, C.; Silverman, R.; Wu, A. An efficient k-means clustering algorithm: Analysis and implementation. *IEEE Trans. Pattern Anal. Mach. Intell.* **2002**, *24*, 881–892. [CrossRef]
26. Shirazi, M.M.H.; Mowla, D. Energy optimization for liquefaction process of natural gas in peak shaving plant. *Energy* **2010**, *35*, 2878–2885. [CrossRef]
27. Moghaddam, M.P.; Abdollahi, A.; Rashidinejad, M. Flexible demand response programs modeling in competitive electricity markets. *Appl. Energy* **2011**, *88*, 3257–3269. [CrossRef]
28. Soroudi, A.; Amraee, T. Decision making under uncertainty in energy systems: State of the art. *Renew. Sustain. Energy Rev.* **2013**, *28*, 376–384. [CrossRef]
29. Khodabakhsh, R.; Sirouspour, S. Optimal Control of Energy Storage in a Microgrid by Minimizing Conditional Value-at-Risk. *IEEE Trans. Sustain. Energy* **2016**, *7*, 1264–1273. [CrossRef]
30. Karaboga, D.; Basturk, B. A powerful and efficient algorithm for numerical function optimization: Artificial bee colony (ABC) algorithm. *J. Glob. Optim.* **2007**, *39*, 459–471. [CrossRef]
31. Zhang, H.; Xie, Z.; Lin, H.-C.; Li, S. Power Capacity Optimization in a Photovoltaics-Based Microgrid Using the Improved Artificial Bee Colony Algorithm. *Appl. Sci.* **2020**, *10*, 2990. [CrossRef]
32. Al-Sakkaf, S.; Kassas, M.; Khalid, M.; Abido, M.A. An Energy Management System for Residential Autonomous DC Microgrid Using Optimized Fuzzy Logic Controller Considering Economic Dispatch. *Energies* **2019**, *12*, 1457. [CrossRef]
33. Habib, H.U.R.; Subramaniam, U.; Waqar, A.; Farhan, B.S.; Kotb, K.M.; Wang, S. Energy Cost Optimization of Hybrid Renewables Based V2G Microgrid Considering Multi Objective Function by Using Artificial Bee Colony Optimization. *IEEE Access* **2020**, *8*, 62076–62093. [CrossRef]
34. Aghajani, G.; Shayanfar, H.; Shayeghi, H. Presenting a multi-objective generation scheduling model for pricing demand response rate in micro-grid energy management. *Energy Convers. Manag.* **2015**, *106*, 308–321. [CrossRef]
35. Faria, P.; Vale, Z. Demand response in electrical energy supply: An optimal real time pricing approach. *Energy* **2011**, *36*, 5374–5384. [CrossRef]
36. Apx Power Spot Exchange. Available online: https://www.apxgroup.com/trading-clearing/apx-power-uk/ (accessed on 1 December 2019).
37. The Solar Power Group Company. Available online: http://thesolarpowergroup.com (accessed on 1 December 2019).
38. Anvari-Moghaddam, A.; Seifi, A.R.; Niknam, T.; Pahlavani, M.R.A. Multi-objective operation management of a renewable MG (micro-grid) with back-up micro-turbine/fuel cell/battery hybrid power source. *Energy* **2011**, *36*, 6490–6507. [CrossRef]
39. Baziar, A.; Kavousi-Fard, A. Considering uncertainty in the optimal energy management of renewable micro-grids including storage devices. *Renew. Energy* **2013**, *59*, 158–166. [CrossRef]
40. Niknam, T.; Golestaneh, F.; Malekpour, A. Probabilistic energy and operation management of a microgrid containing wind/photovoltaic/fuel cell generation and energy storage devices based on point estimate method and self-adaptive gravitational search algorithm. *Energy* **2012**, *43*, 427–437. [CrossRef]

Publisher's Note: MDPI stays neutral with regard to jurisdictional claims in published maps and institutional affiliations.

Article

A Decentralized Informatics, Optimization, and Control Framework for Evolving Demand Response Services

Sean Williams [1,*], Michael Short [1], Tracey Crosbie [1] and Maryam Shadman-Pajouh [2]

1 School of Computing, Engineering and Digital Technologies, Teesside University, Middlesbrough, Tees Valley TS1 3BX, UK; m.short@tees.ac.uk (M.S.); t.crosbie@tees.ac.uk (T.C.)

2 Teesside University Business School, Teesside University, Middlesbrough, Tees Valley TS1 3BX, UK; m.shadmanpajouh@tees.ac.uk

* Correspondence: sean.williams@tees.ac.uk

Received: 1 May 2020; Accepted: 11 August 2020; Published: 13 August 2020

Abstract: This paper presents a decentralized informatics, optimization, and control framework to enable demand response (DR) in small or rural decentralized community power systems, including geographical islands. The framework consists of a simplified lumped model for electrical demand forecasting, a scheduling subsystem that optimizes the utility of energy storage assets, and an active/pro-active control subsystem. The active control strategy provides secondary DR services, through optimizing a multi-objective cost function formulated using a weight-based routing algorithm. In this context, the total weight of each edge between any two consecutive nodes is calculated as a function of thermal comfort, cost (tariff), and the rate at which electricity is consumed over a short future time horizon. The pro-active control strategy provides primary DR services. Furthermore, tertiary DR services can be processed to initiate a sequence of operations that enables the continuity of applied electrical services for the duration of the demand side event. Computer simulations and a case study using hardware-in-the-loop testing is used to evaluate the optimization and control module. The main conclusion drawn from this research shows the real-time operation of the proposed optimization and control scheme, operating on a prototype platform, underpinned by the effectiveness of the new methods and approach for tackling the optimization problem. This research recommends deployment of the optimization and control scheme, at scale, for decentralized community energy management. The paper concludes with a short discussion of business aspects and outlines areas for future work.

Keywords: decentralized; demand response; optimization; community energy management

1. Introduction

1.1. Context and Motivation

The effectiveness of modern technologies continues to improve energy efficiency. However, this does not necessarily translate to a fall in energy demand [1]. Reduction in energy consumption due to technology improvements, somewhat paradoxically, causes energy actors to consume more energy [2]. There is evidence that ongoing trends in energy consumption exist on both the production and consumption side [3]. While policy interventions are advancing technology and economic growth, this is causing environmental stress [4]. Therefore, it is important to improve energy access that is sustainable to help mitigate the risks associated with one of the most extraordinary growth paths in modern times. The ever-increasing presence of sustainable energy supply is lessening harmful emissions from fossil fuel power plants, which contribute to a rise in greenhouse gases [5]. However, the intensified uncertainties associated with modern power systems operating close to their stability boundaries means operators are facing acute challenges when maintaining continuity of supply [6]. Demand response (DR) is an important tool in the energy systems of many developed and industrialized countries. In a future power system, where the contribution of inertia alone can no longer provide resilience during sudden changes in frequency, DR provides an effective mechanism to help balance supply and demand [7].

Traditionally, electricity markets have evolved on the assumption that electric utilities and system network operators will supply all power demands whenever they occur [8]. However, centralized generation and distribution through an ageing infrastructure of high voltage distribution networks to regional system operators are becoming more vulnerable to energy security [9,10]. In 2015, circa 80% of global energy consumption was generated using fossil fuel [11]. Delivery of low carbon, energy-efficient solutions have become more prevalent in recent years [12]. The move away from large fossil fuel power plants operating on a centralized configuration is motivated by greater digitization, the drive for decarbonization and a need for more customer control in energy management [13]. Therefore, to achieve carbon reduction goals, an obvious decarbonization strategy is to extend fuel mix diversity in the electricity sector while displacing the highest polluting power plants [14].

Energy systems are undergoing disruptive change. In the UK, the number of decentralized energy operations is on the increase [15]. These changes are motivated in part by an increasing political drive in response to environmental policy priorities. Consequently, this is provoking a shift towards decentralized energy systems, business models that involve community energy groups and emerging new regulations simultaneously [16]. Innovations in energy evolution are characterized in part by industrial strategy and relations to decarbonization [14]. The fall in the cost of renewables has been significant in the last ten years, which means generating electrical energy from renewables is more economically viable [17]. Nevertheless, when combined with an increased burden on present-day centralized services, risks associated with long-term supply security and the drive to be carbon neutral by 2050 are exposed. While market signals and shifts in government policy are guiding the energy sector transformation, system operators have developed many control strategies to preserve equilibrium in grid frequency during periods of peak demand, including DR.

The UK government has set ambitious targets for electric cars and electrification of heating [18]. These bold steps are accelerating the decarbonization of vehicles and encouraging innovation in electrification technologies, which will further increase the demand for electrical power. The recent emergence of smart cities and communities helps population clusters to become more efficient and their energy infrastructures more sustainable [19,20]. By integrating smart technologies, coupled with a network of sensors and intelligent algorithms, it is often reported that urban smartness is at the forefront of the sustainability transition [21]. However, the realization of smart cities is dependent on concerns about data protection, digital health of interconnected communities, and the reliability of services being addressed [22]. In sustainable development scenarios, a transition towards low carbon energy will operate on different geographical scales. Increased customer participation and increased demand require the decentralization of energy supply [23]. Smart (energy) cities should not only support local needs in terms of energy demands but also feature broader regional or national network demands. However, while the development of smart grids is necessary to modernize the electricity market, many of the reported environmental and security benefits are only realized when they are combined with decentralized energy generation [24]. Besides this, demand for new building stock continues to accelerate, driven in part by renewed industrialization and economic growth [25].

1.2. Previous Work

Studies have highlighted building energy consumption, and contribution to greenhouse gases is significant [26]. In the context of smart energy developments, regulatory control of heating, ventilating, and air conditioning (HVAC) processes in buildings and other thermostatically controlled loads make then exceptionally suitable candidates for providing energy flexibility to the grid [27]. Many control strategies that aim to improve the operation of heating systems have been proposed (e.g., see Reference [28–30]). The slow thermal dynamics and rather stochastic characteristics of buildings (including occupants) mean their power consumption can be easily shifted as part of a DR mechanism without causing a significant short-term impact on space temperatures [31].

Developing energy efficiency in energy systems is perhaps the most sustainable way to reduce carbon emissions [32]. Providing access to electricity brings many socio-economic benefits. Various studies have shown how small-scale distributed renewables are changing people's lives. But many island energy communities fall behind mainland energy network developments when it comes to securing affordable and sustainable supplies. Community energy networks that comprise a small number of distributed renewable electricity generators (DREG) are often more exposed to system vulnerabilities due to the intermittent nature of their energy production [33]. Still, for population clusters that are dependent on conventional diesel generators, decentralized developments offer an alternative sustainable clean energy transition pathway. More recent studies show low carbon smart energy systems offer interconnected islands new opportunities for energy independence [34,35]. With this in mind, harvesting energy from natural resources to achieve specific targets of decarbonization can be realized using smart energy systems combined with efficient control strategies aimed at balancing energy demand and energy production [36,37].

The energy market is moving from a linear centralized system to a more flexible, sophisticated and decentralized system. A decentralized approach can deliver electricity in a controlled environment, providing network operators access to frequency regulation and balancing services [38–40]. Flexibility in energy generation and utility become more prevalent in small geographical areas. Here, a smart grid approach provides technology infrastructure opportunities that enable intermittent DREG to connect with local battery energy storage systems. However, distributed energy installations require coordination mechanisms, especially when network operators request flexibility in consumer behavior to secure operation of the power system. In the context of small island communities, optimization and control of decentralized energy systems may bring economic reward, improve energy security and open opportunities for the end-users to become more active in energy management [41]. Even so, one of the main challenges of integrating several intermittent DREG is the power systems ability to respond to a change in demand. In the absence of robust communication networks, or negative impact due to latency, the ability to react quickly enough is problematic [42].

In contrast, local direct control DR processes may offer a more reactive approach by redistributing energy consumption in response to changes in grid frequency measured at source. However, motivations for decentralization are not universally consistent, and embracing a carbon reduction pathway through decarbonization initiatives is not always the main priority for instigating change [43]. Therefore, these schemes must not be to the detriment of the end customers, such as adversely affecting the thermal comfort of building occupants or loss of essential services [44]. With this in mind, it is important to note that substituting energy from fossil fuels with suitable sustainable energy sources to meet the needs and expectations of the community will help improve the quality of human life [45].

The achievement of a decentralized energy system requires the integration of multiple natural resources, often supplemented by some form of reserve capacity (e.g., electricity storage systems for providing ancillary services or diesel generators for backup power). Furthermore, if the benefits of low carbon power systems within a decentralized setting are to be achieved, then energy management mechanisms must be capable of coordinating and managing a flexible set of services, each characterized by local resources [46]. Alongside the physical transformations, demand side management becomes the most important dimension, especially when there is a tendency to empower consumers to generate electricity [47]. A recent study highlights that prosumers are likely to play a crucial and enabling role in a decentralized system [48]. Ultimately, efficiency improvements established using optimization and control algorithms (demand side management) will help lower emissions and supply energy needs.

As a general proposition, the objective for energy planning is to develop a system that satisfies a dynamic energy forecasting requirement for community energy needs and is consistent with sustainable development scenarios. In contrast, the objective of the optimization procedure will be formulated during the analysis of energy potentials and their geographical location. Such expositions suggest optimization problems may be categorized as either one-dimensional or multi-dimensional depending on the predefined objectives [49]. However, in the case of energy efficiency, there is ample evidence that shows most optimization problems are defined by at least two objectives: time and energy. In practice, many real-world problems are defined as a process of finding a minimal value of an n-dimensional function subject to a set of constraints that may or may not be related [50]. The control of power demand in response to variations in grid frequency is an essential part of the smart grid vision. In the context of DR, the existing research methods are broadly divided into two types, where one approach focuses on classical demand response programs, such as direct control, as well as initiatives that aim to curtail energy consumption during peak times, usually through financial incentives. Furthermore, robust communication protocols are needed to supervise interaction between network operators. A recent feasibility study was conducted on the Italian Pelagie Islands proposed a control system that incorporated DR services [51]. Here, besides cost and usability having been the main features of the DR solution, a telecommunication infrastructure was fundamental to ensure effective regulatory control and exchange of information. Islands have often served as test platforms for distributed smart energy systems [52]. However, most remote communities do not attract this level of energy technology innovation; therefore, such an architecture is out of reach.

In sum, technology innovation, guided by indicators, such as greenhouse gas emissions, is helping policymakers understand the energy transition. Decarbonization pathways are transforming ageing energy (electrical) infrastructures into more flexible decentralized systems. Installation of more remote small-scale renewables means prosumers are more active in energy management. Studies show thermal inertia means community buildings have an important role in demand response. However, although there is a growing amount of research about smart cities, there have been few investigations into the impacts of similar technology insertions in more remote or islanded communities. To fill this gap, this work offers a novel optimization and control technique that supports primary and secondary DR services using pro-active/active control, respectively. Furthermore, the multi-objective optimization algorithm is formulated to optimize the use of thermostatically controlled loads; in this scenario, this is space heating in community-level buildings.

1.3. Contribution

The main objective of this paper is to introduce a decentralized, optimization, and control framework for community energy management. The methodology considers local environmental conditions, user feedback, and economic impacts at the same time as providing flexible primary and secondary demand side response. The contributions of this paper include:

- An integrated, flexible real-time optimization and control framework based on a weight-based routing algorithm, significantly improving efficiency by removing communication network constraints usually associated with centralized control schemes.
- A detailed computational study considering technical and environmental parameters.
- Applying the proposed optimization and control algorithm using prototype hardware in an experiment designed to evaluate interaction with real-world data.

1.4. Structure

The remainder of this paper is structured as follows. Section 2 introduces a generic framework before presenting a detailed description of a real-case study optimization and control framework, including the computer simulation model and its components. Section 3 discusses the results of extensive simulation studies. Experimental tests that validate the optimizer application in real-world conditions are presented in Section 4. Finally, Section 5 summarizes the main conclusions that can be drawn from the work presented and provides insights into what these suggest for future work.

2. Optimization and Control Framework Technical Development

2.1. Generic Framework

A generic decentralized, optimization, and control framework can be used as part of an evolving demand response service; this means both curtailment and generation. This general arrangement will support primary and secondary DR services through frequency regulation and optimal control mechanisms, respectively, and tertiary DR events (Figure 1). Here, optimal performance might be described in terms of energy cost (ec), thermal comfort (tc), and predicted future energy demands (dv). A multi-objective cost function formulated using a weight-based routing algorithm automatically regulates the control of heating to create a meaningful energy demand reduction by shifting energy consumption to out of peak demand periods. Thermostatically controlled loads (TCL) can provide auxiliary services [53]. In this approach, the proposed scheme offers a pro-active control mechanism that changes the TCL operating setpoint proportionally to measured grid frequency. Following this approach avoids synchronization problems that bound the coupling between frequency excursions and load dynamics that switch when prescribed frequency thresholds are exceeded [40]. An optimization algorithm that responds to the real thermal needs of the building occupants is proposed. To achieve this, individual occupants can report their thermal comfort needs using smartphone technology. The feedback reports are processed, and a consensus determined, which is in turn used to influence the room temperature.

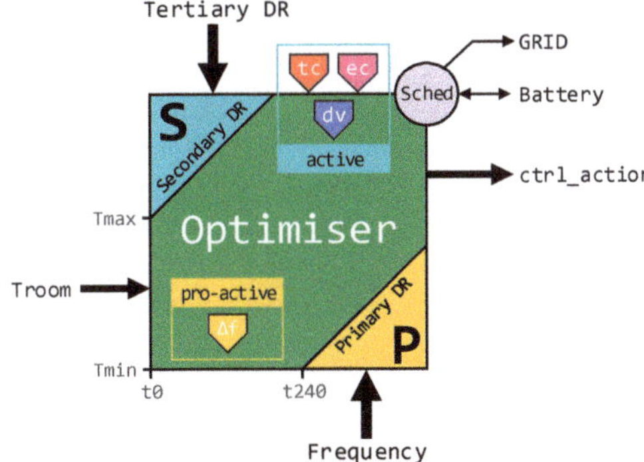

Figure 1. A demand response (DR) framework block diagram.

The inclusion of building occupant feedback is crucial. Recent research has illustrated that engineers tend to assume occupants will not feel small changes in temperature [44]. This oversight can cause a performance gap between the expected and actual results from technologies intended to reduce or shift energy consumption in buildings. The inclusion of occupant feedback ensures that this issue will be avoided in the case of the solution presented in this paper.

This work provides a reference basis for further DR applications in decentralized community-based environments. It is particularly relevant to microgrids that are isolated from the grid as it offers potential for reducing the amount of energy storage required to balance the power fluctuation on those isolated microgrids. Current research has shown that even in the case of a single consumer, a microgrid option could be more economical than network renovation (e.g., provision of underground cabling) to increase the grid reliability [54]. Therefore, the ability to reduce the costs further by utilizing the approach described in this paper could offer real potential for the development of islanded and semi-islanded microgrids in many contexts.

2.2. General Description

The proposed decentralized, informatics, optimization, and control simulation model has been developed to optimize space heating, schedule utility of energy storage assets, and provide pro-active/active control for primary and secondary DR services. Two groups define the simulation model data that aims to replicate the trajectory of the physical systems under consideration so that system configuration parameters can be differentiated from local preferences. Ultimately, the simulation model is designed to assess our understanding of the optimizer and control components in the context of decentralized energy management. The applicability of the optimizer and control component is further demonstrated in hardware-in-the-loop simulation.

The following outline is provided as an overview of the proposed optimization and control strategy. The approach is based on the idea that when the demand for electricity on the distribution network is high, then the system attempts to reduce the local rate of energy consumption by reducing the space heating temperature setpoint. Similarly, during periods of low electricity demand the constraints that govern the temperature setpoint are relaxed, which, in turn, allows, not mandates, an increase in energy consumption by increasing the space heating temperature setpoint.

When we add a measured response from occupants that describes their collective relative thermal comfort, the perception is the rate of energy consumption shifts towards being self-regulatory. For example, if the demand for electricity increases, the system attempts to reduce the local energy consumption at a rate that is inversely proportional to the predicted demand. If space remains void of occupants, this approach is satisfactory and local settings ensure a minimum space temperature is maintained. However, during periods of occupancy, individuals become eligible participants in the optimization algorithm. Subsequently, when individuals report they are feeling cold, and their collective measured response satisfies a set threshold, then the resultant action is to issue a command that counters the instruction to reduce the space temperature further. Conversely, this self-regulatory behavior works equally well during periods of low demand. Consider now introducing a third data type. Incentivizing energy reduction through financial gain aims to reduce or shift energy consumption during periods of high demand [55].

Including information about the cost of energy into the mix introduces an interesting dynamic to the optimization and control strategy. Given a time of use tariff that increases at times when demand is known to peak, the net contribution to the optimizer is to automatically adjust the energy consumption when the cost of electricity exceeds a user-defined threshold. Furthermore, the system can be configured to automatically switch to an alternative power source if demand exceeds a set limit or during periods when the cost of energy makes utilizing an alternative power source more attractive, e.g., energy storage assets.

The immediate outcome attributed to the interaction between the three data types becomes even more attractive if their behaviors can be predicted over a finite time horizon. The opportunity to participate in tertiary DR services by making ready the system in response to a network operator DR instruction becomes feasible. The proposed control algorithm alters the demand profile trajectory such that it adds bias to the tri-data mix in a way that promotes a rise in space temperature. The net effect is to provide optimal space pre-heating in advance of commencing the scheduled DR event. Furthermore, a switching mechanism denies use of a local energy storage asset for a period leading up to the DR event. Instead, resources ensure the energy storage asset is set to recharge. Thus, when the DR event period commences, the system power source automatically switches to the energy storage asset. Previous interventions ensure the energy storage asset capacity is sufficiently charged to enable it to remain the primary source for the duration of the event or until the asset can no longer meet the power demand for continued operation. In this instance, the grid becomes the systems primary power source, and recharging of the energy storage asset is initiated.

The remainder of the section describes the development of individual systems that contribute to the optimization and control framework. Real-time computer simulations that aim to model the behavior of physical systems and the mathematical model of the proposed optimization and control algorithms are performed using the MATLAB/Simulink® environment. Level-2 MATLAB System functions have been used extensively during the design and implementation, providing access to create custom blocks that support multiple input and output ports. Furthermore, this section describes how desktop simulations are reconfigured to validate the optimization and control algorithm using hardware-in-the-loop (HIL) simulation techniques.

The desktop simulation model is shown in Figure 2. In addition to the optimization and control block, the model is composed of a catalogue of supporting subsystems: energy, building, scheduler, date-time (dt), and demand event signal (des).

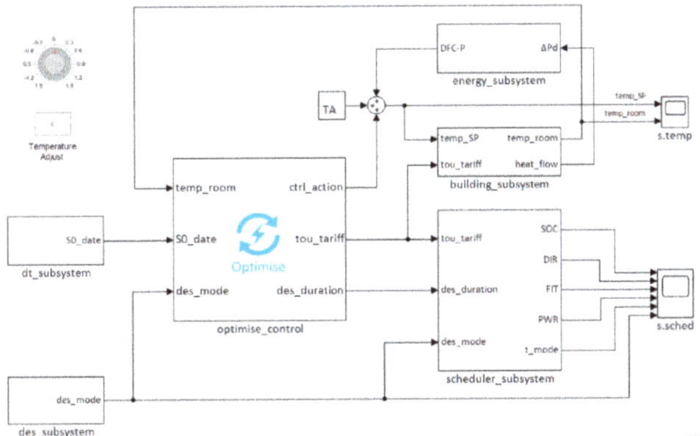

Figure 2. Simulink® model of energy optimization framework.

2.3. Technical Development

The simulation optimizer is constructed in a piecemeal fashion, progressing sequentially by solving problems centered on three data types: (1) thermal comfort, (2) electricity demand forecast, and (3) cost (tariff). In brief, during periods when the system is not responding to a tertiary DR activity, the active control process begins by calculating a predicted or actual value for each data type over a 4-h horizon window at 10 min intervals. Values are mapped onto a multi-dimensional array with a fixed number of rows (magnitude) and columns (time). A Dijkstra's algorithm is then used to project the predicted values over the 4-h horizon window [56,57]. The contribution of each data type is then combined before *k*-means clustering (see Reference [58,59]) is applied iteratively at each 10 min interval. The result yields a new path that follows the optimal temperature setpoint trajectory over the 4-h horizon window. For demand response applications, a model for building design can be successfully implemented using a simplified first-order plus dead time model [60]. Time constants of 10 to 30 min and dead-times between 0 to 5 min are typical [61]. Avoiding complex calculations is achieved by taking a pragmatic approach when determining model control actions. For example, the proposed optimizer has been configured to update the control action at a sample time 10 min.

Since the control objective is to minimize the deviations from a temperature setpoint, according to the system and user-defined rules, at discrete points in time, the optimal cost (shortest path) can be obtained by formulating a Dynamic Programming algorithm that proceeds backwards in time. The algorithm takes a sequence of *k*-means centroid points, where each centroid represents a value that minimizes the total intra-cluster variance of all objects in each cluster. In simple terms, given a time horizon of 240 min, this equates to 24 stages, each separated by a 10-min interval. At each stage, there are 11 objects. A *k*-means algorithm is applied to find the centroid of the 11 objects, at each stage. These calculations result in a series of 24 centroids that contribute to formulating the shortest path.

The objects that belong to each cluster are derived from a series of functions that calculate occupants' relative thermal comfort cost (*tc*), rate of energy consumption (demand forecast value) cost (*dv*), and energy cost (*ec*). Given this, a deterministic problem can be formulated in a finite space

G, which can be equivalently represented by a gridmap of fixed dimension; the problem starts from a source node κ_s, where $\kappa_s = \kappa_0 = G_{(j,S_0)}$, proceeds to $\kappa_1 \in S_1$, and progresses to the final node $\kappa_t = \kappa_n = G_{(j,S_n)}$. An important characteristic of this activity is highlighted. In solving the shortest path problem, the source node κ_s and target node κ_t are revealed to the optimizer just before the first transition from $S_0 \rightarrow S_1$ begins. The trajectory of the shortest path from $S_0 \xrightarrow{\eta} S_n$ will follow a series of weighted edges η that interconnect successive pairs of nodes, i.e., (κ_0, κ_1), $(\kappa_1, \kappa_2), \ldots, (\kappa_{n-1}, \kappa_n)$.

In the framework of the fundamental problem, minimizing the cost in a finite state space G can be translated into mathematical terms:

$$J_n(i) = \min_{\kappa \in S_{n+1}} \left[c^n_{i\kappa} + J_{n+1}(\kappa) \right], \quad i \in S_n, \quad n = 0, 1, 2, \ldots, 24, \tag{1}$$

where the cost of transition at $c^n_{i\kappa}$ is the centroid in a cluster of objects at stage S_n from node $i \in S_n$ to node $\kappa \in S_{n+1}$.

For the problem to have a solution, each object centroid is constructed with k-means++ algorithm. Here, after initially assigning a random object within a cluster as the first centroid, we compute the distance from each remaining object. Based on the square of these distances, a new centroid is defined. The process repeats until k centroids are chosen. We formulate the objects in the following sections.

In addition, when the network operator issues an explicit DR instruction, the optimizer initiates a pre-programmed control strategy that changes the trajectory of subsequent control actions in a period leading up to and during the event window. However, it remains useful if the control actions continue to respond to facility or occupant needs during this mode of operation.

2.4. Optimize and Control Subsystem

The optimize and control subsystem (optimize_control) is a user-defined block written using the MATLAB S-Function application programming interface (API). The proposed optimization algorithm calculates the optimal space heating temperature according to the rate at which electricity is consumed (demand) and cost (tariff). Furthermore, the final temperature value is impacted by the occupants' thermal responses to the combined thermal effect of the environment and physiological variables that influence the relative thermal comfort.

Figure 2 shows the Simulink® optimize and control block includes three input signals: (1) room temperature (temp_room), (2) current date and time (S0_date), and (3) a demand event signal that indicates the status of a tertiary DR service (des_mode). The block output signals provide: (1) a control signal (ctrl_action) that will alter the space heating temperature setpoint, (2) the current cost of energy usage (tou_tariff), and (3) an indication of the tertiary DR event duration (des_duration). The internal architecture of the optimize and control subsystem is shown in Figure 3.

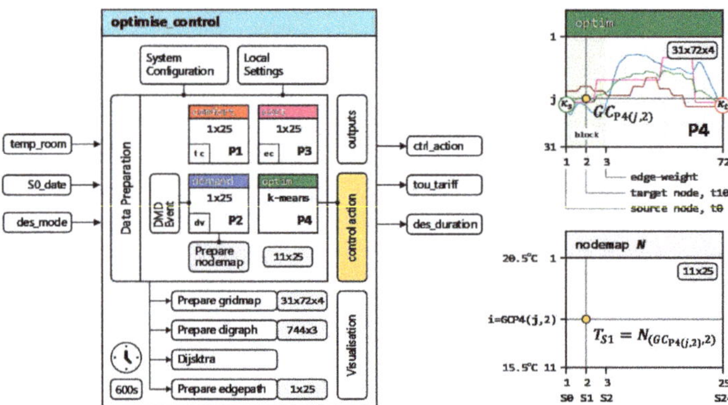

Figure 3. Optimize and control internal block diagram.

The design presented in this article is configured to operate within a custom-built temperature range between T_{min} = 15.5 °C and T_{max} = 20.5 °C. Exception handling ensures temperature values measured outside this range are mapped to either 15.5 °C or 20.5 °C. Default system configuration parameters set the forecast horizon window to 4 h, a demand response temperature step (T_{step}) that instructs the control action to increase the space temperature by 2 °C over the duration of the forecast horizon window, and the duration of a demand event to 40 min. Additional system parameters specific to thermal comfort, electricity demand forecasting and cost (tariff) are described in the corresponding subsections that follow.

2.4.1. Thermal Comfort

The energy demand of buildings is influenced by the presence and behavioral patterns of occupants [62]. The thermal comfort element impacts the temperature setpoint by analyzing the measured room temperature (T_{room}) and occupants' feedback collated at a sample time of 10 min. Weekdays are divided into 7 time intervals $\tau_{(n)}$, configured to mirror a typical teaching timetable, whereas a weekend day consists of only 1 time interval. Changing the weekend day interval pattern to replicate a weekday is straightforward. By considering occupant presence is inhomogeneous, for each $\tau_{(n)}$ we choose an algorithm for the simulation of occupants to be used as an input for current occupant level, u_k. In practice, not all individuals will report their relative thermal comfort; therefore, the model automatically creates several feedback reports u_f, where $u_f \le u_k$. An individual's response is measured using a unipolar Likert scale [63,64]. The question has a five-scale response: too warm, warm, okay, cold, too cold; this is scored mathematically using a scale $u_f \in \{-2, -1, 0, 1, 2\}$. In order to imitate perceived behavior patterns, for each time interval, the following model parameters are defined: $u_{min} = \min u_f$, $u_{max} = \max u_f$ and response threshold u_{th} (%). The thermal model weekday parameters are reported in Table 1.

Table 1. Thermal model parameters.

$\tau_{(n)}$	u_{min}	u_{max}
1	0	0
2	10	40
3	5	20
4	15	70
5	3	12
6	7	30
7	0	0

For any given weekday time, the thermal comfort model output is calculated by the following expression:

$$tc_{\tau_n} = \mathrm{Mo}\left(\sum_{i=1}^{u_k} u_{f(i)}\right), \ u_f \in \{-2\,,-1,\ 0,\ 1,\ 2\},\ n = 1,\ 2,\ldots,\ 7; \tag{2}$$

with respect to:

$$\begin{aligned} \tau_{(1)} &= u_f(3:5); \\ \tau_{(2)(7)} &= u_f(2:5); \\ \tau_{(3)(6)} &= u_f(2:4); \\ \tau_{(4)} &= u_f; \\ \tau_{(5)} &= u_f(1:3); \end{aligned} \tag{3}$$

s.t. constraints:

$$(u_k u_f) \times 100 > u_{th}; \tag{4}$$

$$u_{min} \le u_k \le u_{max}. \tag{5}$$

For weekend days, we assume $u_f = 0$; hence, the model returns a value $tc_{\tau(1)} = u_f(3)$. The variation in $\tau_{(n)}$ represents a bias that is configured to reflect a change in outside temperature over a 24-h period.

It is noted that the seven-time intervals $\tau_{(n)}$ are bounded by a start and stop clock time $\tau_{(n)}(t_1, t_2)$ such that $\tau_{(1)}(t_1, t_2) = \tau_{(1)}(00:00,\ 2n+7)$, $\tau_{(n)}(2n+5,\ 2n+7)$; and terminating at $\tau_{(N)}(2N+5,\ 23:59)$. In practice, if a date and time are specified (e.g., S0_date = Fri, 05-February-2020 07:23:14), then the task to determine if the date-time element occurs on a weekday or weekend day is straightforward. Given a date-time S0_date, it is possible to formulate an algorithm that returns a 1×25 array $\delta_{\text{tc}} = [tc_0,\ tc_1, \ldots, tc_{24}]$, where tc_n represents a thermal comfort value over a 4-h period at $10n$ min. It should be noted that because the optimizer is designed to take into consideration occupants' feedback in real-time at a sample time of 10 min $\delta(2:25) = tc_0 = tc_{\tau(1)}$. However, if during the 4-h horizon window the system identifies a time interval where $u_{max} = 0$, i.e., there are no planned occupants, the model starts a pre-programmed sequence that sets the thermal comfort on a downward trajectory reducing at a rate of 0.5 °C per 10 min interval until a minimum temperature threshold value T^{th}_{min}.is reached. We have by the definition of the 11×25 nodemap δ_{tc} completed the data preparation of thermal comfort shown in Figure 3. It must be remembered that the thermal comfort model is prepared for operation within the simulated environment only. In practice, the implementation proposes occupants' report thermal comfort to the system using a smartphone app. This concept is elaborated further in Section 3.

2.4.2. Electricity Demand Forecasting

A data-driven methodology for modeling electricity demand forecasting is proposed [65]. The implication of this novel semi-autonomous simplified lumped model has the potential to offer decentralized electricity network operators' knowledge of the more extensive aggregated rate of future energy consumption. Thus, enabling decentralized energy management systems to proactively reduce load demand on small island electricity grids or distributed grid-edge systems as part of an evolving DR service. In this paper, we integrate the electricity demand forecasting model as part of the optimize and control framework. Initially, analysis of a chronological sequence of 245,424 discrete observations reveals the composition of the one-dimensional time series is characterized by three seasonal patterns: weekday, weekend day and month. These findings motivate an effort to reduce the dimensionality using piecewise aggregated approximation (PAA). Subsequently, calculating a cubic polynomial that interpolates points of interest yields a $13 \times 4 \times 2$ multi-dimensional array, which in turn helps restore the shape of the original demand forecast profile. The polynomial coefficient structure for weekday and weekend day are listed in the array page 1 and 2, respectively. Given both weekday and weekend day demand profiles recur every 24 h, it turns out using Equation (6) a normalized demand forecast value $M_i(x)$ can be tagged to a specific time in any 24-h period.

$$M_i(x) = a_i + b_i(x - i_{lo}) + c_i(x - i_{lo})^2 + d_i(x - i_{lo})^3, \tag{6}$$

where $i = 0,\ 1, \ldots, n;\ x \in [lo,\ hi]$, lo and hi correspond to the minimum and maximum data points of each PAA 2h segment, respectively, and the cubic polynomial coefficient parameters are a_i, b_i, c_i, and d_i. Moreover, we will show how the demand forecasting model can be used to compute a credible demand forecast value for any given date and time.

There are 12 equidistant segments, which equates to 13 periods (ρ) bounded by minimum and maximum points lo and hi, i.e., $\rho_n(lo, hi)$ where the number of periods $n = 0,\ 1, \ldots,\ N$. In the first period, $\rho_0(lo, hi) = \rho_0(0,\ 4n+2)$, after that $\rho_n(4n-2,\ 4n+2)$; and terminating at $\rho_N(4N-2,\ 4N)$. If we adopt the convention that makes 13-time intervals τ_n bounded by a start and stop clock time $\tau_n(t_1, t_2)$ then $\tau_0(t_1, t_2) = \tau_0(00:00,\ 2n+1)$, after that $\tau_n(2n-1,\ 2n+1)$; and terminating at $\tau_N(2N-1,\ 23:59)$. Thus, it can be seen, given a date-time S0_date it is possible to formulate an algorithm that returns a 1×25 array $\delta_{\text{dv}} = [dv_0,\ dv_1, \ldots, dv_{24}]$ where dv_n represents a normalized demand forecast value over a

4-h period at $t = 10n$ min starting from the specified date-time. This approach works equally well for both weekdays and weekend days.

The normalized demand forecast value dv_n is defined as:

$$dv_n = N_{min} + \left(\frac{M_i(x) - DV_{min}}{DV_{max} - DV_{min}}\right) \times (N_{max} - N_{min}),\ dv_n \in [1,\ 11],\ n = 0,\ 1, \ldots,\ 24, \tag{7}$$

where $N_{min} = \min_{m \in [n]} N_{(m,25)}$, $N_{max} = \max_{m \in [n]} N_{(m,25)}$, $DV_{min} = \min_{i \in [n]} M_i(x)$, $DV_{max} = \max_{i \in [n]} M_i(x)$, noting that a nodemap N is a $m \times n$ two-dimensional array.

2.4.3. Cost (Tariff) Model

A key consideration when taking part in a predefined energy reduction strategy must empower customers to use energy in the lowest price period accessible, at the same time as offering participation in DR initiatives. The cost (tariff) model is configured to integrate a typical static time of use (TOU) tariff [66]. As shown in Figure 4, these tariffs charge cheaper rates when demand is low but increases for electricity consumption at peak times.

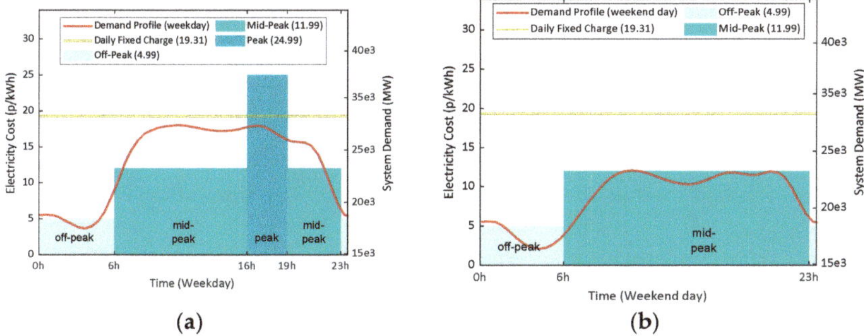

Figure 4. Model static time of use (TOU) tariff: (**a**) Weekday; (**b**) Weekend day.

Given a date-time S0_date, the cost (tariff) model returns a 1×25 array $\delta_{ec} = [ec_0,\ ec_1, \ldots, ec_{24}]$ where ec_n represents a normalized cost (tariff) value over a 4-h period at $t = 10n$ min starting from the specific date-time.

The normalized cost (tariff) value is defined:

$$ec_n = N_{min} + \left(\frac{EC(n) - EC_{min}}{EC_{max} - EC_{min}}\right) \times (N_{max} - N_{min}),\ ec_n \in [3,\ 9],\ n = 0,\ 1,\ \ldots,\ 24, \tag{8}$$

where $EC(n)$ is the cost (tariff) at $t = 10n$ min, $N_{min} = 3$, $N_{max} = 9$, $EC_{min} = 4.99$, and $EC_{max} = 24.99$. The scaling factors are set by design to position δ_{ec} values in the subsequent optimize stage such that a change in price to either off-peak or peak has maximum influence during the optimization outcome. Furthermore, it will be shown δ_{ec} impacts the operation of system assets managed by the scheduler subsystem.

2.4.4. Optimization

The optimization cycle (Figure 3) starts on receipt of the input signal S0_date. Subsequent cycles commence at a block sample time of 10 min (600 s). Previously, data preparation for occupants' thermal comfort, electricity demand forecast, and cost (tariff) each returned a 1×25 array $\delta = [x_0,\ x_1, \ldots, x_{24}]$ where x_n represents a normalised data type (tc, dv and ec) value over a 4-h period at $t = 10n$ min intervals starting from a specific date-time (S0_date). Before each data type array can be processed, it must be homogenized in a way that makes it accessible to the optimizer. The data is transformed into a $m \times n$ two-dimensional nodemap $N(m, n)$ such that $\delta(x_n) \to N(12 - x, t_{10n})$. Accordingly m represents a temperature $T = [T_{max} : -0.5 : T_{min}]$ and n defines 25 stages $(S_n \mid n \in \{0, 1, \ldots, 24\})$ each separated by a 10 min time interval for the duration of the 4-h forecast horizon window, e.g., $S_0 = t_0$ and S_{24} is linked to the 10 min time interval $t_{230} \to t_{240}$. The 11×25 nodemap N is then transformed to a 31×72 gridmap G by the following function:

$$N(\delta(x_n), n) \mapsto G(i, 3n)_{\kappa_s}, \; G(j, 3n+1)_{\kappa_t}, \;\; n = 1, 2, \ldots, 24, \tag{9}$$

where:

$$i = 3\delta(x_n) - \Delta, \;\; \Delta \in \{1, 2, 3\}; \tag{10}$$

s.t. constraints:

$$\Delta = \begin{cases} 1 & \text{if } \delta(x_{n+1}) > \delta(x_n); \\ 2 & \text{if } \delta(x_{n+1}) = \delta(x_n); \\ 3 & \text{if } \delta(x_{n+1}) < \delta(x_n) \end{cases} \tag{11}$$

$$j = \begin{cases} \text{i} + 3 & \text{if } \Delta = 1; \\ \text{i} & \text{if } \Delta = 2; \\ \text{i} - 3 & \text{if } \Delta = 3; \end{cases} \tag{12}$$

$$2 \leq \delta(x_n) \leq 10; \tag{13}$$

when constraint (13) is not satisfied $\Delta = 2$.

The temperature from $t_0 \to t_{10} = T_{S_1}$, where $T_{S_1} \in \{T_{S_0}, T_{S_0} \pm 0.5°\text{C}\}$ s.t. $T_{min} < T_{S_0} < T_{max}$; however, if $T_{S_0} = T_{min}$, then $T_{S_1} \in \{T_{S_0}, T_{S_0} + 0.5°\text{C}\}$; furthermore, if $T_{S_0} = T_{max}$, then $T_{S_1} \in \{T_{S_0}, T_{S_0} - 0.5°\text{C}\}$. Based on this information, this equates to 31 permissible temperature changes between t_n and t_{n+10}. If we continue to record the change in temperature ΔT from $S_n \to S_{n+1}$ using blocks of three columns for each cycle, then it is clear a gridmap of size 31×72 is created. We refer to the three columns in each block as the source node κ_s, target node κ_t, and edge-weight $\lambda_\eta : \kappa_s \xrightarrow{\eta} \kappa_t$, respectively.

The Dijkstra's algorithm computes the shortest path between a specified temperature point given at S_0 and S_{24}. This deterministic problem follows the principle of optimality which suggests if the path taken transits from one legitimate node to the next minimizes the cost-to-go from t_n to t_{n+10}, then the transition between the collective nodes must be optimal [67]. For the Dijkstra's algorithm to solve the shortest path, the 31×72 gridmap is first subjected to a series of simple transformations. The first instruction reshapes the gridmap into a 744×3 matrix referred to as the edgelist. Here, following the same convention to identify columns in the gridmap, the edgelist provides a listed description of all source nodes κ_n, legitimate target nodes κ_{n+1} and their respective connecting edge-weights $\lambda_\eta : \kappa_n \xrightarrow{\eta} \kappa_{n+1}$, i.e., its associated cost. A second instruction creates a digraph object that generates an *Edges* variable (744×2 table) based on the number of source and target nodes extracted from the 744×3 edgelist, and a *Nodes* variable (275×1 table). The 275 value represents the total number of nodes (κ_{275}) in the fixed 11×25 nodemap. Finally, an equivalent sparse adjacency matrix representation of the digraph, which includes the edge-weights, is created. Since the graph object we have constructed is a directed graph, the sparse adjacency matrix is not symmetric. However, we can overcome this by converting the sparse adjacency matrix to a full storage matrix. In this instance, the conversion generates a 275×275 full storage matrix.

The data type shape is now in a format required by the Dijkstra's algorithm. Executing the Dijkstra's algorithm will compute the optimal cost which is equivalent to the summation of all edge weights $\lambda_\eta : \kappa_s \xrightarrow{\eta} \kappa_t$ on the shortest path from κ_s to κ_t between time t_0 and t_{240}.

This process is repeated for each data type. At the end of each transformation the results are assigned to a specific page of a multi-dimensional array where page 1 (P1) is reserved for data type comfort, page 2 (P2) demand, and page 3 (P3) cost (tariff). The fourth page (P4) is reserved for the final stage in the optimization process, which combines the contributions assigned to P1 to P3. Here, every third column in the 31×72 P4 gridmap is allocated a grid centroid value $GC_{P4(j,s)} = 1$, where $j \in \{1, 2, \ldots, 31\}$ and $s \in \{3, 6, \ldots, 72\}$, and assigned to row index j that is equivalent to the k-means cluster centroid index that partitions the observations in the corresponding column s on P1 to P3. Note, for each data type $c_{i\kappa}^{n} = GC_{P4(j,s)}$; see Equation (1). The remaining values in each column are incremented by one until the row index j has reached its boundary limit, i.e., 1 or 31. When the Dijkstra's algorithm subsequently computes the shortest path between the source node $\kappa_s = GC_{P4(j,1)} = 1$ and target node $\kappa_t = GC_{P4(j,71)}$ where $j = GC_{P4(j,72)} = 1$, the results yield the optimal path that transits from $S_0 \rightarrow S_{24}$. The control action $T_{S_1} = N\big(GC_{P4(j,2)}, 2\big)$. Simply stated, the control action is a fixed temperature value that is linked to the 11×25 nodemap $N(m,n)$ at row index $m = GC_{P4(j,2)}$, where $N(1,n) = 20.5$ °C, $N(2,n) = 20.0$ °C, $\ldots, N(11,n) = 15.5$°C, where $n \in \{1, 2, \ldots, 25\}$. The relationship between the gridmap and nodemap is highlighted in Figure 3. The pseudocode describing the operating principle of the optimize and control algorithm is listed in Appendix A.

2.5. Demand Event Signal Subsystem

The demand event signal subsystem (des_subsystem) simulates actions in response to a network operator instigated instruction. These signals are sent to individual customers enrolled in a campaign designed to deliver aggregated tertiary DR. The Simulink® model itself is trivial (Figure 5); however, the subsequent sequence of events requires further explanation. Firstly, the objective shifts to making the system ready for a DR event; this includes setting the control action to increase the room temperature in a measured approach by a pre-set value T_{step}(°C) within the 4-h horizon window. Secondly, there is the objective to ensure the battery energy storage system (BESS) is available with enough charge at the start of the DR event.

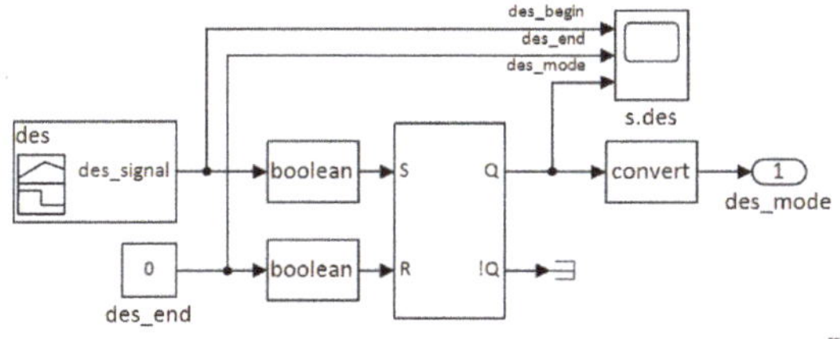

Figure 5. Simulink® model of demand event signal subsystem.

The period of pre-heating is regulated by altering the demand forecast profile. By default, $T_{step} = 2$ °C. Therefore, the normalized demand forecast value $\delta_{dv} = [dv_0, dv_1, \ldots, dv_{24}]$ is recast to $\epsilon_{dv} = [dv_{\epsilon 0}, dv_{\epsilon 1}, \ldots, dv_{\epsilon 24}]$, where $\epsilon_{dv}(0:4) = dv_0$, $\epsilon_{dv}(i:j) = \epsilon_{dv}(i-5, j-5) - 1$, where $i \in \{5, 10, \ldots, 20\}$, $j \in \{9, 14, \ldots, 24\}$ s.t. $dv_0 - 1 \geq 1$. This new trajectory increases the last recorded room temperature by 2 °C at a rate of 0.5 °C every 50 min. At the beginning of each subsequent optimization cycle, the trajectory leading up to the DR event is maintained, i.e., it advances closer to the plus 2 °C temperature at each iteration and towards the DR projected start time. However, before ϵ_{dv} reverts to δ_{dv}, the trajectory is modified further, this time by reducing the temperature setpoint 2 °C less than the temperature recorded immediately before the start of the tertiary DR event. The system reinstates δ_{dv} immediately after the DR event terminates.

The des_mode signal triggers the scheduler subsystem to start charging the BESS. The energy storage asset will continue to charge until the start of the DR event. The battery will then start to work from this time, reducing the stored charge of the battery while it continues to provide primary power to the heating system. The heating system will continue to be supplied from the battery until a state of charge (SOC) minimum threshold has been reached. The scheduler switches primary power to the grid and the battery to charge.

2.6. Energy Subsystem

Decentralized DR frequency regulation, when used in building stock, can regulate short-term frequency excursions in demanded electrical energy [68]. The contribution of a decentralized frequency regulator has been analyzed [68]. Results presented suggest that small excursions in measured temperature from a TCL setpoint value will not compromise indoor comfort temperatures but can contribute to the restoration of frequency equilibrium during network stress events. In this paper, we integrate the implied linear power system and frequency regulator as part of the optimize and control framework. The model (energy_subsystem) shown in Figure 6 replicates a power system rating of 300 MVA. Initial conditions assume the balance in supply and demand is at equilibrium, measured frequency is 50 Hz and the steady-state frequency error is zero. The energy subsystem model parameters are reported in Table 2.

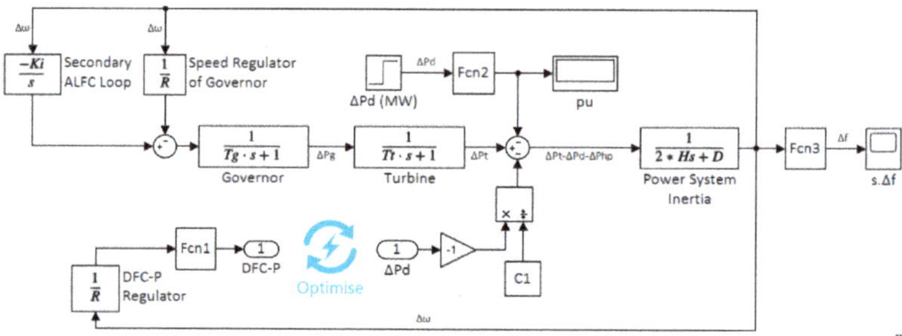

Figure 6. Simulink® model of energy subsystem.

Table 2. Energy subsystem parameters.

Parameter	Description	Value
Ki	Secondary ALFC integral gain	1.667×10^{-3}
R	Governor speed regulator	0.05 Hz/pu MW
Tg	Governor time constant	0.25 s
Tt	Turbine time constant	0.60 s
H	Inertia time constant	5 s
D	Load damping constant	0.8 s
C1	Constant	10×10^6
ΔPd	Continency load	75 MW

2.7. Building Subsystem

The building subsystem model (building_subsystem) (Figure 7) is a simplified thermostatically controlled (on/off) space heating system with feedback loops which typically maintains the air temperature at a set level. The model emulates building thermodynamics (building), calculating variations in temperature based on heat flow, $H(t)$, and heat losses, $H_{loss}(t)$.

$$H_{loss}(t) = \frac{T_{room} - T_{out}}{R_{th}}, \tag{14}$$

$$\frac{\Delta T_{heater}}{\Delta t} = \frac{1}{\dot{M}c}(H(t) - H_{loss}(t)). \tag{15}$$

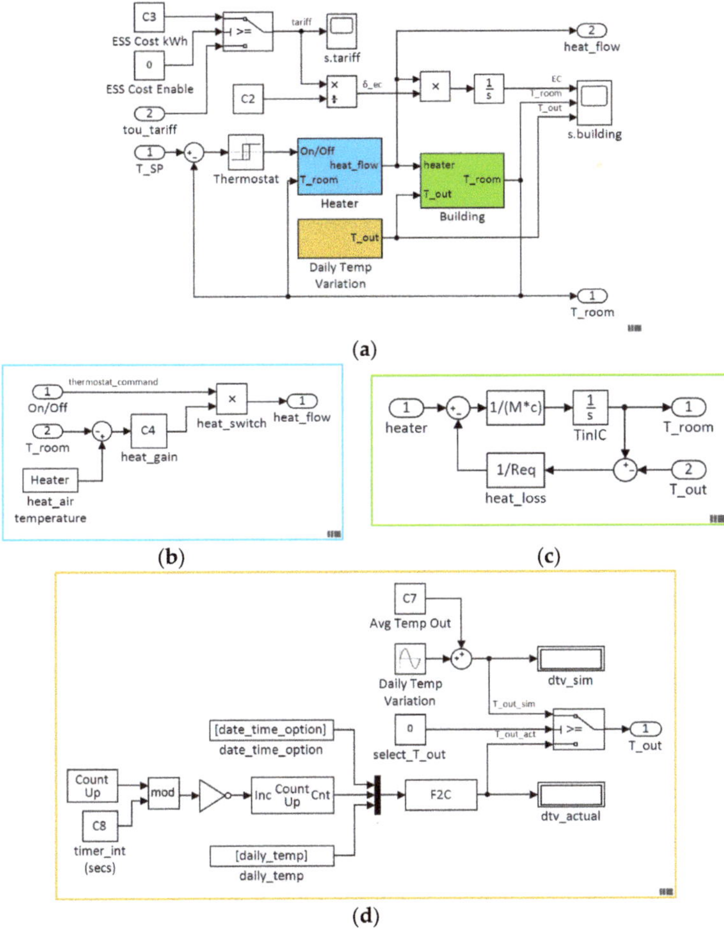

Figure 7. Simulink® model: (**a**) Building subsystem; (**b**) heater; (**c**) building; (**d**) daily temperature variation.

A series of embedded lookup tables representative of seasonal variation are used to model outside air temperature over a 24-h period at a sample rate of 30 min [69]. In practice, the local outdoor temperature is measured using sensors and input into the system. Energy cost (EC [p/kWh]) is calculated as a function of time and heat flow and is expressed in following equation:

$$EC = \int_{t_0}^{t_n} \left((T_{heater} - T_{room})\dot{M}c \right) \delta_{ec}\left(x_{t(n)} \right), \tag{16}$$

where $\dot{M}$ [kg/hr] denotes air mass flow rate through the heater; c specific heat at constant air pressure, and $\delta_{ec}(x_{tn})$ [p/kWh] is the energy price at time t_n. The building subsystem model parameters are reported in Table 3.

Table 3. Building subsystem parameters.

Parameter	Value
c	1005.4
Rth	0.0015
C2	3.6×10^3
C3	0.0199
C7	15
C8	1800

2.8. Scheduler Subsystem

The scheduler subsystem primary job is to monitor several signals and direct the operation of an automatic transfer switch between a grid and an alternative backup source of power. To ensure the appropriate power source is selected, the scheduler requires knowledge of the current cost (tariff) of electrical energy, whether a tertiary DR event is in progress including information of the event duration and BESS state of charge. The Simulink® model of the scheduler subsystem is shown in Figure 8 and includes three input signals and six output signals. The output signals are provided for visual indication of various signal status. A simplified BESS element (ess_subsystem) simulates a battery SOC using a first-order transfer function. Locally defined parameters SOC_hi and SOC_lo set maximum and minimum state of charge values (expressed as a percentage), which determine when the BESS is declared available for use. In this context, initial values are defined in Section 3. The model also includes a self-discharge rate (SDR) which reduces the stored charge of the battery naturally over time.

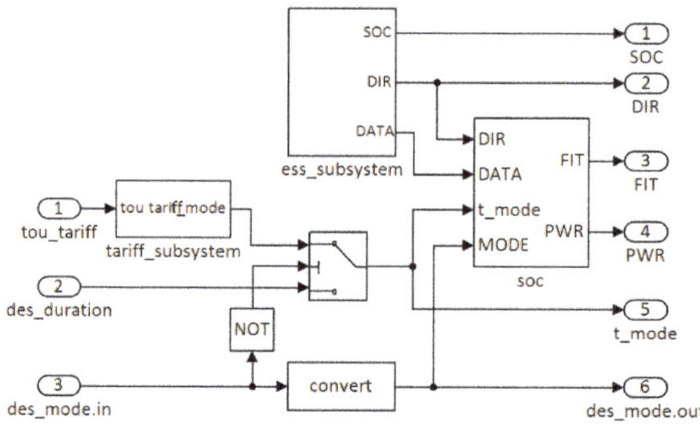

Figure 8. Simulink® model of scheduler subsystem.

The BESS availability function is represented by Equation (17), where SOC_lo is a low-level SOC threshold (locally defined parameter).

$$\text{FIT} = \begin{cases} 0 & \text{if SOC} \leq \text{SOC\_lo} \\ 1 & \text{otherwise} \end{cases}. \tag{17}$$

Control rules that determine when the primary power source is set to grid or BESS is illustrated in Figure A1 (Appendix B). The decision variable t_mode is the cost (tariff) threshold and automatically switches the power source to BESS when the cost (tariff) is high s.t. Equation (17). Furthermore, when signal des_mode =1 (0 = normal, 1 = tertiary DR event), t_mode = 0, thus preventing a control action that switches the power source to BESS during the period leading up to the start of the DR event (nominally 4 h). Signal CDir (change direction) reports if the battery is in charge or discharge (0 = discharge, 1 = charge); PWR denotes primary power source (0 = grid, 1 = BESS); SOC_EC denotes cost (tariff) in use, (0 = TOU, 1 = BESS).

2.9. Date-Time Subsystem

For completeness, the Simulink® model of the date-time subsystem (dt_subsystem) is shown (Figure 9). Its primary function is to provide a date-time element at a sample time of 10 min. The model has been configured to run in real-time during experimental evaluation. By default, dt is set to the current date and time, using format dd-mmm-yyyy hh:mm:ss, with the option to set to any data time during model analysis. The date-time model parameters are reported in Table 4.

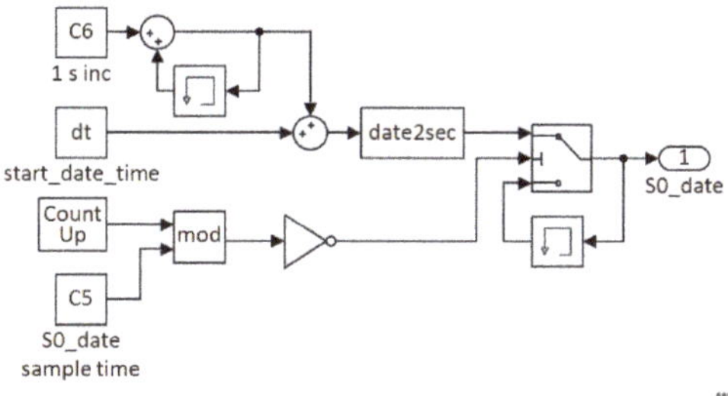

Figure 9. Simulink® model of date-time subsystem.

Table 4. Date time subsystem parameters.

Parameter	Value
C5	600
C6	$1.157412771135569 \times 10^{-5}$
dt	dd-mmm-yyyy hh:mm:ss

3. Computational Study

In this section, we report the findings from a computational study (desktop simulation). By design, the computational study validates the functionality of critical services. In contrast, the experimental evaluation (Section 4) is explicitly directed on proving the interaction of proposed data types within the optimization subsystem. A simulated tertiary DR event is considered in both scenarios.

The interaction between decision variables and control actions of individual subsystems is complex. Accordingly, the computational study validates the functionality of the following vital services:

- Thermal comfort model
- Electrical demand forecasting model
- Cost (tariff) model
- Optimizer
- Tertiary DR activity
- Pro-active frequency control

To begin, we evaluate the data input models. Individual charts created using nodemap data, and corresponding gridmap data validate the optimization and control behavior. In the second study, the results obtained from a simulated tertiary DR event are discussed. Finally, we monitor the system behavior during an imbalance between supply and demand. Here, the pro-active frequency control reacts to a simulated load disturbance causing a frequency excursion from the nominal 50 Hz steady-state. The model is initialized using the values reported in Table 5.

Table 5. Computational model initialization parameters.

Parameter	Value	Description
SO_date	10-Oct-2019 16:00	Stage 0 date time
des_begin	10-Oct-2019 16:40	Notification of DR event
T^{th}_{min} (°C)	16.5	Minimum temperature threshold
T_{step} (°C)	3	Temperature step increase
T_{room} (°C)	18	Room temperature
Horizon (h)	4	Forecast horizon
DR_t (min)	40	Tertiary DR event duration
SOC_hi	0.8	State of charge maximum threshold
SOC_lo	0.2	State of charge minimum threshold

Occupant thermal comfort feedback is shown in Figure 10a. At 16:40, the model reports the aggregated occupant thermal comfort is "too warm". This consensus triggers the optimization algorithm to set the comfort level gridmap trajectory on a path that reduces the measured room temperature by 0.5 °C, i.e., $S_0 \xrightarrow{\eta} S_1$, where $\eta = T_{S_0} - 0.5$°C. In addition, according to local settings, the timetable sets the number of occupants in a space to zero at 19:00. A 'no occupancy' status has clearly defined adaptive triggers. Firstly, the comfort signal values (occupants, response, and comfort) are held at a constant zero, while the number of occupants present in a space is zero. Secondly, at 19:00, the optimizer begins to alter the comfort level gridmap trajectory by reducing the temperature to a minimum temperature threshold T^{th}_{min}.(local setting) at a rate of 0.5 °C every 10 min. This behavior is confirmed in the optimizer gridmap visualization and subsequent optimizer nodemap shown in Figure 11.

The price in the three-tier TOU tariff is translated visually in Figure 10b. Initially, from 16:00 to 19:00 the TOU signal value is set to 9, which represents cost 24.99 p/kWh (peak), reducing to 6 (11.99 p/kWh mid-peak price) at 19:00. The energy cost nodemap data (δ_{ec}) transformation to the optimizer gridmap is shown in Figure 11. During peak periods, when the cost of energy is highest, the gridmap interpretation is to influence the control variable by reducing the temperature setpoint, which in turn reduces the cost of energy. Similarly, at 19:00 (mid-peak), the gridmap tou signal is set at mid-scale (nominally 18 °C). The electricity demand forecast is shown in Figure 10c. To help interpret the demand signals shown, Figure 10d illustrates the calculated weekday demand profile over a 24-h period. The red circle marks the start of the 4-h horizon window (shaded area). The dv (gridmap) signal is reconstructed within the optimization algorithm. The results are consistent with the modified layout of corresponding digraph object node coordinates, which describes the relationship between directional edges and connecting nodes shown in Figure 11a. The optimal temperature path is calculated at a sample rate of 10 min. Figure 11b highlights the optimal temperature value over a 4-h horizon window commencing 16:40. The control action for the continuing 10 min cycle shown is the temperature value specified at 16:50, that is $T_{S_1} = 16.5$ °C. This accords with our earlier occupant thermal comfort feedback report, which registered a consensus to reduce the room temperature by 0.5 °C.

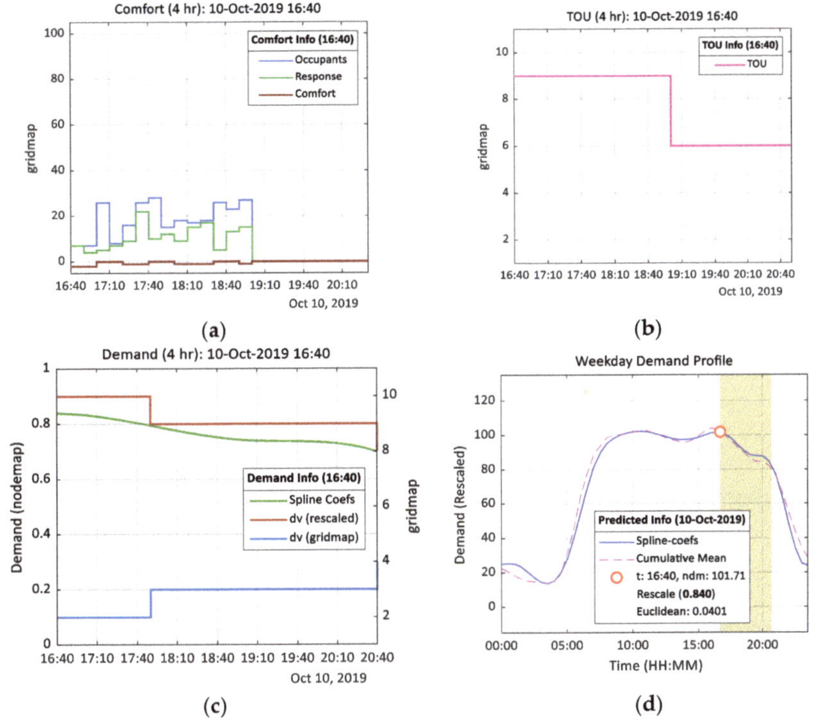

(a) (b) (c) (d)

Figure 10. Gridmap visualization of data type function response at 10-Oct-2019 16:00 over a 4-h horizon window: (**a**) Occupant thermal comfort feedback response; (**b**) electricity demand forecast: weekday 24 h; (**c**) cost (tariff); (**d**) electricity demand forecast.

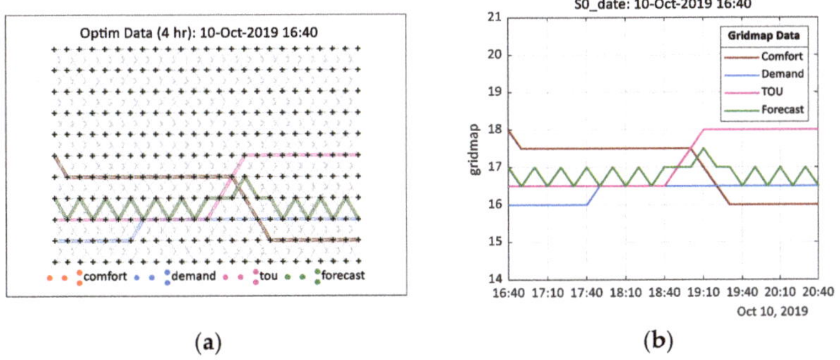

(a) (b)

Figure 11. Optimization response showing individual data types and forecast response: (**a**) $31 \times 72 \times 4$ gridmap visualization; (**b**) gridmap projected onto 11×25 nodemap.

On receipt of a DR event notice (16:40) the normalized demand forecast value δ_{dv} is recast to ϵ_{dv}. The modified demand profile trajectory is defined by the Dijkstra's shortest path algorithm $\kappa_s \xrightarrow{\eta} \kappa_t$, where $\eta = T_{S_0} + T_{step}(°\text{C})$. As can be observed in Figure 12, the change in demand profile at 16:50 increases from 16 °C (T_{S_0}) to 19 °C ($T_{S_{24}}$). A sample rate of 600 s accounts for the slight delay from the start of the DR preparatory window to the change in demand profile trajectory. Although the supposed outcome is to promote an increase in temperature equivalent to T_{step} (°C) leading up to the start of the DR event, the projected value is offset by the continued influence of the thermal comfort (ϵ_{tc}) and energy cost (ϵ_{ec}) (tou) decision variables. Consequently, in this instance, the optimization algorithm sets the 4-h ahead optimal temperature value slightly less than the anticipated 19 °C. The layout of individual digraph objects and their corresponding nodemap representation, shown in Figures 11 and 12, respectively, serve to provide a snapshot of the optimizer outputs over a 4-h horizon window at any given time. The benefit of the optimizer is now translated into Figure 13, which plots several decision variables and control actions over a 24-h period. Between Figure 13a,b, we observe the impact of demand and tariff data on the temperature setpoint (TS1). Furthermore, the outside temperature (Tout) as no impact on the measured room temperature during this simulation.

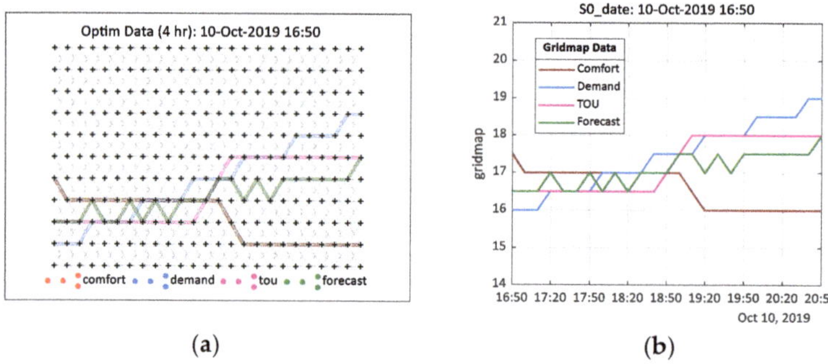

Figure 12. Optimization response showing individual data types and forecast response on receipt of demand events signal, $T_{step} = 3$ °C: (**a**) $31 \times 72 \times 4$ gridmap visualization; (**b**) gridmap projected onto 11×25 nodemap.

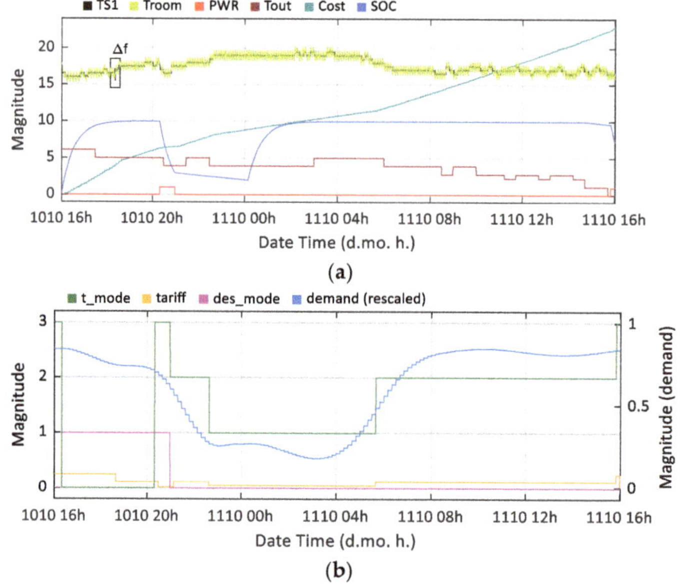

Figure 13. Simulation study at 10-Oct-2019 16:00 for 24 h with DR event: (**a**) shows temperature setpoint (°C) (TS1), room temperature (°C) (Troom), primary power switch signal (PWR), outdoor temperature (°C) (Tout), cost (p/kWh) (Cost), and battery energy storage (BESS) state of charge (SOC) (%) (rescaled) (SOC) profiles; (**b**) shows tariff mode (t_mode), TOU tariff (tariff), demand event signal mode (des_mode), and demand (rescaled) (demand) profiles.

The start of the DR preparatory window is recorded at 16:40 and subsequently sets and holds des_mode = 1 for 4 h and 40 min (the time leading up to and including the DR event). The BESS is seen to start a charge period in readiness to the start of the DR event. A tariff mode signal (t_mode) automatically restricts the use of the BESS until the DR event starts. At 16:40, the power signal (PWR) switches the primary power source from the grid to BESS. If the cost of energy is peak tariff immediately after the DR event (t_mode = 3), then the BESS would continue as the primary power source. However, as can be observed, the BESS SOC signal (SOC) indicates the BESS starts a discharge phase at from the start of the DR event and continues, in this scenario, to the end of the DR event. At 20:20, the primary power source reverts to the grid, but the BESS remains available ($SOC > SOC_{lo}^{th}$).

The rate at which the energy source naturally discharges has been magnified to evaluate control actions when SDR exceeds low and high charge threshold values (local settings). In practice, SDR parameters should be set accordingly. The simulation results show the calculated electricity demand forecast profile (demand). Its impact on the optimization algorithm is clear when demand is high (06:00 to 22:00) the aggregated effect is to limit the temperature setpoint (reducing the demand for electricity on the distribution network). Conversely, when demand is low (22:00 to 06:00), the constraints that govern the temperature setpoint are relaxed. Here, the optimizer allows, not mandates, an increase in energy consumption by increasing the space heating temperature setpoint. This finding, while preliminary, suggests the proposed control strategy has the potential to deliberately lessen peaks in demand (electrical) and fill in the period of low demand.

At 18:20.36, the impact of a simulated load disturbance ΔPd (Table 2) within the power subsystem is highlighted. The large and rapid decreasing frequency excursion shown in the box highlight, signifying an imbalance between supply and demand, is observed more clearly in Figure 14a. The proposed system immediate response is to lower the temperature setpoint (T_{S_1}), reducing the on-site heat source energy consumption and thus providing a pro-active response to the stability of the electrical distribution network [68]. As can be observed in Figure 14b, and in the broader context in Figure 13a, these immediate interventions have minimal impact on measured room temperature (T_{room}), hence minimizing occupant thermal discomfort.

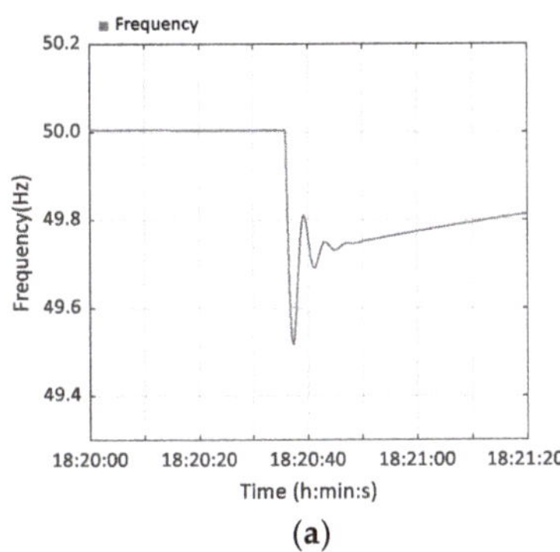

(a)

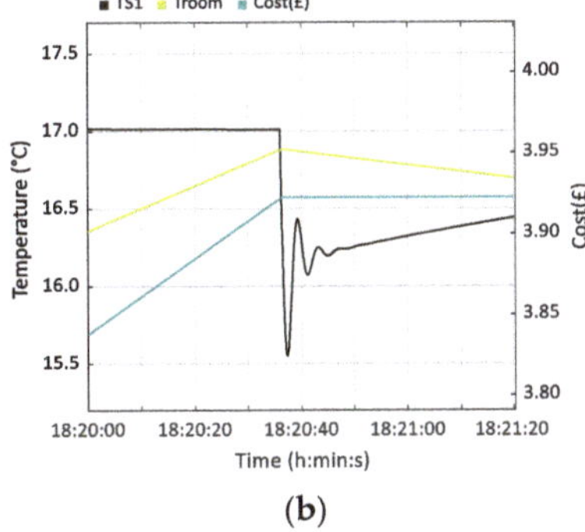

(b)

Figure 14. Frequency response 10-Oct-2019: (**a**) impact on mains grid frequency due to simulated load disturbance; (**b**) proposed model response.

4. Experimental Evaluation

Testing cannot be expected to catch every error in the software, and system complexity makes it difficult to evaluate every branch. A traditional approach to software testing during earlier development and subsequent simulation testing provided a satisfactory level of acceptance. However, as the hardware-in-the-loop test environment is not entirely under the control of the tester, an element of nondeterminism is introduced in the test. Furthermore, because of observations documented during early simulation testing, new features are added to help eliminate transitions to deadlock states.

These situations arise when a stalemate between two or more processes occurs, and the process is unable to proceed because each is waiting for the other to respond [70]. Therefore, it is considered helpful to outline an appropriate test and level of abstraction of the software and hardware devices for testing.

One of the significant objectives of testing is to assess the integration of the optimizer software code by connecting other software and hardware components. Therefore, a test environment was designed to evaluate and report as accurately as possible on the proposed optimization algorithm interaction with real-world data. An image of prototype equipment, including an industrial controller and sensor equipment, is shown in Figure 15a.

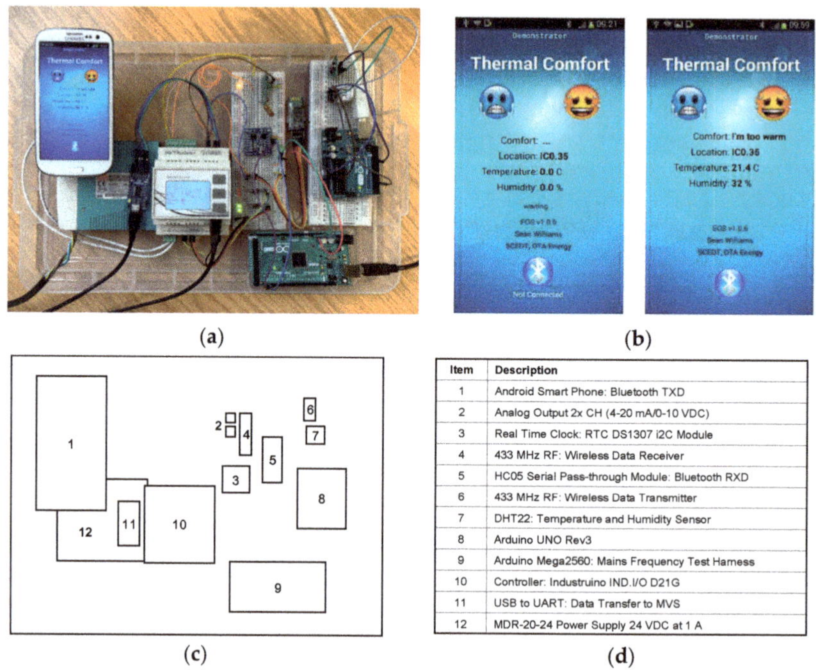

Item	Description
1	Android Smart Phone: Bluetooth TXD
2	Analog Output 2x CH (4-20 mA/0-10 VDC)
3	Real Time Clock: RTC DS1307 I2C Module
4	433 MHz RF: Wireless Data Receiver
5	HC05 Serial Pass-through Module: Bluetooth RXD
6	433 MHz RF: Wireless Data Transmitter
7	DHT22: Temperature and Humidity Sensor
8	Arduino UNO Rev3
9	Arduino Mega2560: Mains Frequency Test Harness
10	Controller: Industruino IND.I/O D21G
11	USB to UART: Data Transfer to MVS
12	MDR-20-24 Power Supply 24 VDC at 1 A

Figure 15. Hardware-in-the-loop test environment: (**a**) Arduino equipment; (**b**) Android smartphone demonstrator app example screen images; (**c**) Arduino equipment block identification map; (**d**) Arduino equipment legend.

The test environment composed of the following main components: (1) a revised Simulink® model designed to send/receive serial data, (2) electronic fan speed controller (EFSC) to regulate the heat transfer through flow, (3) a 240 VAC 3 kW box fan portable heater, (4) an Industruino IND.I/O 32u4 Arduino-compatible industrial controller, which includes 2 CH 0 to 10 VDC/4-20 mA 12bit output, and (5) Arduino-compatible remote sensors and communication equipment, including Android smartphone pre-loaded with an app, developed using MIT App Inventor 2 version nb183c. In addition to streaming data into the software environment, the industrial controller on-board liquid crystal display (LCD) panel was codified to visualize the data from remote sensors and user thermal preferences (registered using the smartphone app). These feedback indicators were supplemented by a series of light emitting diodes (LEDs) reporting the status of several decision-making variables.

Dedicated values for energy demand forecasting and price indicators are embedded in the computer model. However, occupant thermal comfort feedback data is input into the system in real-time using Bluetooth technology. This experiment utilizes the thermal comfort feedback data reported by a single occupant. A technology update to manage multiple users is a relatively straightforward task. Images of the App demonstrator designed to capture occupants thermal comfort report is shown in Figure 15b. Remote sensors monitor room temperature data, which is communicated to the optimizer in real-time using low power device 433 MHz (LPD433) equipment. A block diagram of the general arrangement is shown in Figure 16. The positioning of the optimizer indicates this configuration has the potential to participate in similar energy management schemes with minimal impact on existing infrastructure.

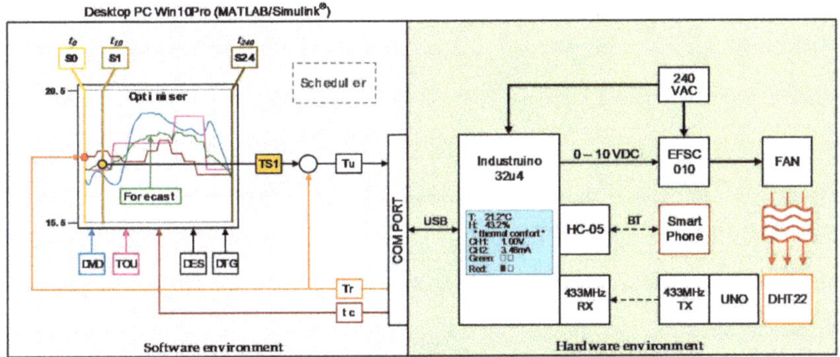

Figure 16. Block diagram of experimental evaluation setup.

This paper reports the results of an experimental test carried out in real-time. The test started on Monday, 6-April-2020, 16:00. At 16:40, the start of a DR preparatory event triggers a pre-set sequence of control actions designed to prepare the heating services in advance of the 40 min DR event, which started at 20:40. The test was run for 5.5 h, finishing 10 min after the DR event. Comparison of the findings shown in Figure 17 with those of earlier computation studies confirms the operation of the optimization algorithm is consistent with our mathematical arguments, which posits that the interaction between declared data types can influence an environment space heating. Increasing the temperature setpoint successively by 0.5 °C at 10-min intervals during the DR preparatory stage increased the space temperature by 2 °C from the start of the DR preparatory window. Figure 18a confirms a temperature value of 18.5 °C was recorded at approximately 19:10. It can be observed the temperature then decreased to 17.4 °C at 20:40, which is the start time of DR event. This behavior may be explained by the fact that the thermal comfort profile (dark red color) reduced to an equivalent of 16 °C (T^{th}_{min}) at 19:00, which is consistent with an expected zero occupancy at the same time.

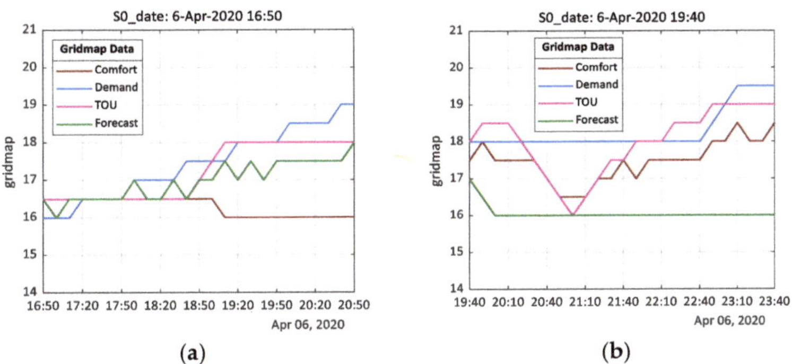

Figure 17. Visual representations of gridmap data showing 4-h horizon window of predicted values of each data type and optimized temperature profile: (**a**) on receipt of a simulated DR event signal at 16:50; (**b**) during the 40 min DR event (20:40 to 21:20).

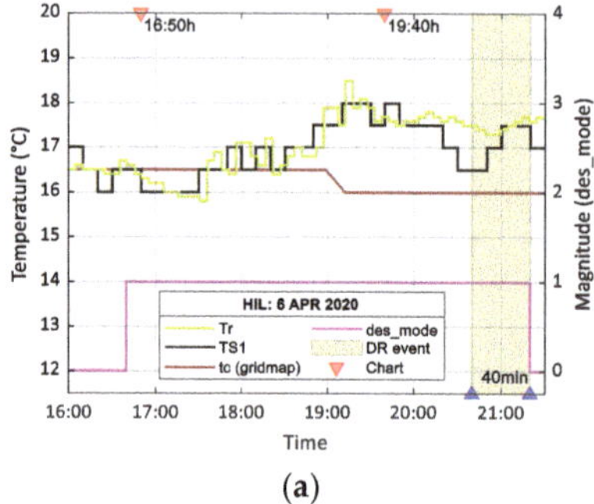

(a)

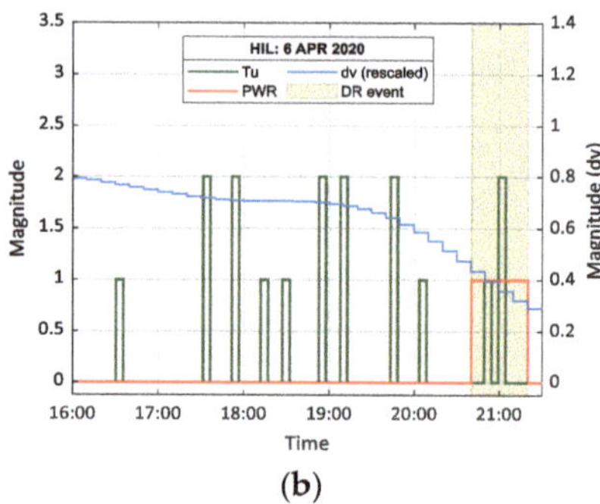

(b)

Figure 18. Experimental evaluation recorded results at 6-April-2020 16:00 for 5.5 h with DR event: (**a**) room temperature (Tr), temperature setpoint (TS1), thermal comfort gridmap data (tc), demand event signal mode (des_mode), and DR event; (**b**) control action signal (Tu), primary power switch signal (PWR), demand (rescaled) (dv), and DR event.

Furthermore, as can be observed in Figure 18b, the control action signal utilized in the earlier computational study has been modified to regulate the physical heat transfer through flow. Here, the control action signal (Tu), which operates a 0 to 10 VDC EFSC, is proportional to the difference between the calculated optimal temperature setpoint (TS1) and the measured temperature (Tr), i.e., $Tu \propto Te$, where $Te = T_{S_1} - T_{room}$. The power switch signal (PWR) shows the virtual energy storage system is activated at 20:40 and continues to operate as the heating system primary energy source for the duration of the DR event (shaded area).

Overall, these results are very encouraging. The experimental evaluation raises the possibility that the proposed optimization algorithm may support small communities in a decentralized environment with limited access to communication networks. Comparison of the findings with other studies confirms the novelty of the proposed framework for energy management. It is encouraging elements of this research are consistent with results found in previous work. Eriksson et al. [71] developed a normalized weighted constrained multi-objective meta-heuristic optimization algorithm to consider technical, economic, socio-political, and environmental objectives. The results emphasized the application of a modified Particle Swarm Optimization (PSO) algorithm to optimize a renewable energy system of any configuration. The implementation of the Dijkstra's algorithm (used in this study) is more prevalent in other applications (e.g., see Reference [72–74]).

Nevertheless, the simplicity Dijkstra's algorithm makes it a versatile heuristic algorithm. The shortest path optimization algorithm was designed to compute an optimal water heating plan based on specific optimality criteria and inputs [75]. The significant feature reported of the proposed algorithm was its low computational complexity, which opens the possibility to deploy directly on low-cost embedded controllers. In a further study, a strong relationship between optimization and space heating has been reported [76]. Here, a neural network algorithm was used to build a predictive model for the optimization of a HVAC is combined with a strength multi-objective PSO algorithm. Although results show satisfactory solutions at hourly time intervals for users with different preferences, demand response mechanisms have not been considered. However, leveraging upon the concepts of Industry 4.0, Short et al. [77] demonstrated the potential to dispatch HVAC units in the presence of tertiary DR program in a distributed optimization problem could deliver satisfactory performances. Finally, a more inclusive study proposed an optimization model which takes total operational cost and energy efficiencies as objective functions [78]. Here, a thermal load is adjusted in the knowledge that a managed change in temperature value has no significant impact on user comfort. An integrated demand response mechanism is also considered. Although the results provide a new perspective for integrated energy management and demand side load management, there is no further exploitation in real-time user engagement or perspectives on decentralization.

5. Conclusions

A real-time DR strategy in a decentralized grid has been formulated with both spatial and temporal constraints. A generic framework established regulatory control of space heating through mechanisms that automatically respond to changes in grid frequency and in response to explicit tertiary DR event signals. Design considerations set decision variables and control actions to illustrate the effectiveness of the novel multi-objective cost function, which is based on a weight-based routing algorithm. A series of embedded lookup tables based on historical operational data calculate an aggregated rate of electricity consumption over a rolling 4-h horizon window. As the techniques for enabling and controlling DR events emerged, extensive simulation studies demonstrated power consumption could be easily shifted without causing any significant short-term impact on space temperatures. Increasing growth of renewable energy resources could reduce system inertia, which means networks are more vulnerable to energy security. The approach offered may benefit rural decentralized community power systems, including geographical islands, seeking to optimize heating services through optimization and collaborative energy management. This operation was validated using experimental testing, which included a response to simulated tertiary DR event signals. The obtained results show an effectiveness of the decentralized, informatic optimization, and control framework for evolving DR services. The energy transition offers small communities' opportunities to meet decarbonization targets. The gradual shift from centralized fossil fuel power networks to more low carbon decentralized sustainable smart energy systems is set to disrupt businesses, policymakers, and system analysts. As energy markets change to meet innovations, the reliance on a single energy source is slowly diminishing. Still, as technology advances and support for an emerging group of consumers that produce energy continues to gather momentum, system operators should embrace a changing market to remain relevant in the future.

The context within which DR operates is important as related initiatives need to support the objectives of DR itself. These are usually encased within a policy objective in response to concerns relating to the environment. Concerning supporting the achievement of such policy goals, organizations seeking to participate in the sector need a business model/strategy that will, in the longer-term support and sustain the success of such policy goals. Therefore, the business model is essential to the objective of the policy. Martin et al. [79] illustrated the dangers of applying the business model/strategy that does not support the overall objective of an organization. Here, the policy goal is the European targets related to energy efficiency and climate change by consumers reducing or shifting their electricity usage during periods of peak electricity demand in response to time-based tariffs or other forms of financial incentives. The success of such policy objectives, therefore, are dependent on several factors, including end-user participation. The opportunities for realizing DR, however, vary across Europe as they are dependent on the particular regulatory, market, and technical contexts in different European countries [80]. It is estimated that the distribution networks share of the overall network investment in energy networks will be 80% by 2050 [81]. Hence, the need for DR solutions to reduce peaks in energy demand is significant. Thus, in recent times, the emphasis is on developing novel solutions which can align the energy demand to the energy supply in real-time [82]. A decentralized approach will support creativity and tailor-made solutions [83,84]. The role of the end-user in successfully delivering the policy objective is essential, and their buying into new usage patterns is critical [85]. Therefore, a business model that encourages end-user participation becomes crucial. Hence, a business model that understands and appreciates the variables of regulatory, market and technical contexts in different European countries, enabling enhanced end-user engagement is required to support the achievement of the policy objectives.

Future research will extend the test and validation work, including the integration of scalable communities and other forms of energy storage and distributed renewable energy generators. In addition, a business case that scrutinizes how the proposed optimization and control framework can be mobilized will be investigated in future work.

Author Contributions: Conceptualization, S.W.; methodology, S.W.; software, S.W.; validation, S.W. and M.S.; formal analysis, S.W. and M.S.; investigation, S.W.; resources, S.W.; data curation, S.W.; writing—original draft preparation, S.W.; writing—review and editing, S.W., M.S., T.C. and M.S.-P.; visualization, S.W.; supervision, M.S. and T.C.; project administration, S.W.; funding acquisition, M.S. and T.C. All authors have read and agreed to the published version of the manuscript.

Funding: Elements of the work presented in this paper was carried out as part of the REACT project (01/01/2019–31/12/2022) which is co-funded by the EU's Horizon 2020 Framework Programme for Research and Innovation under Grant Agreement No. 824395.

Acknowledgments: The first author wishes to acknowledge the financial support provided by Teesside University and the Doctoral Training Alliance (DTA) scheme in Energy.

Conflicts of Interest: The authors declare no conflict of interest. The funders had no role in the design of the study; in the collection, analysis, or interpretation of data; in the writing of the manuscript, or in the decision to publish the results.

Appendix A

Algorithm A1. Optimize and control algorithm

```
inputs:
  temp_room = {n|n is pos, and 15.5 ≤ n ≤ 20.5}      ▷ T_room (°C)
  S0_date ← now()
  des_mode = {n|n is int, and n ∈ {0, 1}}      ▷ 0=normal, 1=event
outputs:
  ctrl_action; tou_tariff; des_duration
initialise:
  visual_mode; gridmap
  horizon = 4                                   ▷ duration (h)
  des_mode = 0
  T_step (°C) = {n|n = 2, and n ∈ {2, 3}}
  des_duration = {n|n = 40, and n ∈ {30, 40, 50}}      ▷ duration (min)
  T_min = {n|n = 16.0, and 15.5 ≤ n ≤ 17.5}
  dt = {tc, dv, ec, optim}
for every 10 min interval do
  S0_date ← S0_date + 10 min
  for each dt do
    if dt = tc then
      T^th_min ← T_min                          ▷ min temp threshold (°C)
      prepare comfort values ∀Sn = {n|n is an integer, and 0 ≤ n ≤ 24}
    else if dt = dv then
      require: des_mode; des_duration; T_step
      prepare demand values ∀Sn = {n|n is an int, and 0 ≤ n ≤ 24}
      prepare node path
    else if dt = ec then
      prepare tou values ∀Sn = {n|n is an int, and 0 ≤ n ≤ 24}
    end if
    get: δ_dt(x_n)
    prepare gridmap
    adjacency matrix ← digraph ← edgelist ← gridmap
    optimize using dijkstra algorithm
    identify edgepath from start to end node ∀Sn = {n|n is an int, and 0 ≤ n ≤ 24}
    if dt = optim then
      prepare control action                    ▷ T_S1 (°C)
    end if
    get: visual_mode
    display: visualization ∈ {horizon, gridmap, bigpath, biggridmap}
  end for
end for
```

Appendix B

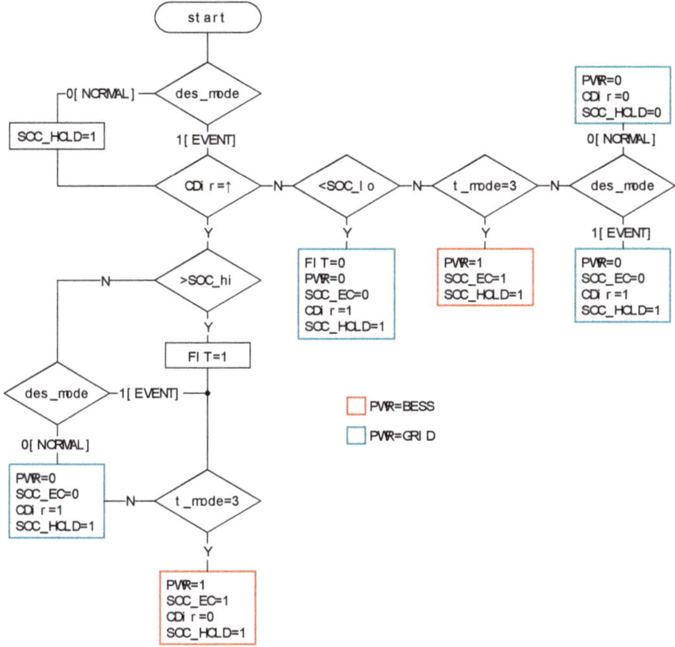

Figure A1. Scheduler control logic flowchart.

References

1. Alcott, B. Jevons' paradox. *Ecol. Econ.* **2005**, *54*, 9–21. [CrossRef]
2. Copiello, S. Building energy efficiency: A research branch made of paradoxes. *Renew. Sustain. Energy Rev.* **2017**, *69*, 1064–1076. [CrossRef]
3. Harris, T.M.; Devkota, J.P.; Khanna, V.; Eranki, P.L.; Landis, A.E. Logistic growth curve modeling of US energy production and consumption. *Renew. Sustain. Energy Rev.* **2018**, *96*, 46–57. [CrossRef]
4. Freire-González, J.; Puig-Ventosa, I. Energy efficiency policies and the Jevons paradox. *Int. J. Energy Econ. Policy* **2015**, *5*, 69–79.
5. Vujanović, M.; Wang, Q.; Mohsen, M.; Duić, N.; Yan, J. Special issue of applied energy dedicated to SDEWES conferences 2018: Sustainable energy technologies and environmental impacts of energy systems. *Appl. Energy* **2019**, *256*. [CrossRef]
6. Abdullah, M.A.; Agalgaonkar, A.P.; Muttaqi, K.M. Assessment of energy supply and continuity of service in distribution network with renewable distributed generation. *Appl. Energy* **2014**, *113*, 1015–1026. [CrossRef]
7. Ratnam, K.S.; Palanisamy, K.; Yang, G. Future low-inertia power systems: Requirements, issues, and solutions—A review. *Renew. Sustain. Energy Rev.* **2020**, *124*, 109773. [CrossRef]
8. Bayod-Rújula, A.A. Future development of the electricity systems with distributed generation. *Energy* **2009**, 377–383. [CrossRef]
9. Neelawela, U.D.; Selvanathan, E.A.; Wagner, L.D. Global measure of electricity security: A composite index approach. *Energy Econ.* **2019**, *81*, 433–453. [CrossRef]
10. Kosai, S.; Unesaki, H. Short-term vs long-term reliance: Development of a novel approach for diversity of fuels for electricity in energy security. *Appl. Energy* **2020**, *262*, 114520. [CrossRef]
11. IEA Statistics The World Bank. Available online: https://data.worldbank.org/indicator/EG.USE.COMM.FO.ZS (accessed on 14 April 2020).
12. Price, J.; Zeyringer, M.; Konadu, D.; Sobral Mourão, Z.; Moore, A.; Sharp, E. Low carbon electricity systems for Great Britain in 2050: An energy-land-water perspective. *Appl. Energy* **2018**, *228*, 928–941. [CrossRef]

13. Hope, A.; Roberts, T.; Walker, I. Consumer engagement in low-carbon home energy in the United Kingdom: Implications for future energy system decentralization. *Energy Res. Soc. Sci.* **2018**, *44*, 362–370. [CrossRef]
14. Barrett, J.; Cooper, T.; Hammond, G.P.; Pidgeon, N. Industrial energy, materials and products: UK decarbonisation challenges and opportunities. *Appl. Therm. Eng.* **2018**, *136*, 643–656. [CrossRef]
15. Foxon, T.; Hammond, G.; Pearson, P. Socio-technical transitions in UK electricity: Part 2 – technologies and sustainability. *Proc. Inst. Civ. Eng. Energy* **2020**, *173*, 123–136. [CrossRef]
16. Johnstone, P.; Kivimaa, P. Multiple dimensions of disruption, energy transitions and industrial policy. *Energy Res. Soc. Sci.* **2018**, *37*, 260–265. [CrossRef]
17. IRENA International Renewable Energy Agency. *Renewable Power Generation Costs in 2018*; IRENA: Abu, Dhabi, 2018.
18. Imperial College London. *Report Prepared for Committee on Climate Change Final*; Vivid Economics Limited: London, UK, 2019.
19. Zheng, C.; Yuan, J.; Zhu, L.; Zhang, Y.; Shao, Q. From digital to sustainable: A scientometric review of smart city literature between 1990 and 2019. *J. Clean. Prod.* **2020**, *258*, 120689. [CrossRef]
20. Sodiq, A.; Baloch, A.A.B.; Khan, S.A.; Sezer, N.; Mahmoud, S.; Jama, M.; Abdelaal, A. Towards modern sustainable cities: Review of sustainability principles and trends. *J. Clean. Prod.* **2019**, *227*, 972–1001. [CrossRef]
21. Mosannenzadeh, F.; Bisello, A.; Vaccaro, R.; D'Alonzo, V.; Hunter, G.W.; Vettorato, D. Smart energy city development: A story told by urban planners. *Cities* **2017**, *64*, 54–65. [CrossRef]
22. Wang, S.J.; Moriarty, P. Energy savings from Smart Cities: A critical analysis. *Energy Procedia* **2019**, *158*, 3271–3276. [CrossRef]
23. Di Silvestre, M.L.; Favuzza, S.; Riva Sanseverino, E.; Zizzo, G. How Decarbonization, Digitalization and Decentralization are changing key power infrastructures. *Renew. Sustain. Energy Rev.* **2018**, *93*, 483–498. [CrossRef]
24. Thellufsen, J.Z.; Lund, H.; Sorknæs, P.; Østergaard, P.A.; Chang, M.; Drysdale, D.; Nielsen, S.; Djørup, S.R.; Sperling, K. Smart energy cities in a 100% renewable energy context. *Renew. Sustain. Energy Rev.* **2020**, *129*, 109922. [CrossRef]
25. Marinova, S.; Deetman, S.; van der Voet, E.; Daioglou, V. Global construction materials database and stock analysis of residential buildings between 1970–2050. *J. Clean. Prod.* **2020**, *247*, 119146. [CrossRef]
26. Qiao, R.; Liu, T. Impact of building greening on building energy consumption: A quantitative computational approach. *J. Clean. Prod.* **2020**, *246*, 119020. [CrossRef]
27. Paridari, K.; Nordström, L. Flexibility prediction, scheduling and control of aggregated TCLs. *Electr. Power Syst. Res.* **2020**, *178*, 106004. [CrossRef]
28. Ibrahim, O.; Fardoun, F.; Younes, R.; Louahlia-Gualous, H. Review of water-heating systems: General selection approach based on energy and environmental aspects. *Build. Environ.* **2014**, *72*, 259–286. [CrossRef]
29. Yang, L.; Yan, H.; Lam, J.C. Thermal comfort and building energy consumption implications—A review. *Appl. Energy* **2014**, *115*, 164–173. [CrossRef]
30. Ascione, F.; Bianco, N.; Mauro, G.M.; Napolitano, D.F.; Vanoli, G.P. Weather-data-based control of space heating operation via multi-objective optimization: Application to Italian residential buildings. *Appl. Therm. Eng.* **2019**, *163*, 114384. [CrossRef]
31. Lin, S.; Liu, D.; Hu, F.; Li, F.; Dong, W.; Li, D.; Fu, Y. Grouping control strategy for aggregated thermostatically controlled loads. *Electr. Power Syst. Res.* **2019**, *171*, 97–104. [CrossRef]
32. Erdogan, S.; Yıldırım, S.; Yıldırım, D.C.; Gedikli, A. The effects of innovation on sectoral carbon emissions: Evidence from G20 countries. *J. Environ. Manag.* **2020**, *267*, 110637. [CrossRef]
33. Kumar, D.; Mathur, H.D.; Bhanot, S.; Bansal, R.C. Modeling and frequency control of community micro-grids under stochastic solar and wind sources. *Eng. Sci. Technol. Int. J.* **2020**. [CrossRef]
34. Loisel, R.; Lemiale, L. Comparative energy scenarios: Solving the capacity sizing problem on the French Atlantic Island of Yeu. *Renew. Sustain. Energy Rev.* **2018**, *88*, 54–67. [CrossRef]
35. Pfeifer, A.; Dobravec, V.; Pavlinek, L.; Krajačić, G.; Duić, N. Integration of renewable energy and demand response technologies in interconnected energy systems. *Energy* **2018**, *161*, 447–455. [CrossRef]
36. Curto, D.; Favuzza, S.; Franzitta, V.; Musca, R.; Navarro Navia, M.A.; Zizzo, G. Evaluation of the optimal renewable electricity mix for Lampedusa island: The adoption of a technical and economical methodology. *J. Clean. Prod.* **2020**, *263*. [CrossRef]

37. Cabrera, P.; Lund, H.; Carta, J.A. Smart renewable energy penetration strategies on islands: The case of Gran Canaria. *Energy* **2018**, *162*, 421–443. [CrossRef]
38. Tindemans, S.; Strbac, G. Low-complexity control algorithm for decentralised demand response using thermostatic loads. In Proceedings of the Proceedings—2019 IEEE International Conference on Environment and Electrical Engineering and 2019 IEEE Industrial and Commercial Power Systems Europe (EEEIC/I and CPS Europe), Genova, Italy, 11–14 June 2019.
39. Schellenberg, C.; Dimache, L.; Lohan, J. Grid-edge technology—Exploring the flexibility potential of a heat pump and thermal energy storage system. In Proceedings of the E3S Web of Conferences, Galway, Ireland, 13 August 2019.
40. Shen, Y.; Li, Y.; Zhang, Q.; Shi, Q.; Li, F. State-shift priority based progressive load control of residential HVAC units for frequency regulation. *Electr. Power Syst. Res.* **2020**, *182*, 106194. [CrossRef]
41. Croce, D.; Giuliano, F.; Bonomolo, M.; Leone, G.; Musca, R.; Tinnirello, I. A decentralized load control architecture for smart energy consumption in small islands. *Sustain. Cities Soc.* **2020**, *53*, 101902. [CrossRef]
42. Esposito, C.; Tamburis, O.; Su, X.; Choi, C. Robust Decentralised Trust Management for the Internet of Things by Using Game Theory. *Inf. Process. Manag.* **2020**, *57*, 102308. [CrossRef]
43. McLellan, B.; Florin, N.; Giurco, D.; Kishita, Y.; Itaoka, K.; Tezuka, T. Decentralised energy futures: The changing emissions reduction landscape. *Procedia CIRP* **2015**, *29*, 138–143. [CrossRef]
44. Breukers, S.; Crosbie, T.; Van Summeren, L. Mind the gap when implementing technologies intended to reduce or shift energy consumption in blocks-of-buildings. *Energy Environ.* **2020**, *31*, 613–633. [CrossRef]
45. Ala-Juusela, M.; Crosbie, T.; Hukkalainen, M. Defining and operationalising the concept of an energy positive neighbourhood. *Energy Convers. Manag.* **2016**, *125*, 133–140. [CrossRef]
46. Crosbie, T.; Broderick, J.; Short, M.; Charlesworth, R.; Dawood, M. Demand response technology readiness levels for energy management in blocks of buildings. *Buildings* **2018**, *8*, 13. [CrossRef]
47. Zapata Riveros, J.; Kubli, M.; Ulli-Beer, S. Prosumer communities as strategic allies for electric utilities: Exploring future decentralization trends in Switzerland. *Energy Res. Soc. Sci.* **2019**, *57*, 101219. [CrossRef]
48. Brown, D.; Hall, S.; Davis, M.E. What is prosumerism for? Exploring the normative dimensions of decentralised energy transitions. *Energy Res. Soc. Sci.* **2020**, *66*. [CrossRef]
49. Martínez Fernández, P.; Villalba Sanchís, I.; Yepes, V.; Insa Franco, R. A review of modelling and optimisation methods applied to railways energy consumption. *J. Clean. Prod.* **2019**, *222*, 153–162. [CrossRef]
50. Borhanazad, H.; Mekhilef, S.; Gounder Ganapathy, V.; Modiri-Delshad, M.; Mirtaheri, A. Optimization of micro-grid system using MOPSO. *Renew. Energy* **2014**, *71*, 295–306. [CrossRef]
51. Zizzo, G.; Beccali, M.; Bonomolo, M.; Di Pietra, B.; Ippolito, M.G.; La Cascia, D.; Leone, G.; Lo Brano, V.; Monteleone, F. A feasibility study of some DSM enabling solutions in small islands: The case of Lampedusa. *Energy* **2017**, *140*, 1030–1046. [CrossRef]
52. Skjølsvold, T.M.; Ryghaug, M.; Throndsen, W. European island imaginaries: Examining the actors, innovations, and renewable energy transitions of 8 islands. *Energy Res. Soc. Sci.* **2020**, *65*, 101491. [CrossRef]
53. Tindemans, S.H.; Trovato, V.; Strbac, G. Decentralized Control of Thermostatic Loads for Flexible Demand Response. *IEEE Trans. Control Syst. Technol.* **2015**, *23*, 1685–1700. [CrossRef]
54. Uski, S.; Rinne, E.; Sarsama, J. Microgrid as a cost-effective alternative to rural network underground cabling for adequate reliability. *Energies* **2018**, *11*, 1978. [CrossRef]
55. Bradley, P.; Coke, A.; Leach, M. Financial incentive approaches for reducing peak electricity demand, experience from pilot trials with a UK energy provider. *Energy Policy* **2016**, *98*, 108–120. [CrossRef]
56. Dijkstra, E.W. A note on two problems in connection with graphs. *Numer. Math.* **1959**, *1*, 269–271. [CrossRef]
57. Williams, S.; Short, M. Electricity Demand Forecasting in Decentralised Demand Side Response for Blocks of Building. In Proceedings of the International Conference on Energy and Sustainable Futures (ICESF), Nottingham, UK, 9–11 September 2019.
58. Hartigan, J.A.; Wong, M.A. Algorithm AS 136: A K-Means Clustering Algorithm. *Appl. Stat.* **1979**, *28*, 100–108. [CrossRef]
59. Cai, W.; Zhao, J.; Zhu, M. A real time methodology of cluster-system theory-based reliability estimation using k-means clustering. *Reliab. Eng. Syst. Saf.* **2020**, *202*. [CrossRef]
60. Mathews, E.; Richards, P.; Lombard, C. A first-order thermal model for building design. *Energy Build.* **1994**, *21*, 133–145. [CrossRef]
61. Bi, Q.; Cai, W.J.; Wang, Q.G.; Hang, C.C.; Lee, E.-L.; Sun, Y.; Liu, K.D.; Zhang, Y.; Zou, B. Advanced controller auto-tuning and its application in HVAC systems. *Control Eng. Pract.* **2000**, *8*, 633–644. [CrossRef]

62. Happle, G.; Fonseca, J.A.; Schlueter, A. A review on occupant behavior in urban building energy models. *Energy Build.* **2018**, *174*, 276–292. [CrossRef]
63. Likert, R. A Technique for the Measurement of Attitudes. *Arch. Psychol.* **1932**, *22*, 5–53.
64. Joshi, A.; Kale, S.; Chandel, S.; Pal, D. Likert Scale: Explored and Explained. *Br. J. Appl. Sci. Technol.* **2015**, *7*, 396–403. [CrossRef]
65. Williams, S.; Short, M. Electricity demand forecasting for decentralised energy management. *Energy Built Environ.* **2020**, *1*, 178–186. [CrossRef]
66. Green Energy UK Green Energy UK. Available online: https://www.greenenergyuk.com/ (accessed on 31 March 2020).
67. Bellman, R.E. The Theory of Dynamic Programming. *Bull. Am. Math. Soc.* **1954**, *60*, 503–515. [CrossRef]
68. Williams, S.; Short, M.; Crosbie, T. On the use of thermal inertia in building stock to leverage decentralised demand side frequency regulation services. *Appl. Therm. Eng.* **2018**, *133*, 97–106. [CrossRef]
69. Weather Underground Weather History & Data Archive. Available online: https://www.wunderground.com/history (accessed on 3 April 2020).
70. Laneve, C. A lightweight deadlock analysis for programs with threads and reentrant locks. *Sci. Comput. Program.* **2019**, *181*, 64–81. [CrossRef]
71. Eriksson, E.L.V.; Gray, E.M.A. Optimization of renewable hybrid energy systems—A multi-objective approach. *Renew. Energy* **2019**, *133*, 971–999. [CrossRef]
72. Wang, H.; Mao, W.; Eriksson, L. A Three-Dimensional Dijkstra's algorithm for multi-objective ship voyage optimization. *Ocean Eng.* **2019**, *186*, 106131. [CrossRef]
73. Rosita, Y.D.; Rosyida, E.E.; Rudiyanto, M.A. Implementation of dijkstra algorithm and multi-criteria decision-making for optimal route distribution. *Procedia Comput. Sci.* **2019**, *161*, 378–385. [CrossRef]
74. Shen, L.; Shao, H.; Wu, T.; Lam, W.H.K.; Zhu, E.C. An energy-efficient reliable path finding algorithm for stochastic road networks with electric vehicles. *Transp. Res. Part C Emerg. Technol.* **2019**, *102*, 450–473. [CrossRef]
75. Kapsalis, V.; Safouri, G.; Hadellis, L. Cost/comfort-oriented optimization algorithm for operation scheduling of electric water heaters under dynamic pricing. *J. Clean. Prod.* **2018**, *198*, 1053–1065. [CrossRef]
76. Kusiak, A.; Xu, G.; Tang, F. Optimization of an HVAC system with a strength multi-objective particle-swarm algorithm. *Energy* **2011**, *36*, 5935–5943. [CrossRef]
77. Short, M.; Rodriguez, S.; Charlesworth, R.; Crosbie, T.; Dawood, N. Optimal dispatch of aggregated HVAC units for demand response: An industry 4.0 approach. *Energies* **2019**, *12*, 4320. [CrossRef]
78. Wang, Y.; Ma, Y.; Song, F.; Ma, Y.; Qi, C.; Huang, F.; Xing, J.; Zhang, F. Economic and efficient multi-objective operation optimization of integrated energy system considering electro-thermal demand response. *Energy* **2020**, *205*. [CrossRef]
79. Martin, M.; Pajouh, M.S. Rebalancing the balance: How the WTO's HR policy impacts on its very objectives for welfare enhancement and development. *J. Int. Trade Law Policy* **2011**, *10*, 243–254. [CrossRef]
80. Crosbie, T.; Short, M.; Dawood, M.; Charlesworth, R. Demand Response in Blocks of Buildings: Opportunities and Requirements. *J. Entrep. Sustain. Issues* **2017**, *4*, 271–281. [CrossRef]
81. Eurelectric Union of the Electricity Industry. *Power Distribution in Europe: Facts Figures*; Eurelectric aisbl: Brussels, Belgium, 2013.
82. Patteeuw, D.; Bruninx, K.; Arteconi, A.; Delarue, E.; D'haeseleer, W.; Helsen, L. Integrated modeling of active demand response with electric heating systems coupled to thermal energy storage systems. *Appl. Energy* **2015**, *151*, 306–319. [CrossRef]
83. Pop, C.; Cioara, T.; Antal, M.; Anghel, I.; Salomie, I.; Bertoncini, M. Blockchain Based Decentralized Management of Demand Response Programs in Smart Energy Grids. *Sensors* **2018**, *18*, 162. [CrossRef]
84. Mengelkamp, E.; Notheisen, B.; Beer, C.; Dauer, D.; Weinhardt, C. A blockchain-based smart grid: Towards sustainable local energy markets. *Comput. Sci. – Res. Dev.* **2018**, *33*, 207–214. [CrossRef]
85. Radenković, M.; Bogdanović, Z.; Despotović-Zrakić, M.; Labus, A.; Lazarević, S. Assessing consumer readiness for participation in IoT-based demand response business models. *Technol. Forecast. Soc. Chang.* **2020**, *150*, 119715. [CrossRef]

MDPI
St. Alban-Anlage 66
4052 Basel
Switzerland
Tel. +41 61 683 77 34
Fax +41 61 302 89 18
www.mdpi.com

Energies Editorial Office
E-mail: energies@mdpi.com
www.mdpi.com/journal/energies

www.ingramcontent.com/pod-product-compliance
Lightning Source LLC
LaVergne TN
LVHW070103170726
843515LV00007B/1602

* 9 7 8 3 0 3 6 5 1 3 3 6 2 *